W9-DDK-172

Factor Analysis in Chemistry
Third Edition

Factor Analysis in Chemistry

Third Edition

EDMUND R. MALINOWSKI

WILEY-INTERSCIENCE

A John Wiley & Sons, Inc., Publication

For ordering and customer service, call 1-800-CALL-WILEY.

Library of Congress Cataloging-in-Publication Data is available.

ISBN 0-471-13479-1

Printed in the United States of America.

10 9 8 7 6 5 4 3 2 1

To the next generation of chemometricians.

CONTENTS

PREFACE

It is now 10 years since publication of the second edition, and 20 years since the first edition. Chemical factor analysis has grown so large that it is impossible to keep pace with the innovations developed during the interim. The third edition is an attempt to decipher and explain many of these exciting developments. This was not an easy task since each edition represents a quantum leap in conceptual as well as mathematical understanding.

Except for minor corrections and literature updates, the introductory chapters, 1 through 5, remain unchanged. Chapter 6 has been edited to focus on evolutionary methods. Window factor analysis, transmutation, DECRA, as well as other techniques, have been incorporated in this chapter. The sections on partial least squares and multimode analysis have been expanded and consolidated into two new chapters, Chapters 7 and 8. Chapter 7 is devoted to multimode analysis, including PARAFAC and related methods. To remove much of the mystery, Chapter 8 explains partial least squares, including kernel algorithms, using a simplified notation, easily translated into the classical notation. Some of the latest advances in a wide variety of fields, such as chromatography, nuclear magnetic resonance (NMR), biomedicine, environmental science, food, and fuels are described in the applications chapters, Chapters 9 through 12.

The true authors of the text are listed in the Author Index, comprising more than 400 prominent scientists. My role in compiling this monogram is simply that of a transcriber inspired by the ingenuity of these investigators.

The author expresses his appreciation to those who supplied copies of their work for inclusion in the third edition. Special thanks go to Rasmus Bro, Roma Tauler, D. L. Massart, Willem Windig, John Kalivas, Barry Lavine, and David Haaland. The author also thanks his wife, Helen, who helped with the myriad of mechanical details associated with the revisions.

EDMUND R. MALINOWSKI
Stuart, Florida

PREFACE TO THE SECOND EDITION

More than 10 years have passed since publication of the first edition. I am pleased with the response to the initial effort. I have followed the field closely and have paid serious attention to the comments, criticisms. and witticisms of reviewers, correspondents, and associates. Exciting advances have been made in the interim, and it is time to compile this new material in the form of a revised edition.

Since the first edition, factor analysis has become prominent in chemistry primarily due to the advances in microcomputer technology. The cost-effectiveness, speed, and ease of the computer to carry out the tedious computations are responsible for the surge in interest, resulting in new and ingenious methods. Many of these methods are included in this revised edition, which emphasizes principles and methodologies, illustrated with examples and applications.

The overall organization of the text is the same as the original edition. The notation has been revised to conform closely to the current practice found in chemical journals. For example. the square bracket notation has been replaced by bold characters and the overhead "hat" and "tilde" are used to designate calculated or estimated quantities. The most difficult task has been to transcribe the rather inconsistent hieroglyphics of various investigators into a uniform notation to be maintained throughout the text while preserving the integrity of the original work.

Except for minor changes, Chapter 1 remains the same. The essence of Chapter 6 has been incorporated into and combined with Chapter 2. To chapter 3 the following topics have been added: Q- and R-mode analysis, singular value decomposition, NIPALS decomposition, and iterative key set factor analysis. To Chapter 4 the following advances are introduced: new methods for rank determination (including submatrix analysis, autocorrelation, frequency distribution of Fourier transformed eigenvectors, cross validation, and the reduced eigenvalue), statistical methods for target testing, errors in the factor loadings, and weighted factor analysis. Some of these advances are illustrated in Chapter 5 and throughout the text wherever possible. Chapter 6 has been completely revised. This new chapter describes special methods of factor analysis, such as classical factor analysis, common factors, communality, partial least squares, modeling and self-modeling methods of evolutionary factor analysis, the latest advances in rank annihilation factor analysis, and multimode factor analysis. Chapters 7 through 10, concerning applications, remain essentially unchanged.

For the record, I would like to thank Professor Darryl Howery, co-author of the original text, for continued use of his contributions. I also extend my loving thanks to my wife, Helen, who catered to my needs and idiosyncracies during the revisions.

EDMUND R. MALINOWSKI
Hoboken, New Jersey
January 1991

PREFACE TO THE FIRST EDITION

The development of new methods for interpreting large data sets ranks as one of the major advances in chemistry during the past decade. Major weapons in this still evolving "chemometric revolution" are high-speed digital computers and a variety of mathematical–statistical methodologies.

Factor analysis, a mathematical technique for studying matrices of data, is one of the most powerful methods in the arsenal of chemometrics. The method has been used more than 50 years in the behavioral sciences, and has been applied successfully to chemical problems during the past 15 years. In this book we cover the theory, practice, and application of factor analysis in chemistry. Whereas previous publications have dealt solely with classical, "abstract" factor analysis, we shall stress the target transformation version of factor analysis. The model-testing capabilities of "target" factor analysis enable the chemist to identify physically significant factors that influence data.

Although chemical applications are emphasized, the book should be of interest to physical scientists in general. In addition, the target transformation approach should prove valuable to behavioral scientists.

Our main objectives are to explain factor analysis and thereby to facilitate greater use of this technique in chemical research. The qualitative material can be followed by persons having a background in physical chemistry, and the theoretical sections can be assimilated if the chemist has some familiarity with matrix algebra. We have four specific objectives:

1. To explain the essentials of factor analysis in qualitative terms.
2. To develop the mathematical formulation of factor analysis in a rigorous fashion.
3. To describe how to carry out a factor analysis.
4. To survey the applications of factor analysis in chemistry.

Readers can acquire a qualitative overview of factor analysis in Chapters 1 and 2. A detailed study of the mathematical formulation presented and illustrated in Chapters 3, 4, and 5 is necessary for a thorough understanding of factor analysis. Users of factor analysis will find practical instructions for carrying out an analysis in Chapter 6. Chapters 7 through 10 constitute a review of the many applications of factor analysis to chemical problems.

For the record, Chapters 1, 2, 6, and 9 are primarily the work of D. G. H.; Chapters 3, 4, 5, 7, and 8 are due almost entirely to E. R. M.; and Chapter 10 is a joint effort. We owe special thanks to Dr. Paul H. Weiner for useful discussions and for contributions to Chapters 8, 9, and 10. We also thank Matthew McCue for his excellent technical suggestions, comments, and corrections. For the tedious task of typing the many drafts of the manuscript, we are indebted to Louise Sasso, Joanne Higgins, and our wives, Helen M. and Yong H.

EDMUND R. MALINOWSKI
DARRYL G. HOWERY
Hoboken, New Jersey
Brooklyn, New York
March 1980

HISTORICAL PERSPECTIVE
OF THE AUTHOR

Factor analysis was devised and developed by the behavioral scientists in the early 1930s. In spite of the fact that the methodology is ideally suited to solving chemical problems, it went unnoticed by the chemical profession until 1970.

My interest in factor analysis was inspired by a chapter in *Applied Group Theoretical and Matrix Methods* (Oxford University Press, 1955), a book written by Higman, which was recommended to me by my thesis advisor, Professor Luigi Pollara. This chapter concerned linear free energy, historically representing the first application of factor analysis to a chemical problem. In 1961, I defended my doctoral dissertation, part of which was devoted to the development and applications of target factor analysis to chemistry. The laborious calculations were carried out with a Marchant hand-operated, electric calculator, a prototype of which is on display in the Science Museum in London, England.

My first publication on factor analysis. "Application of Factor Analysis to the Prediction of Activity Coefficients of Non-Electrolytes," appeared in *Separation Science*, Vol. 1, p. 661 (1966), representing a joint effort with Phil Funke, Dan Martire, and Lui Pollara, my colleagues at Stevens Institute. This work required the use of an IBM 1620 computer, which often devoured the computer cards.

In 1975, I began to compile a list of publications from the chemical literature that could, in some sense, be classified as involving factor analytical principles. In collaboration with Professor Howery of Brooklyn College, and with the helpful suggestions and contributions from Paul Weiner, this effort produced the first edition of *Factor Analysis in Chemistry*, published by Wiley-Interscience in 1980 and reprinted by Krieger Publishing in 1989.

Thus the seeds of factor analysis were planted in the field of chemistry. The story that followed took place so quietly and in such small steps that it was hardly noticeable, but it has steadily unfolded. The seeds of the early investigations are just beginning to blossom and bear fruit. So many new, inventive, and ingenious developments and applications of factor analysis have recently appeared in the chemical literature that it is impossible to keep abreast of the field. Exciting contributions have been made by investigators from a wide range of scientific disciplines, including geologists, pharmacologists, biochemists, food specialists, chemical engineers, and statisticians, as well as chemists. The rain of information precipitated by these efforts will undoubtedly produce new flowers in the fertile field of factor analysis.

PROLOGUE

ANOTHER DIMENSION

Marlo slid over to Sprig, peeked across his shoulder at the paper Sprig was so intently viewing. "What are you reading?" he asked inquisitively.

Sprig was startled by the unexpected voice. "Oh. . .a science-fiction story. . .about another world that has a third dimension."

"Three dimensions? What's a third dimension?" Marlo puzzled.

"Something called depth," Sprig replied.

Marlo shrugged. "Depth? What's depth?" he asked.

"Well, its very complicated," Sprig mulled, "but let me try to explain it to you. . .A point in space represents the zero dimension. . .By moving the point forward we form a line. . . . The line exists in one dimension. . .Do you understand that?"

Marlo nodded.

Sprig continued, "Now let's move the line sideways forming a plane. . .The plane lies in two dimensions. . .Just like the two-dimensional world we live in today. . .A world with only length and width. . ." He paused, carefully contemplating his thoughts.

"Yes, yes, go on," urged Marlo.

"Let me try to demonstrate."

Sprig spun in planar circles while he tried to think of a simple explanation, a most challenging task.

"First we draw a straight line like this," he said, gliding forward smoothly. Then he spun completely around and glided back over the line to the middle.

Reprinted with permission from E. R. Malinowski, *Tales Abstracted*, copyright 1993.

"Now, we draw a second line...But this line we draw perpendicular to the first line...Like this..."

He continued. "Although each line has only one dimension, the two lines, together, lie in a two-dimenension world..."

"Yes, yes." Marlo interrupted. "I understand. But what is depth?"

"You must use all the power of your imagination to visualize the third dimension," cautioned Sprig. "Imagine another plane, different from the plane we live in...Now turn this plane so it is perpendicular to our planar world...Each plane has only two dimensions...but, together, they lie in a three-dimensional world.

The extra dimension is called...depth."

Marlo shrugged his head. "How fantastic!"

"It's only science fiction," added Sprig as he darted away in spiraling circles.

1

Every problem has a solution.

INTRODUCTION

The use of computers to solve chemical problems has increased tremendously during the past 20 years. Indeed, the discipline of chemistry called chemometrics is flourishing as a result of this revolutionary marriage of chemistry and computer science. Chemometrics is the use of mathematical and statistical methods for handling, interpreting, and predicting chemical data. Powerful methodologies have opened new vistas for chemists and have provided useful solutions for many complex chemical problems.[1-29] Factor analysis has proved to be one of the most potent techniques in the chemometric arsenal.

Chapters 1 and 2 provide a simplified, qualitative introduction to factor analysis. These two chapters furnish an overview of factor analysis, written especially for readers who do not wish to wrestle with mathematical details. Chapter 1 explores the purposes and advantages of factor analysis, while avoiding detailed methodology and mathematical derivations. For the mathematically inclined reader and for all who wish to acquire an in-depth understanding of factor analysis, a rigorous development is presented in Chapters 3 and 4.

1.1 DEFINITION OF FACTOR ANALYSIS

Factor analysis (FA) was founded by the behavioral scientists. Its early history has been recorded by Harman,[30] who ascribes its origin to a paper on orthogonal regression lines published by Pearson[31] in 1901. The first real development was

1

accomplished by Hotelling[32] in 1933. Although an ideal tool for solving chemical problems, the method went unnoticed by the chemical profession until the birth of chemometrics in the 1970s.

In the early years factor analysis suffered a tarnished reputation because many variations of the method did not yield the same results. This was due to the different simplifying assumptions and auxiliary conditions needed during the precomputer era. Many of these ingenious methods have fallen by the wayside. During the years, the computer greatly influenced the development of factor analysis. Today, chemists, and scientists in general, are familiar with computers, mathematics, and statistics, the prerequisites for factor analysis. The field has grown so large that it is impossible for us to examine all the methodologies. In this text, we concentrate on those techniques that have become popular in chemistry.

If you ask for a definition of *factor analysis*, you will obtain a different answer from each practitioner. The definition has changed over the years, encompassing a much wider selection of techniques than originally intended. As an attempt to provide a global definition, we offer the following:

Factor analysis is a multivariate technique for reducing matrices of data to their lowest dimensionality by the use of orthogonal factor space and transformations that yield predictions and/or recognizable factors.

1.2   PHILOSOPHICAL BASIS

The nature and objectives of factor analysis can be illustrated with an example from academic life. As we know, the same laboratory report will receive different grades from different professors because of variations in marking criteria. The assigned grade represents a composite of a variety of factors (i.e., subjects), such as chemistry, physics, mathematics, grammar, and organization. Each factor is weighted in importance according to the personal judgment of each professor. The various grades a given report receives are due to the differences in importance conferred upon each factor by each professor.

In the technique of factor analysis, the grade is viewed as a linear sum of factors, each factor being weighted differently. Grade d_{ik} received by student i from professor k is assumed to have the form

$$d_{ik} = s_{i1}l_{1k} + s_{i2}l_{2k} + \cdots + s_{in}l_{nk} = \sum_{j=1}^{n} s_{ij}l_{jk} \qquad (1.1)$$

where s_{ij} is the true *score* of student i in factor j, l_{jk} is the relative *loading* (importance) given by professor k to factor j, and the sum is taken over the n *factors* or subjects.

Factor analysis deals with a battery of grades involving a number of students and professors. The grades are arranged in a *matrix* such that each row concerns a particular student and each column concerns a particular professor. Such a data matrix, involving four students and three professors, is shown in (1.2). If only two factors, such as chemistry and English, were considered important in the grading, each data point could be broken down into a sum of two factors, as shown in the first equality:

$$
\begin{array}{c} \text{Students} \end{array}
\begin{array}{c} \\ 1 \\ 2 \\ 3 \\ 4 \end{array}
\overset{\begin{array}{ccc} & \text{Professors} & \\ 1 & 2 & 3 \end{array}}{
\begin{bmatrix} d_{11} & d_{12} & d_{13} \\ d_{21} & d_{22} & d_{23} \\ d_{31} & d_{32} & d_{33} \\ d_{41} & d_{42} & d_{43} \end{bmatrix}}
=
\begin{bmatrix} s_{11}l_{11}+s_{12}l_{21} & s_{11}l_{12}+s_{12}l_{22} & s_{11}l_{13}+s_{12}l_{23} \\ s_{21}l_{11}+s_{22}l_{21} & s_{21}l_{12}+s_{22}l_{22} & s_{21}l_{13}+s_{22}l_{23} \\ s_{31}l_{11}+s_{32}l_{21} & s_{31}l_{12}+s_{32}l_{22} & s_{31}l_{13}+s_{32}l_{23} \\ s_{41}l_{11}+s_{42}l_{21} & s_{41}l_{12}+s_{42}l_{22} & s_{41}l_{13}+s_{42}l_{23} \end{bmatrix}
$$

$$
=
\begin{array}{c} \text{Students} \end{array}
\begin{array}{c} 1 \\ 2 \\ 3 \\ 4 \end{array}
\overset{\begin{array}{cc} \text{Factors} \\ 1 & 2 \end{array}}{
\begin{bmatrix} s_{11} & s_{12} \\ s_{21} & s_{22} \\ s_{31} & s_{32} \\ s_{41} & s_{42} \end{bmatrix}}
\overset{\begin{array}{ccc} & \text{Professors} & \\ 1 & 2 & 3 \end{array}}{
\begin{bmatrix} l_{11} & l_{12} & l_{13} \\ l_{21} & l_{22} & l_{23} \end{bmatrix}}
\begin{array}{c} 1 \\ 2 \end{array} \text{Factors}
$$

(1.2)

The second equality in (1.2) is the result of applying standard rules of matrix multiplication. In matrix notation (1.2) becomes

$$
\underset{\text{data}}{\mathbf{D}} = \underset{\text{scores}}{\mathbf{S}} \ \underset{\text{loadings}}{\mathbf{L}}
$$

(1.3)

Here $\mathbf{D}$ is the *data matrix* that consists of the grades; $\mathbf{S}$ is the matrix of the students' true scores in each subject, called the *score* matrix; and $\mathbf{L}$ is the matrix of importance conferred upon each subject by each professor, called the *loading matrix*.

The purpose of factor analysis, as visualized by psychologists, is to extract the student score matrix from the data matrix in order to determine the students' true abilities in each subject, in effect removing the professors' prejudices (the loadings) from the grades. Such reasoning forms the philosophical basis of factor analysis.

The form of the data matrix discussed above is analogous to many types of data matrices encountered in chemistry, where, for example, molecules emulate the students and chemical measurements emulate the professors. Physical observations, such as boiling points, melting points, spectral intensities, and retention values, are data analogous to grades. In a chemical problem, a row of data may concern a particular molecule and a column may concern a particular measurement. Factor analysis yields a molecule score matrix, which depends solely on the characteristics of the molecules, and a measurement loading matrix, which depends solely on the nature of the measurements. Such a separation of the features of the molecules from the features of the measurements provides the chemist with a better insight into the true nature of the chemical phenomenon involved.

1.3 GENERALIZATIONS

This section concerns notation and terminology, which can be used in a general manner. Data matrix **D**, consisting of r rows and c columns, is written as

$$
\text{Column designee}
$$

$$
\mathbf{D} = \text{Row designee}
\begin{bmatrix}
d_{11} & d_{12} & \cdots & d_{1c} \\
d_{21} & d_{22} & \cdots & d_{2c} \\
\vdots & \vdots & & \vdots \\
d_{r1} & d_{r2} & \cdots & d_{rc}
\end{bmatrix}
\tag{1.4}
$$

The row and column headings of the matrix are called *designees*. Each measured data point in **D** is specified by a subscript denoting its row and column position in the matrix. The symbol d_{ik} represents the data point associated with the ith row and kth column of the matrix.

Abstract Model. The first objective of factor analysis is to obtain a mathematical, "abstract" solution wherein each point in the data matrix is expressed as a linear sum of product terms. The number of terms in the sum, n, is called the *number of factors*. Specifically, we seek solutions of the form

$$
d_{ik} = \sum_{j=1}^{n} r_{ij} c_{jk}
\tag{1.5}
$$

In this equation r_{ij} and c_{jk} are called *factors*. For the jth factor in the sum, *row factor* r_{ij} is associated with the ith row of the data matrix, and the corresponding *column factor* c_{jk} is associated with the kth column of the matrix. In classical abstract factor analysis, the row factors are called *scores* and the column factors are called *loadings*.

For data modeled by (1.5), the data matrix can be decomposed into two matrices:

$$
\underset{\substack{\text{data} \\ \text{matrix}}}{\mathbf{D}} = \underset{\substack{\text{row} \\ \text{matrix}}}{\mathbf{R}_{\text{abstract}}} \underset{\substack{\text{column} \\ \text{matrix}}}{\mathbf{C}_{\text{abstract}}}
\tag{1.6}
$$

where

$$
\mathbf{R}_{\text{abstract}} = \underset{\text{Row designee}}{\overset{\text{Factor}}{\begin{bmatrix} r_{11} & r_{12} & \cdots & r_{1n} \\ r_{21} & r_{22} & \cdots & r_{2n} \\ \vdots & \vdots & & \vdots \\ r_{r1} & r_{r2} & \cdots & r_{rn} \end{bmatrix}}} \tag{1.7}
$$

$$
\mathbf{C}_{\text{abstract}} = \underset{\text{Factor}}{\overset{\text{Column designee}}{\begin{bmatrix} c_{11} & c_{12} & \cdots & c_{1c} \\ c_{21} & c_{22} & \cdots & c_{2c} \\ \vdots & \vdots & & \vdots \\ c_{n1} & c_{n2} & \cdots & c_{nc} \end{bmatrix}}} \tag{1.8}
$$

Since this solution is purely mathematical and is devoid of physical meaning, these matrices are called *abstract matrices*. The columns of $\mathbf{R}_{\text{abstract}}$ are called *abstract factors*. *Row matrix* $\mathbf{R}_{\text{abstract}}$ contains a row for each of the r row designees and a column for each of the n factors, while *column matrix* $\mathbf{C}_{\text{abstract}}$ has a column for each of the c column designees and a row for each factor. The factor analytical solution isolates the row-designee factors from the column-designee factors.

Methodologies for determining the number of factors and for calculating the abstract row and column matrices are discussed in Chapter 3. Since the abstract solution should involve a physically meaningful number of factors, determination of n, the correct factor "size," is a particularly important step. As a result of this step, an estimate of the complexity of the data space, information normally lacking even for the simplest chemical problems, is obtained.

Interpreting Factors. The ultimate objective of factor analysis is to develop a complete, physically meaningful model for the data. Hence the second objective of factor analysis is to convert the abstract solution into a real solution. To do this, the abstract factors are mathematically "transformed" into physically significant, "real" factors. Transforming the abstract solution into a real solution is a difficult, but realizable, goal of factor analysis.

To carry out the transformations, an appropriate information matrix, $\mathbf{T}$, is required. Postmultiplying $\mathbf{R}_{\text{abstract}}$ by $\mathbf{T}$ and premultiplying $\mathbf{C}_{\text{abstract}}$ by the inverse of the transformation $\mathbf{T}^{-1}$, the data matrix in (1.6) can be expressed as

$$
\begin{aligned}
\mathbf{D} &= \{\mathbf{R}_{\text{abstract}}\mathbf{T}\}\{\mathbf{T}^{-1}\mathbf{C}_{\text{abstract}}\} \\
&= \mathbf{R}_{\text{transformed}}\mathbf{C}_{\text{transformed}}
\end{aligned} \tag{1.9}
$$

If the transformed solution can be shown to have physical significance, a real solution to the problem will have been found so that

$$\mathbf{D} = \mathbf{X}_{real}\mathbf{Y}_{real} \tag{1.10}$$

where $\mathbf{X}_{real} = \mathbf{R}_{transformed}$ and $\mathbf{Y}_{real} = \mathbf{C}_{transformed}$. This equation summarizes the ultimate objective of factor analysis.

How such magical transformations can be carried out is one of the main topics of Chapters 2 through 6. Using factor analysis, we may be able to ascribe meanings to chemical data that initially appear to be exceedingly complicated because of the myriad of uncontrollable factors at play. The potential for modeling data with real factors is the most exciting feature of factor analysis.

A technique called *target testing* is especially valuable for achieving meaningful transformations. Suspected parameters (such as physical properties or structural features of molecules) can be tested individually as possible factors, and complete models of real factors can be systematically pieced together. This individual testing ability is one of the most valuable features of the target factor analysis method.

1.4 CHEMICAL EXAMPLE

To illustrate how factor analysis can be applied to chemical problems, let us consider the following data matrix, **A**, involving the ultraviolet absorbances of five different mixtures of the same absorbing components measured at six wavelengths:

$$
\mathbf{A} =
\begin{array}{c}
\\
\text{Wavelength} \\
278\text{ nm} \\
274\text{ nm} \\
270\text{ nm} \\
266\text{ nm} \\
262\text{ nm} \\
258\text{ nm}
\end{array}
\begin{array}{ccccc}
& & \text{Mixture} & & \\
1 & 2 & 3 & 4 & 5 \\
0.005 & 0.031 & 0.063 & 0.091 & 0.046 \\
0.040 & 0.172 & 0.356 & 0.444 & 0.218 \\
0.103 & 0.283 & 0.484 & 0.471 & 0.208 \\
0.116 & 0.323 & 0.562 & 0.548 & 0.241 \\
0.125 & 0.318 & 0.516 & 0.450 & 0.185 \\
0.104 & 0.267 & 0.430 & 0.376 & 0.154
\end{array}
\tag{1.11}
$$

Such special information can be obtained from a liquid chromatograph where samples are collected at five different elution time intervals. Similar data can be collected from a chemical kinetics study if samples of the reaction mixture are collected at different times during the experiment. The problem here is to determine the number of components, to identify the chemical constituents, and to ascertain their concentrations.

According to (1.5), factor analysis will automatically furnish an abstract solution for each absorbance datum, A_{ik}, in the form

$$A_{ik} = \sum_{j=1}^{n} w_{ij} m_{jk} \qquad (1.12)$$

Here w_{ij} and m_{jk} are the jth abstract row and column factors associated with the ith wavenumber and the kth mixture, respectively. To account for the absorbances within experimental error, n factors are included in the sum. According to (1.12), the absorbance data matrix has an abstract factor analytical solution expressed by

$$\mathbf{A} = \mathbf{W}_{\text{abstract}} \mathbf{M}_{\text{abstract}} \qquad (1.13)$$

where $\mathbf{W}_{\text{abstract}}$ and $\mathbf{M}_{\text{abstract}}$ are wavenumber-factor and mixture-factor matrices, respectively.

The most important feature of the abstract solution is that it reveals the number of factors responsible for the absorbance data. Ultimately, we search for an appropriate transformation matrix that will convert the abstract solution into a physically significant real solution:

$$\mathbf{A} = \mathbf{W}_{\text{real}} \mathbf{M}_{\text{real}} \qquad (1.14)$$

Going from (1.13) to (1.14) is not automatic. On the contrary, this step presents the most difficult challenge to the chemist, requiring a great deal of effort, knowledge, and intuition. If theoretical speculations can be invoked, the transformation has a better chance of being successful.

If the absorbance data obey Beer's law, the factors can be interpreted chemically. For a mixture containing n absorbing components, Beer's law models each absorbance datum by the equation

$$A_{ik} = \sum_{j=1}^{n} \varepsilon_{ij} c_{jk} \qquad (1.15)$$

Here ε_{ij} is the molar absorptivity per unit path length of component j at wavelength i, and c_{jk} is the molar concentration of component j in the kth mixture. Equation (1.15) involves a linear sum of products analogous to (1.12); therefore, data that obey Beer's law should have meaningful factor analytical solutions. To solve the problem completely, we must find a transformation matrix that will convert the abstract solution into the real solution. When this is done correctly, (1.14) will take the form

$$\mathbf{A} = \mathbf{E}_{\text{real}} \mathbf{C}_{\text{real}} \qquad (1.16)$$

Each column of the molar absorptivity matrix, $\mathbf{E}_{\text{real}}$, corresponds to the absorbances of one of the pure components at the five wavelengths, essentially tracing out the spectrum of the pure component. Each row of the molar concentration matrix, $\mathbf{C}_{\text{real}}$, corresponds to the concentrations of one of the n components in each of the four mixtures.

Let us now summarize the kinds of information that might be furnished by factor analyses of absorbance data. First and quite important, determining the number of factors is tantamount to finding the total number of absorbing components in the mixtures. Second, astute transformations of the abstract factor analytical solution furnishes a good factor analytical representation of the real situation. The molar absorptivity matrix and the molar concentration matrix are the desired, physically significant transformations of $\mathbf{W}_{abstract}$ and $\mathbf{M}_{abstract}$. Successful transformation to $\mathbf{E}_{real}$ identifies each component chemically via its spectrum. The concomitant transformation to $\mathbf{C}_{real}$ furnishes the concentrations of the components in each sample mixture.

In summary, the ultimate payoff from factor analysis in this type of problem might be to determine:

1. The number of absorbing components
2. The concentration of each component in each mixture
3. The spectrum of each component

The factor analytical approach is far more useful than the popular determinant method for finding the concentrations of components in multicomponent mixtures, since the spectra of all components must be specified initially in the latter approach. By contrast, factor analysis can furnish the number of components, the concentrations, and the spectral information via a purely mathematical route. Describing exactly how factor analysis can accomplish these and even other, more difficult tasks is the primary objective of this book.

1.5   ATTRIBUTES

Factor analysis often allows us to answer the most fundamental questions in a chemical problem: How many factors influence the observable? What are the natures of these factors in terms of physically significant parameters? Factor analysis enables chemists to tackle problems that in the past had to be avoided because too many uncontrollable variables influenced the data. Factor analysis not only enables us to correlate and explain data, but also to fill in gaps in our data store.

In this section we list some of the virtues of the factor analytical approach. In particular, the following five general attributes illustrate why a chemist might want to use factor analysis:

1. *Data of great complexity can be investigated.* Factor analysis, being a method of "multivariate" analysis, can deal with many factors simultaneously. This feature is of special importance in chemistry, since interpretations of most chemical data require multivariate approaches.

2. *Large quantities of data can be analyzed.* Factor analyses can be carried out efficiently using standard factor analytical computer programs. Methods such as factor analysis are needed to properly use the voluminous data sets of chemistry.

3. *Many types of problems can be studied.* Factors analysis can be applied regardless of the initial lack of insight into the data. Although, ideally, factor analysis is used in conjunction with theoretical constructs, the approach can yield valuable predictions based on empirical applications.

4. *Data can be simplified.* Matrices can be modeled concisely with a minimum of factors, and generalizations that bring out the underlying order in the data can be obtained. Huge volumes of data can be compressed into small packages without loss of accuracy.

5. *Factors can be interpreted in useful ways.* The nature of the factors can be clarified and deciphered, and data can be classified into specific categories. Complete physically significant models can be developed systematically, and these models can be employed to predict new data.

In general, factor analysis provides a means to attack those problems that appear to be too difficult to solve. Such problems are bountiful in chemistry, making factor analysis an ideal probe for exploration. Finding the controlling factors is akin to an engineer drilling for oil. To increase the chances for success, the engineer must use every scientific resource available; blind drilling can be extremely expensive, time-consuming, and fruitless. A great deal of scientific input and intuition are required in the factor analytical approach.

Thousands of publications bear witness to the power and utility of factor analysis in chemistry. Howery,[14,15] Weiner,[16] and Llinas and Ruiz[17] have reviewed the role of factor analysis in chemistry during the early years. Its importance in mixture analysis has been reviewed by Gemperline[18] and by Hamilton and Gemperline.[19] A tutorial on target transformation factor analysis has been written by Hopke.[20] Bro[21] has published an exposition of multiway analysis, including parallel factor analysis (PARAFAC). Multiway analysis has been reviewed by Bro et al.[22] Geladi[23] has compiled the history of partial least squares (PLS). Classical PLS methodology is explained in the studies of Geladi and Kowalski.[24] Various chemical applications are described in Chapters 6 through 12.

REFERENCES

1. B. R. Kowalski (Ed.), *Chemometrics: Theory and Applications*, ACS Symp. Ser., 52, American Chemical Society, Washington, DC, 1977.

2. R. F. Hirsch (Ed.), *Statistics*, Franklin Institute Press, Philadelphia, 1978.

3. B. R. Kowalski (Ed.), *Chemometrics and Statistics in Chemistry*, Reidel, Dordrecht, Holland, 1983.

4. M. A. Sharaf, D. L. Illman, and B. R. Kowalski, *Chemometrics*, Wiley-Interscience, New York, 1986.

5. G. L. McClure (Ed.), *Computerized Quantitative Infrared Analysis*, American Society for Testing and Materials, Philadelphia, 1987.

6. D. L. Massart, B. G. M. Vandeginste, S. N. Deming, Y. Michotte, and L. Kaufman, *Chemometrics: A Textbook*, Elsevier Science, Amsterdam, 1988.

7. H. Martens and T. Naes, *Multivariate Calibration*, Wiley, New York, 1989.

8. J. E. Jackson, *A User's Guide to Principal Components*, Wiley, New York, 1991.

9. R. Q. Yu, *Introduction to Chemometrics*, Hunan Educational Press House, Changsha, 1991.

10. Z. X. Pang, S. Z. Si, S. Z. Nie, and M. Z. Zhang, *Chemical Factor Analysis*, Publishing House of Science and Technology University of China, Hehui, 1992.

11. J. H. Kalivas and P. M. Lang, *Mathematical Analysis of Spectral Orthogonality*, Dekker, New York, 1994.

12. E. J. Karjalainen and U. P. Karjalainen, *Data Analysis for Hyphenated Techniques*, Elsevier Science, Amsterdam, 1996.

13. R. Kramer, *Chemometric Techniques for Quantitative Analysis*, Marcel Dekker, New York, 1998.

14. D. G. Howery, *Am. Lab.*, **8**(2), 14 (1976).

15. D. G. Howery, in R. F. Hirsch (Ed.), *Statistics*, Franklin Institute Press, Philadelphia, 1978, p. 185.

16. P. H. Weiner, *Chem Tech.*, **1977**, 321.

17. J. R. Llinas and J. M. Ruiz, in G. Vernin and Chanon (Eds.), *Computer Aids to Chemistry*, Wiley, New York, 1986, Chap. V.

18. P. J. Gemperline, *J. Chemometrics*, **3**, 549 (1989).

19. J. C. Hamilton and P. J. Gemperline, *J. Chemometrics*, **4**, 1 (1990).

20. P. K. Hopke, *Chemometrics Intell. Lab. Syst.*, **6**, 7 (1989).

21. R. Bro, Multi-Way Analysis in the Food Industry: Models, Algorithms and Applications, Doctoral Thesis, Royal Veterinary and Agricultural University, Denmark, 1998.

22. R. Bro, J. J. Workman, Jr., P. R. Mobley, and B. R. Kowalski, *Appl. Spectrosc. Rev.*, **32**, 237–261 (1997).

23. P. Geladi, *J. Chemometrics*, **2**, 231 (1988).

24. P. Geladi and B. R. Kowalski, *Anal. Chim. Acta*, **185**, 1 (1986); **185**, 18 (1986).

25. O. Matthias, *Chemometrics*, Wiley-VCH, New York, 1999.

26. D. Livingstone, *Data Analysis for Chemists: Applications to QSAR and Chemical Products Design*, Oxford University Press, Oxford, 1995.

27. K. R. Beebe, R. J. Pell, and M. B. Seasholtz, *Chemometrics: A Practical Guide*, Wiley-Interscience, New York, 1998.

28. E. K. Kemsley, *Discriminant Analysis and Class Modeling of Spectroscopic DataI*, Wiley, Chichester, 1998.

29. A. Höskuldsson, *Prediction Methods in Science and Technology*, Thor Publishing, Arnegaards Alle 7, Copenhagen, Denmark, 1998.

30. H. H. Harmon, *Modern Factor Analysis*, 2nd ed., University of Chicago Press, Chicago, 1967.

31. K. Pearson, *Philos. Mag.*, Series 6, **2**, 559 (1901).

32. H. Hotelling, *J. Educ. Psych.*, **24**, 417 (1933).

2

Every problem has more than one solution.

MAIN STEPS

This chapter concerns notation and summarizes the methodology of factor analysis in a qualitative manner. The chapter serves as a prelude to the rigorous development of factor analysis presented in Chapters 3 and 4.

2.1   NOTATION

Factor analysis involves the use of matrices, vectors, and scalars. Throughout this text the following mathematical notation is followed. Scalar quantities (i.e., numbers) are represented by lowercase letters—$a, b, c, x, y,$ and z. Vectors (i.e., one-dimensional arrays of numbers) are symbolized by bold, lowercase letters—

$\mathbf{s}, \mathbf{t}, \mathbf{u}, \mathbf{x}, \mathbf{y}$, and $\mathbf{z}$. In particular, all vectors are considered to be column vectors unless otherwise indicated. Row vectors are denoted by a prime—$\mathbf{s}', \mathbf{t}', \mathbf{u}'$ and $\mathbf{z}'$. Lowercase letters are used to designate components of a vector. Subscripts are used to characterize matrices, vectors, and scalars. Numbers and lowercase letters are used as subscripts. Bold, uppercase letters or enclosures in square brackets [ ] signify matrices. Matrix transformation, whereby rows and columns are interchanged, are denoted by a prime, consistent with the vector notation.

Based on this notation:

b_{ik} is a scalar.

$$\mathbf{b}_k = \begin{bmatrix} b_{1k} \\ b_{2k} \\ b_{3k} \end{bmatrix} \text{ is the } k\text{th column vector.}$$

$\mathbf{b}'_i = [b_{ik} \quad \cdots \quad b_{ik}]$ is the ith row vector.

$$\mathbf{B} = [\mathbf{B}] = \begin{bmatrix} b_{11} & b_{12} \\ b_{21} & b_{22} \\ b_{31} & b_{32} \end{bmatrix} \text{ is a matrix.}$$

$$\mathbf{B}' = [\mathbf{B}]' = \begin{bmatrix} b_{11} & B_{21} & b_{31} \\ b_{12} & b_{22} & b_{32} \end{bmatrix} \text{ is the transposed matrix.}$$

The $\wedge$, called "hat," or a $\sim$, called "tilde," above a quantity signifies an estimated (or calculated) quantity. A bar above a quantity indicates an estimation based on the reduced factor space.

$\mathbf{D}$ is the matrix containing the measured data, d_{ik}.

$\hat{\mathbf{D}}$ and $\tilde{\mathbf{D}}$ represent different estimations of $\mathbf{D}$.

$\bar{\mathbf{D}}$ is an estimation of $\mathbf{D}$ based on the reduced factor space.

Other specialized notation that will be used consistently throughout the text is:

$\mathbf{X}^+ = (\mathbf{X}'\mathbf{X})^{-1}\mathbf{X}'$ is the pseudoinverse of $\mathbf{X}$.

$\|r\| = (\sum r_i^2)^{1/2} = (\mathbf{r}'\mathbf{r})^{1/2}$ is the norm of vector $\mathbf{r}$.

$\|\mathbf{R}\| = (\sum \sum r_{ij}^2)^{1/2}$ is the norm of matrix $\mathbf{R}$.

$\mathbf{U}$ = normalized row-factor matrix, composed of eigenvectors of the row-factor space.

$\bar{\mathbf{U}} = \mathbf{U}$ in the reduced factor space.

$\mathbf{V}$ = normalized column-factor matrix, composed of eigenvectors of the column-factor space.

$\mathbf{S}$ = diagonal matrix composed of the square roots of the eigenvalues.

$\mathbf{I}$ = identity matrix composed of ones along the diagonal, zeros elsewhere.

2.2 SUMMARY OF STEPS

Factor analysis involves the following main steps: preparation, reproduction, transformation, combination, and prediction. Figure 2.1 shows the sequencing of the steps and the most important information resulting from each step. The essence of each step is as follows. In the *preparation* step the data to be factor analyzed are selected and pretreated mathematically. The *reproduction* step, the foundation of the analysis, furnishes an abstract solution based on the reduced factor space. These two

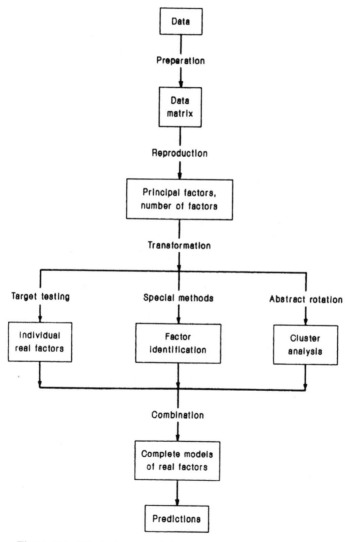

Figure 2.1 Block diagram of the main steps in factor analysis.

steps are common to all factor analyses. Three kinds of mathematical transformations—target testing, special methods, and abstract rotation—are employed to obtain more useful solutions. *Target testing* is a special technique for identifying individual, real factors. *Special methods* take advantage of known chemical information, placing important constraints on the transformation; these techniques are used when targets cannot be accurately or properly formulated. *Rotation* converts the abstract factors into easier-to-interpret abstract factors. The *combination* step furnishes complete models that can be used to calculate new data in the *prediction* step.

2.3   PREPARATION

The objective of the data preparation step is to obtain a data matrix best suited for factor analysis. This step involves formulating the problem, selecting the data, and mathematically pretreating the data to conform to appropriate theoretical or statistical criteria. The ultimate success or failure of factor analysis depends strongly on the preparation step. The kinds of information sought from a data analysis should carefully be considered before applying factor analysis. Factor analysis is just one of a host of chemometric methods,[1,2] each of which furnishes different kinds of information.

2.3.1   Problem Selection

Some general aspects of problem selection are discussed in this section. Thoughtful evaluation of the problem within the context of factor analysis is the first essential consideration in a well-designed research plan.

Factor Analyzability. One asks first and foremost whether the data might be modeled as a linear sum of product terms as required for factor-analyzable solutions. The decision to apply factor analysis should be based on theoretical or chemical concepts rather than on intuition. Results of factor analysis will be easier to evaluate if the researcher has prior insight into the problem. If the applicability is not known a priori, the abstract reproduction step can be used to judge whether to continue or to terminate the analysis.

Of the many kinds of data that might have factor analytical solutions, two kinds are especially suitable. Matrices involving spectral intensities at several wavenumbers for several multicomponent mixtures will usually have factor analytical solutions because, according to Beer's law, absorbance can be expressed as a linear sum of product terms, consistent with the factor analytical model. Solute–solvent problems are often anticipated to be factor analyzable when the observable is suspected to arise from a sum of chemical interactions involving solute–solvent terms. For example, Hammett-like functions, based on models consistent with factor analysis, explain large quantities of chemical data.

Types of Problems. Many kinds of data can be expressed as a matrix. Problems suited to factor analysis can be classified according to the types of rows and columns of the data matrix. Two types of designees—objects and variables—can be used to describe chemical phenomenon. The word *object* encompasses any sample of matter, from subatomic particles to stellar galaxy. Molecules, mixtures, instruments, and persons are examples of objects. Variables characterize and distinguish the chemical systems. Boiling point, melting point, and density are examples of variables that distinguish chemicals. Spectral interval, chromatographic interval, and pH are examples of variables that are somewhat controllable.

Based on the preceding classification, three types of data matrices can be formed: variable–object, object–object, and variable–variable. The matrix expressed in Eq. (1.11) is an example of a variable–object matrix involving absorbances at various wavelengths for different mixtures. A matrix of chromatographic data, involving solutes and solvents, forms an object–object matrix. Fluorescent excitation–emission matrices, described in Chapter 7, represent variable–variable matrices.

2.3.2 Data Selection

After selecting an appropriate problem, the next step is to assemble the data. Data matrices for factor analysis should meet the criteria discussed in this subsection. Careful data selection is a crucial prerequisite for successful factor analyses.

Reliable data should be chosen, especially since an accurate determination of the factor size usually hinges on the quality of the data. If data were collected in several laboratories, self-consistent data should be chosen and dubious data should be deleted. The uniqueness test (see Chapter 3) is useful for detecting points having gross errors.

For mathematical reasons, the data must be complete. When points in a matrix are missing, an analysis can be conducted on the smaller, but complete, submatrix that can be constructed by deleting the rows and columns associated with such points.

The size of the data matrix depends on the availability of data, the objectives of the research, and the computational facilities at hand. To ensure general solutions, the largest possible matrix should be used. However, large matrices may be too complex to analyze initially. In order to make some progress, we may purposely work with a selection of submatrices. At a later stage, successively larger matrices can be analyzed to obtain more general solutions.

2.3.3 Data Pretreatment

Data pretreatment should be based on sound theoretical or statistical criteria. The decision to factor analyze the raw data (covariance), or to normalize or standardize the data before factor analysis (correlation), is based on the type of data. Standardization should always be applied when the columns (or rows) in the matrix involve different units. Logarithmic transformations of the data are quite common in chemistry. For example, the logarithm of equilibrium constants and rate constants, rather than the raw data, should be factor analyzed because their logarithms are related to the thermodynamic free energy, which is an additive property.

2.4   REPRODUCTION

The abstract reproduction step is the mathematical underpinning of factor analysis. Reproduction involves two procedures: obtaining the "principal" factor solution and determining the correct number of factors. The abstract solution is expressed as principal factors (see Figure 2.2). Since all subsequent procedures are based on the abstract solution, the reproduction step should be considered with exceptional care.

Principal factor eigenanalysis can be carried out routinely with many readily available computer programs. Determining the size of the factor space, a not-so-routine procedure, entails the use of nonstatistical as well as statistical criteria.

Examples of various strategies employed to accomplish this important task are presented in Chapter 4. These strategies should be reviewed carefully.

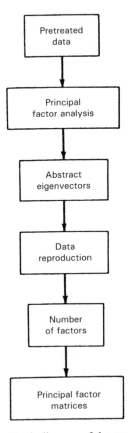

Figure 2.2   Block diagram of the reproduction step.

2.4.1 Principal Factor Analysis

The procedure for calculating the abstract solution involves a mathematical method called *eigenanalysis*, yielding eigenvalues and associated eigenvectors. The four most commonly used methods are the power method, the Jacobi method, singular value decomposition (SVD), and NIPALS. An excellent introduction to the first three methods, including FORTRAN and Pascal source codes, is found in *Numerical Recipes*, a book written by Press et al.[3] A detailed description of NIPALS has been presented by Wold.[4] Although these methods differ in their mathematical approaches, they achieve the same results, extracting eigenvectors and eigenvalues.

The power method and the NIPALS method are derived in Chapter 3. The numerical stability and ruggedness of SVD have been discussed by Lawson and Hanson[5] and by Shrager.[6] SVD is the preferred algorithm because it is most stable under the widest range of applications. Because of this preference, the notation associated with SVD has become increasingly popular. In the following chapters, we amalgamate this notation with other popular notation to take advantage of the best aspects of each, as needed.

Principal component analysis (PCA) is another popular name for *eigenanalysis*. To chemists the word "component" conjures up a different meaning and therefore the terminology *principal factor analysis* (PFA) is recommended as an alternative. Because eigenanalysis, PCA, PFA, and SVD produce essentially the same results, we use these terms interchangeably to signify the decomposition of a data matrix into eigenvalues and eigenvectors.

Principal factor analysis yields an abstract solution consisting of a set of abstract eigenvectors and an associated set of abstract eigenvalues. Each principal eigenvector represents an abstract factor. In factor analysis the terms "factor" and "vector" are used interchangeably. Each eigenvalue measures the relative importance of the associated eigenvector. A large eigenvalue indicates a major factor, whereas a very small eigenvalue indicates an unimportant factor.

If the data were free of experimental error, PFA would yield exactly n eigenvectors, one for each of the controlling factors. Because of experimental error, PFA solutions in chemical problems invariably generate s eigenvectors, where s is equal to either the number of data rows, r, or the number of data columns, c, whichever is smaller. However, only n of this set of s eigenvectors, associated with the n largest eigenvalues, have physical meaning.

As shown in Section 1.3, factor analysis decomposes the data matrix into the product of an abstract row matrix and an abstract column matrix. Standard factor analytical computer programs are used to calculate the complete principal factor solution:

$$\underset{r \times c}{\mathbf{D}} = \underset{r \times s}{\mathbf{R}} \ \underset{s \times c}{\mathbf{C}} \tag{2.1}$$

Because there are s factors, the row matrix contains s columns and the column matrix contains s rows. The complete solution overspans the true factor space, involving more eigenvectors than necessary.

In SVD notation, (2.1) has the form

$$\mathbf{D}_{r\times c} = \mathbf{U}_{r\times s}\ \mathbf{S}_{s\times s}\ \mathbf{V}'_{r\times s} \tag{2.2}$$

where

$$\mathbf{U} = \mathbf{RS}^{-1} \tag{2.3}$$

$$\mathbf{V}' = \mathbf{C} \tag{2.4}$$

In these expressions $\mathbf{S}$ is a diagonal matrix whose elements are the square roots of the eigenvalues. PFA separates the data into two sets of eigenvectors, one spanning the row space and the other spanning the column space. More specifically, each column of $\mathbf{U}$ is an abstract, orthonormal eigenvector that spans the row space; and each column of $\mathbf{V}$ is an abstract, orthonormal eigenvector that spans the column space. Both sets of eigenvectors lie completely inside the s-dimensional factor space defined by s abstract factor axes. The elements in each row of $\mathbf{U}$ represent the coordinate positions of the row designee on the factor axes. Similarly, the elements in each row of $\mathbf{V}$ represent the coordinate positions of the column designee on the factor axes.

The jth eigenvector of $\mathbf{U}$ and the jth eigenvector of $\mathbf{V}$ share the same jth eigenvalue. Each eigenvalue represents a portion of the total variation in the data. Factors are ranked according to their ability to account for the variation in the data. Hence, eigenvectors of the abstract matrices are arranged in decreasing order of importance. The first pair of eigenvectors, representing the first factor, is associated with the largest, most important eigenvalue. The sth pair, associated with the smallest eigenvalue, is least important. The first factor accounts for the greatest percentage of variation in the data, the second for the next greatest percentage, and so forth, so that the complete set of s abstract factors accounts exactly for all the data, including experimental error.

2.4.2   Number of Factors

Having calculated the complete PFA solution, we seek to discover how many of the s factors are physically important. The abstract factors can be divided into two sets: a primary set of n factors, which account for the real, measurable features of the data, and a secondary set of $s - n$ factors, called the *null* set, which are associated entirely with experimental error. By eliminating the secondary factors from the initial solution, we "compress" the factor model, incorporating only the physically significant factors. After compression, (2.1) and (2.2) become

$$\overline{\mathbf{D}}_{r\times c} = \overline{\mathbf{R}}_{r\times n}\ \overline{\mathbf{C}}_{n\times c} = \overline{\mathbf{U}}_{r\times n}\ \overline{\mathbf{S}}_{n\times n}\ \overline{\mathbf{V}}'_{n\times c} \tag{2.5}$$

where the overbar indicates the deletion of the error factors. Equation (2.5) concisely expresses the properly dimensioned abstract solution. All subsequent computations in factor analysis are based on this result.

Determination of the correct factor size is the first practical dividend that we receive from the use of factor analysis. The number of real factors serves as a measure of the true complexity of the data, information seldom obtainable by other methods. If the error in the data is known, a variety of methods, based on data reproduction, can be employed to find the number of factors. If the error in the data is not known, special mathematical techniques can be used to estimate the factor size. Empirical, theoretical, and statistical methods for determining the number of real factors are explained in Chapter 4.

Stepwise procedures are used to deduce the correct number of factors. Each stage of the reproduction involves the following computation and comparison:

$$\overline{\mathbf{R}}_{r \times j} \overline{\mathbf{C}}_{j \times c} = \overline{\mathbf{D}}_{r \times c} \stackrel{?}{=} \mathbf{D}_{r \times c} \tag{2.6}$$

Here $\overline{\mathbf{R}}$ and $\overline{\mathbf{C}}$ are abstract matrices based on the j most important eigenvectors, $\overline{\mathbf{D}}$ is the data matrix reproduction using the first j abstract factors, and $\mathbf{D}$ is the original data matrix. In the first reproduction attempt, only the single most important factor ($j = 1$) is used; in the second stage of reproduction, the first and second most important factors ($j = 2$) are employed simultaneously; and so forth, until all c factors are used together in the final reproduction.

As additional factors are incorporated into (2.6), data reproduction becomes more accurate, since, cumulatively, a greater fraction of the total variation in data is accounted for. When the correct number of factors is employed ($j = n$), the reproduced data matrix, $\mathbf{D}$, should equal the original data matrix within experimental uncertainty, as determined by some statistical criterion. A certain number, n, of the complete set of s factors are required to reproduce the data within the uncertainty criterion. If too few factors are employed in the abstract factor analytical model, the data will not be reproduced with sufficient accuracy. If too many factors are used, the extra factors will reproduce experimental error and will therefore serve no useful purpose, unless one is interested in modeling error.

To illustrate the reproduction step, let us return to the absorbance problem described in Section 1.4. Carrying out a PFA on the hypothetical data matrix in (1.11), we find that two factors are required to reproduce the data within the built-in error of ± 0.002 absorbance unit. Statistical tests, based on the Fisher variance ratio and the chi-squared criterion, confirm this conclusion. Thus the factor size in this problem is two, indicating that the number of absorbing components in the mixtures is two. This result is consistent with the fact that the hypothetical data were generated from a two-component model.

When the only information required in a chemical problem is the number of factors, the preparation and reproduction steps constitute the complete factor analysis. However, in most chemical problems, the main objective is to gain insight into the nature of the factors. For this purpose the abstract solution must be transformed into a more meaningful solution.

2.5 TRANSFORMATION

Transformation of the principal factors into recognizable parameters is the most important dividend of factor analysis. As explained in Section 1.3, a transformation matrix, $\mathbf{T}$, and its inverse, denoted by $\mathbf{T}^{-1}$, are employed to carry out a transformation. The mathematical basis of the transformation is summarized by the sequence of equations

$$\overline{\mathbf{D}} = \overline{\mathbf{R}}\overline{\mathbf{C}}$$
$$= \{\overline{\mathbf{R}}\mathbf{T}\}\{\mathbf{T}^{-1}\overline{\mathbf{C}}\}$$
$$= \hat{\mathbf{X}}\hat{\mathbf{Y}} \tag{2.7}$$

Here $\hat{\mathbf{X}}$ and $\hat{\mathbf{Y}}$ correspond to the transformed row and column matrices. If the transformation is successful, these matrices will correspond to $\mathbf{X}_{real}$ and $\mathbf{Y}_{real}$ in Eq. (1.10).

Three distinctly different approaches (abstract rotation, target testing, and special methods) are employed to transform the PFA solution:

1. *Abstract factor analysis* (AFA), involving mathematical rotations.
2. *Target factor analysis* (TFA), involving target testing.
3. *Special methods* (such as evolving factor analysis (EVA), rank annihilation factor analysis (RAFA), and iterative key set factor analysis (IKSFA)), involving known chemical constraints.

Abstract rotation is the name given to a host of techniques for transforming PFA abstract matrices into other abstract matrices. Those interested in detailed discussions of abstract rotations should consult the readable monograph of Rummel.[7] *Target testing* is a unique method for testing potential factors one at a time. *Special methods* are those transformation techniques that impose known chemical or physical constraints on the system.

Transformation involves five types of factors:

1. *Principal* (eigenvectors obtained from PFA)
2. *Latent* (hidden factors common to two data matrices)
3. *Rotated* (abstract factors obtained by rotations)
4. *Typical* (rows or columns of the original data matrix)
5. *Basic* (chemically recognizable factors)

Principal, latent, and rotated factors are abstract factors, whereas typical and basic factors are real factors identified by transformation. *Principal* factors are those obtained directly from PFA solutions. Rotations of PFA matrices produce *rotated* factors in a new abstract form. By transforming the principal factors, we seek factors that will have physical meaning. *Typical* factors are simply rows or columns of the original data matrix which can be employed as factors.[8] *Basic* factors, which describe the properties of the designees in the data matrix, are the most fundamental of the four types of factors.[8] *Latent* factors are those intrinsic factors that relate one data set to another. *Latent* factors describe the subset of abstract factors that are common to two different matrices.[4] Such factors are used in partial least-squares computations, a special method, where their existence, but not their significance, is important.

2.5.1 Abstract Rotation

Principal factor solutions can be transformed into other abstract solutions by rotation of the principal factor axes. More specifically, the column matrix obtained from PFA can be transformed mathematically into a new abstract matrix by multiplication with a transformation matrix according to

$$\overline{\mathbf{R}}\mathbf{T} = \mathbf{X} \tag{2.8}$$

Here $\mathbf{T}$ is the transformation matrix required to carry out the desired rotation, $\overline{\mathbf{R}}$ is the principal factor matrix given in (2.5), and $\mathbf{X}$ is the new abstract row-factor matrix resulting from the rotation.

There are many types of rotational criteria, each involving a different mathematical principle, yielding a different solution. A number of these methods are discussed in Chapters 16 and 17 of Rummel's monograph.[7] The theory behind several, popular, abstract rotational methods (varimax, quartimax, equimax, and oblimin) is discussed in Section 3.4.2. Packages of standard computer programs, such as the *Statistical Analysis System* (SAS)[9] and the *Statistical Package for the Social Sciences* (SPSS),[10] can be employed to carry out several types of rotations.

Rotated solutions are easier to interpret than are principal factor solutions. In particular, by bringing out "clusterings" in the data, rotated factors are valuable for classifying the *objects* or *variables* in a data matrix. If a group of objects have similar scores on the abstract factors, these objects form a *cluster*. More so than in the principal factor model, specific features of the data are indicated in the rotated factors. Varimax rotation, the most popular rotation, tends to yield a few large scores, many small scores, and very few intermediate-valued scores, thereby emphasizing clusters of objects.

Because abstract factors are linear combinations of basic factors, it is difficult to relate rotated factors directly to real factors. Should one of the rotated factors be similar to a known basic factor, one is tempted to associate the abstract factor with the basic factor. However, such interpretations are fraught with uncertainty since abstract rotations cannot adequately isolate individual basic factors. Basic factors can be identified with greater assurance using target testing.

2.5.2 Target Testing

Target transformation, involving target testing, serves as a mathematical bridge between abstract and real solutions. By target testing, we can evaluate ideas concerning the nature of the factors and thereby develop physically significant models for the data.

Regardless of the complexity of the problem, we can test potential factors *individually*. Tests can be associated with either the row designees or the column designees of the data matrix by focusing attention either on the row-factor matrix or the column-factor matrix. The procedure for target testing row factors is summarized by

$$\overline{R}t = \hat{x} \overset{?}{=} x \qquad (2.9)$$

Here $\overline{R}$ is the row matrix from PFA based on n factors [see (2.5)] and the other three quantities are vectors. The target transformation vector, t, results from a least-squares operation involving the PFA solution and the individual "target" being tested, designated as x. If the test vector is a real factor, the predicted vector, $\hat{x}$, obtained from (2.9) will be reasonably similar to the test vector, confirming the idea embodied in the test vector. If the tested and predicted vectors are sufficiently dissimilar, the parameter tested is not a real factor, thus leading to rejection of the tested idea.

Target testing is carried out using special computer programs developed for this purpose.[11,12] Statistical criteria for evaluating the success or failure of a target test are available (see Section 4.6).

To illustrate the power of target testing, consider the three tests shown in Table 2.1. These tests concern the absorbance problem discussed in Section 1.4. TFA is used to identify the components in the mixtures by testing the absorbance spectra of various pure components. For each suspected component being tested, the absorptivities measured at the six wavelengths constitute the test vector, as shown in the table. Each test is carried out separately.

Since abstract reproduction indicated two factors in this problem, the target tests involved only the two most important principal factors. The least-squares vectors, obtained from TFA via (2.9), are listed as predicted vectors in the table. In tests 1 and 3, concerning *ortho-* and *para*-xylene, the point-by-point agreement between the test

TABLE 2.1   Examples of Target Testing

Wavelength (nm)	Spectrum 1 o-xylene		Spectrum 2 m-xylene		Spectrum 3 p-xylene	
	Test	Predicted	Test	Predicted	Test	Predicted
278	0.015	0.014	0.037	0.066	0.112	0.112
274	0.143	0.144	0.397	0.366	0.512	0.511
270	0.400	0.397	0.404	0.519	0.416	0.417
266	0.452	0.452	0.638	0.599	0.489	0.490
262	0.495	0.497	0.581	0.560	0.344	0.337
258	0.416	0.416	0.501	0.468	0.275	0.281
%SL	81.5		0.0		34.1	

vectors and their predicted vectors is quite good, signifying that these compounds are likely components. In test 2 the agreement between the test and predicted vectors is poor, indicating that the second test compound is not present. These conclusions are verified statistically by the %SL (the percent significance level) shown at the bottom of the table. As described in Section 4.6.3, a %SL greater than 5% (or 10%) is evidence for a valid test vector. Values of 81.5 and 34.1% for o-xylene and p-xylene, respectively, are highly significant, whereas the 0.0% value for m-xylene is not significant. These conclusions are in accord with the known facts, since the data in (1.11) were obtained from simulated, chromatographic mixtures containing compounds 1 and 3 only.

Target testing is a unique method for deciphering the chemical factors responsible for the data. Three important features of target testing are the following:

1. Each factor can be evaluated independently, even if a multitude of other factors simultaneously influence the data. The vector of interest is tested without the need to identify or specify any of the other factors. This feature, in particular, distinguishes TFA from multiple regression analysis (MRA) and from principal component regression (PCR), which require specification of all components simultaneously. The overall complexity of the data does not deter the mathematical isolation of single real factors.

2. A test vector need not be complete. Missing points in a successful test vector are automatically calculated, affording a unique way for predicting data.

3. Target factor analysis serves as both a theoretical tool and an empirical tool. Target testing can be used not only to confirm theory but also to extend and modify theoretical models. When insight into the data is lacking, target testing can be used as a guide to search, term by term, for the best empirical model.

2.5.3 Special Methods

The third classification for transforming abstract factors into chemically meaningful information is categorized as special methods. These techniques are employed when target vectors cannot be formulated properly because of a lack of information. Instead, they make use of known chemical constraints that are indigenous to the specific problem under study. Key set factor analysis (KSFA), partial least squares (PLS), evolving factor analysis (EFA), rank annihilation factor analysis (RAFA), iterative factor analysis (IFA), evolutionary factor analysis (EVOLU), and the variance diagram (VARDIA) are just a few of the methods that fall into this category.

Key set factor analysis searches for the most orthogonal set of typical rows or columns of the data matrix. Partial least squares is designed to find a linear relation between two matrices, a calibration matrix and a measurement matrix, which are related by a latent set of factors that are mutually orthogonal. The method is particularly useful for problems that have too many uncontrollable factors and for problems that lack theoretical foundations. Evolutionary factor analysis takes advantage of the constraints imposed on data matrices, which are formulated from chemical processes where each chemical species evolves and devolves in a systematic way as dictated by the chemical equilibria or kinetics involved. Although exact expressions for the target vectors may not be known, the known constraints are used to extract the chemical factors. Such valuable methods are described in detail in Chapter 6.

2.6   TARGET COMBINATION

Complete models of real factors are tested in the target combination step. By comparing the results for different sets of real factors, the best TFA solution to a problem can be determined. The combination step is mathematically equivalent to principal component regression (PCR). In the combination step the data matrix is reproduced from real factors rather than from abstract factors. A row matrix, $\mathbf{X}$, is formed from the selected set of n real factors that successfully passed the target test:

$$\mathbf{X} = [\mathbf{x}_1, \mathbf{x}_2, \ldots, \mathbf{x}_n] \tag{2.10}$$

For this combination of factors, a target transformation matrix, $\mathbf{T}$, can be constructed using the target transformation vectors for each test factor from (2.9):

$$\mathbf{T} = [\mathbf{t}_1, \mathbf{t}_2, \ldots, \mathbf{t}_n] \tag{2.11}$$

The inverse of this matrix, denoted by $\mathbf{T}^{-1}$, is then used in accord with (1.9) to transform $\overline{\mathbf{C}}$ into a new column matrix, $\hat{\mathbf{Y}}$. The combination procedure is summarized by the following equations and comparison:

$$\mathbf{X}\{\mathbf{T}^{-1}\overline{\mathbf{C}}\} = \mathbf{X}\hat{\mathbf{Y}}$$

$$= \hat{\mathbf{D}} \overset{?}{=} \mathbf{D} \tag{2.12}$$

If $\mathbf{X}$ represents all the factors in the problem, the combination-reproduced matrix, $\hat{\mathbf{D}}$, will be reasonably similar to the original data matrix, thus confirming the reliability of the tested combination of real factors.

For successful combination tests, the factors in $\mathbf{X}$ are called *key factors*. Sometimes a best set of factors is dictated from theoretical principles. Otherwise, the best set of real factors can be found by examining all or a selected combination of real factors, *n* at a time.

By testing combinations of basic factors, we try to model the data in a fundamental way, seeking a solution of the form

$$\mathbf{X}_{\text{basic}}\mathbf{Y}_{\text{basic}} = \mathbf{D} \tag{2.13}$$

where each of the factors in $\mathbf{X}_{\text{basic}}$ and $\mathbf{Y}_{\text{basic}}$ are basic factors. Equation (2.13) expresses the high point of TFA. Solutions of this type are the crown jewels of factor analysis. Although the effort required to develop basic solutions can be extensive, TFA can yield detailed models that cannot be obtained by other methods..

Key combination solutions should be justified in terms of theoretical and empirical knowledge. Carefully planned factor analyses based on sound conceptual models pay handsome dividends at this stage of the analysis. Correlating combination TFA models with chemical insight is the most satisfying end product of target factor analysis.

Although it may be too difficult to find key sets of basic factors in some problems, combination TFA of typical factors furnishes a routine procedure for compressing the data into a few key typical factors. Such combination of factors can be found by methods such as iterative key set factor analysis (IKSFA).

2.7 PREDICTION

The target prediction step furnishes practical bonuses of TFA. Missing data in test vectors as well as new data rows and new data columns can be obtained from target prediction.

Data can be predicted routinely via target testing. Test vectors need not be complete; instead, missing values for test points can be left blank, that is, *free-floated*. Since predicted vectors are always complete because of the nature of target testing, values for the free-floated points are predicted automatically in successful target tests. In this way, basic data that have not been measured can be estimated from TFA. For example, consider the first target test in Table 2.1. If the molar absorptivity of the test compound at one of the wavenumbers had not been known, that test point could have been free-floated in the target test, and its value would have been predicted accurately since the test vector is a real factor.

The combination TFA procedure described in Section 2.6 can be adapted to add new rows and new columns to the original data matrix, thus expanding the matrix. For example, to predict a datum associated with a new row designee, i, and an original column designee, k, a modified form of the basic equation is employed; namely,

$$\hat{d}_{ik}(\text{predicted}) = \sum_{j=1}^{n} x_{ij}(\text{known})\hat{y}_{jk}(\text{combination}) \qquad (2.14)$$

Column factors $\hat{y}_{jk}$ are taken from $\hat{\mathbf{Y}}$ and calculated from (2.12), but the values of the row factors x_{ij} must be known from sources independent of the factor analysis. The key sets of real factors from which $\hat{\mathbf{Y}}$ is calculated can contain either basic or typical factors.

2.8 SYNOPSIS

The main features of each step in a factor analysis are summarized in Table 2.2. Factor analysis can contribute unique answers to several of the fundamental questions that are asked in most chemical problems. Specifically, factor analysis can:

TABLE 2.2 Features of the Main Steps in Factor Analysis

Step	Purpose	Procedure	Result
1. Preparation	To obtain a matrix best suited for factor analysis	Data selection, data pretreatment	Complete data matrix in suitable form
2. Reproduction	To generate an abstract model	Principal factor analysis, abstract reproduction	Principal factor matrices, number of factors
3. Transformation			
Rotation	To interpret abstract model	Transformation into new abstract matrices	Clusterings of data
Target testing	To evaluate test factors individually	Transformation into real factors	Identification of real factors
Special	To transform into real factors	Invoke known chemical constraints	Extraction of real factors
4. Combination	To develop models from sets of real factors	Simultaneous transformation into a set of real factors	Key sets of real factors
5. Prediction	To calculate new data	Free-float missing points, employ key combination step	New target data, new data rows and columns

1. Furnish the number of factors that influence the data.
2. Identify the real factors.
3. Detect clustering of data.
4. Identify key sets of real factors.
5. Predict new data.

The specific dividends discussed in this chapter coupled with the general attributes discussed in Section 1.5 testify to the power of factor analysis.

REFERENCES

1. M. A. Sharaf, D. L. Illman, and B. R. Kowalski, *Chemometrics*, Wiley-Interscience, New York, 1986.
2. D. L. Massart, B. G. M. Vandeginste, S. N. Deming, Y. Michotte, and L. Kaufman, *Chemometrics: A Textbook*, Elsevier, Amsterdam, 1988.
3. W. H. Press, B. P. Flannery, S. Teukolsky, and W. T. Vetterling, *Numerical Recipes: The Art of Scientific Computing*, Cambridge University Press, Cambridge, 1986.
4. H. Wold, in P. Krishnaiah (Ed.), *Multivariate Analysis*, Academic, Orlando, 1966, p. 391.
5. C. L. Lawson and R. J. Hanson, *Solving Least Squares Problems*, Prentice-Hall, Englewood Cliffs, NJ, 1974.
6. R. I. Shrager, *Chemometrics Intell. Lab. Syst.*, **1**, 59 (1986).
7. R. J. Rummel, *Applied Factor Analysis*, Northwestern University Press. Evanston, IL, 1970.
8. R. W. Rozett arid E. M. Petersen, *Anal. Chem.*, **47**, 2377 (1975).
9. A. J. Barr, J. H. Goodnight, J. P. Sall, and J. T. Hewig, *Statistical Analysis System*, Statistical Analysis Institute, Raleigh, NC, 1976.
10. N. H. Nie, C. H. Hull, J. G. Jenkins, K. Steinbrenner, and D. H. Bent, *Statistical Package for the Social Sciences*, 2nd ed., McGraw-Hill, New York, 1975.
11. E. R. Malinowski, D. G. Howery, P. H. Weiner, J. R. Soroka, P. T. Funke, R. S. Selzer, and A. Levinstone, "FACTANAL," Program 320, Quantum Chemistry Program Exchange, Indiana University, Bloomington, IN, 1976.
12. E. R. Malinowski, "TARGET 90," Stevens Institute of Technology, Hoboken, NJ, 1989.

3

Every solution generates two or more new problems.

MATHEMATICAL FORMULATION OF TARGET FACTOR ANALYSIS

This chapter concerns mathematical derivations and techniques pertinent to solving chemical problems by target factor analysis.[1-3] The treatment is different from the more widely known forms of classical factor analysis developed for psychological and sociological studies.[4-9] However, the mathematics is applicable in a quite general sense.

3.1 INTRODUCTION

3.1.1 Criteria for Factor Analysis

Consider a matrix of experimental data with elements symbolized as d_{ik}, where subscript i refers to a particular row and subscript k refers to a particular column. Factor analysis is applicable whenever a measurement can be expressed as a linear sum of product terms of the form

$$d_{ik} = \sum_{j=1}^{n} r_{ij} c_{jk} \tag{3.1}$$

where r_{ij} is the jth factor associated with row i, and c_{jk} is the jth factor associated with column k. The number of terms in the summation is n, the number of factors that adequately account for the measurement in question. This equation has the same mathematical form as that used in multiple linear regression analysis where d_{ik} is the dependent variable, the c_{jk}'s are the regression coefficients, and the r_{ij}'s are the independent variables. In comparison with regression analysis, the factor analytical method has many inherent advantages, which will become apparent during the development.

Attention is focused on either the row designees or the column designees. Where attention is focused is called the *scores*; the counterpart is called the *loadings*. For example, if r_{ij} is the score, then c_{jk} is the loading. Alternatively, if c_{jk} is the score, then r_{ij} is the loading.

We write **D**, **R**, and **C** as matrices whose elements are d_{ik}, r_{ij}, and c_{jk}, respectively. The pair of subscripts designates the exact location of the elements in the matrix, specifically, the respective row and column. From (3.1) and the definition of matrix multiplication, we may write

$$\mathbf{D} = \mathbf{RC} \tag{3.2}$$

where **D** is the experimental data matrix. In general, **R** is called the row matrix since this matrix is associated with the row designees of the data matrix. Similarly, **C** is called the column matrix. In general, **R** and **C** are not square matrices.

Using language developed by the pioneers of factor analysis, if we focus attention on **R**, we would call **R** the score matrix and **C** the loading matrix. If we focus attention on **C**, we would call **C** the score matrix and **R** the loading matrix. Because this terminology can be quite confusing, we tend to favor the simplest and most specific designations: row matrix and column matrix.

3.1.2 Mathematical Synopsis

A brief overview of the mathematical steps of abstract factor analysis (AFA) and target factor analysis (TFA) is presented in this section. The key steps are shown in

Figure 3.1. The problem to be solved by factor analysis is simply this: From a knowledge of **D**, find various sets of **R**'s and **C**'s that reproduce the data in accord with (3.2). This is accomplished by the procedure described below.

The first step in FA involves decomposing the data matrix into eigenvectors and eigenvalues. This can be done directly as described in Section 3.3.8. However, for pedagogical reasons, we make use of the covariance or correlation matrix. (The difference between covariance and correlation is discussed in Section 3.2.1.) By standard mathematical techniques the covariance or correlation matrix is decomposed into a set of "abstract" factors, which, when multiplied together, reproduce the original data. These factors are called abstract because, although they have mathematical meaning, they have no real physical or chemical meaning in their present forms. Target transformation enables us to convert these factors into physically significant parameters that reproduce the experimental data.

The covariance matrix **Z** is constructed by premultiplying the data matrix by its transpose:

$$\mathbf{Z} = \mathbf{D}'\mathbf{D} \tag{3.3}$$

This matrix is then diagonalized by finding a matrix **Q** such that

$$\mathbf{Q}^{-1}\mathbf{Z}\mathbf{Q} = [\lambda_j \delta_{jk}] = \mathbf{\Lambda} \tag{3.4}$$

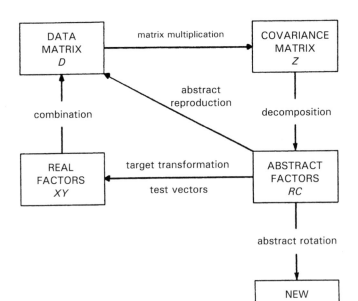

Figure 3.1   Key steps in factor analysis.

Here δ_{jk} is the well-known Kronecker delta,

$$\delta_{jk} = \begin{cases} 0 & \text{if } j \neq k \\ 1 & \text{if } j = k \end{cases}$$

and λ_j is an eigenvalue of the set of equations

$$\mathbf{Z}\mathbf{q}_j = \lambda_j \mathbf{q}_j \tag{3.5}$$

where $\mathbf{q}_j$ is the jth column of $\mathbf{Q}$. These columns, called *eigenvectors*, constitute a mutually orthonormal set. Hence

$$\mathbf{Q}^{-1} = \mathbf{Q}' \tag{3.6}$$

The following shows that $\mathbf{Q}'$ is identical to $\mathbf{C}$:

$$\begin{aligned} \mathbf{Q}^{-1}\mathbf{Z}\mathbf{Q} &= \mathbf{Q}^{-1}\mathbf{D}'\mathbf{D}\mathbf{Q} \\ &= \mathbf{Q}'\mathbf{D}'\mathbf{D}\mathbf{Q} \\ &= \mathbf{P}'\mathbf{P} \end{aligned} \tag{3.7}$$

where

$$\mathbf{P} = \mathbf{D}\mathbf{Q} \tag{3.8}$$

Solving (3.8) for $\mathbf{D}$ yields

$$\mathbf{D} = \mathbf{P}\mathbf{Q}' \tag{3.9}$$

By comparing this with (3.2), we see that

$$\mathbf{Q}' = \mathbf{C} \tag{3.10}$$
$$\mathbf{P} = \mathbf{R} \tag{3.11}$$

Thus the transpose of the matrix that diagonalizes the covariance matrix represents the column matrix. Since each row of this matrix is an eigenvector, this matrix is generally called the *eigenvector matrix*. Furthermore, because the eigenvectors are orthonormal

$$\mathbf{C}^{-1} = \mathbf{C}' \tag{3.12}$$

The row matrix can be calculated from (3.8) and (3.11). Finally, the data matrix can be reproduced from $\mathbf{R}$ and $\mathbf{C}$. Thus the abstract reproduction (see Figure 3.1) is readily accomplished.

The problem, however, is to reproduce the data within experimental error with a minimum of eigenvectors. In general, not all eigenvector columns of $\mathbf{Q}$ are required. Section 3.3 shows that the magnitude of the eigenvalue is a gauge of the importance of the corresponding eigenvector. Eigenvectors associated with the largest eigenvalues are most important, whereas eigenvectors associated with the smallest eigenvalues are least important. In practice, the least important eigenvectors are negligible and are dropped from the analysis. In fact, their retention would simply regenerate experimental error. Unnecessary reproduction of error by the inclusion of too many eigenvectors is contrary to the original intent of factor analysis, which is to reproduce the data within experimental error using the minimum number of eigenvectors. Nothing is gained by retaining an excessive number of eigenvectors. The dropping of unwanted eigenvectors is called *factor compression*.

As a first trial, the data matrix is recalculated using only the most important eigenvector, $\mathbf{q}_1$, associated with the largest eigenvalue, λ_1. In accord with (3.10), $\mathbf{c}_1 = \mathbf{q}_1$. Premultiplying $\mathbf{c}_1'$ by $\mathbf{r}_1$, outer-product-wise, yields a recalculated data matrix that is compared to the original data matrix. If the agreement is within experimental error, only one factor is important. If not, a second trial is made using the two most important eigenvectors (associated with the two largest eigenvalues, λ_1 and λ_2), letting $\mathbf{R} = [\mathbf{r}_1 \quad \mathbf{r}_2]$ and $\mathbf{C} = [\mathbf{c}_1 \quad \mathbf{c}_2]'$. Again we ascertain whether or not the reproduced data adequately represent the original data. We continue adding eigenvectors associated with the largest eigenvalues, sequentially, until the data are reproduced satisfactorily:

$$[\mathbf{r}_1 \quad \mathbf{r}_2 \quad \cdots \quad \mathbf{r}_n] \begin{bmatrix} \mathbf{c}_1' \\ \mathbf{c}_2' \\ \vdots \\ \mathbf{c}_n' \end{bmatrix} = \overline{\mathbf{R}}\,\overline{\mathbf{C}} = \overline{\mathbf{D}} \simeq \mathbf{D} \qquad (3.13)$$

The overbar indicates that only n important (primary) eigenvectors are employed in constructing the row-factor and column-factor matrices. $\overline{\mathbf{D}}$ is the reproduced data matrix, based on n eigenvectors. The minimum number of eigenvectors, n, required to reproduce the data within experimental error represents the number of factors involved. This number also represents the "dimensionality" (rank or size) of the factor space.

At this stage of the scheme, the abstract reproduction (see Figure 3.1) is properly completed. However, in their present forms the row and column matrices represent mathematical solutions to the problem and are not recognizable in terms of any known chemical or physical quantities. Nevertheless, in their abstract mathematical forms, these factors contain useful information. Examples of the utility of the abstract reproduction scheme are presented in detail later in the book.

Most often the chemist is not interested in the abstract factors produced by AFA but is interested in finding physically significant factors that are responsible for the data. Factor analysis can be used in a powerful and unique fashion for such purposes. This is achieved by transforming the abstract eigenvectors into vectors that have physical meaning. A method called *target transformation* allows us to test and develop theoretical and empirical hypotheses concerning the fundamental nature of the variables at play.

Mathematical transformation is accomplished by performing the following matrix multiplication:

$$\hat{\mathbf{X}} = \overline{\mathbf{R}}\mathbf{T} \tag{3.14}$$

Here $\mathbf{T}$ is the appropriate transformation matrix and $\hat{\mathbf{X}}$ is the row matrix in the new coordinate system. The corresponding column matrix (the eigenvector matrix) is obtained as follows:

$$\hat{\mathbf{Y}} = \mathbf{T}^{-1}\overline{\mathbf{C}} \tag{3.15}$$

This is true since

$$\overline{\mathbf{D}} = \overline{\mathbf{R}}\mathbf{C}$$
$$= \overline{\mathbf{R}}\mathbf{T}\mathbf{T}^{-1}\overline{\mathbf{C}}$$
$$= \hat{\mathbf{X}}\hat{\mathbf{Y}} \tag{3.16}$$

Transformation matrices can be obtained by many different methods. The transformation of special interest to chemists is target transformation. This method consists of finding a set of transformation vectors by means of a least-squares method. Each transformation vector is a column of the transformation matrix. Target transformation refers to the procedure of testing a single physical or structural concept. By means of target transformation, we can test theoretical concepts and obtain physically significant parameters that reproduce the data (see Figure 3.1). Details of this feature are presented in Section 3.4.

3.2 PRELIMINARY CONSIDERATIONS

3.2.1 Constructing the Covariance Matrix

The covariance matrix is obtained by premultiplying the data matrix by its transpose:

$$\mathbf{Z} = \mathbf{D}'\mathbf{D} \tag{3.17}$$

Although the data matrix is not necessarily square, the covariance matrix is square. If the data matrix has r rows and c columns, the covariance matrix will be of size $c \times c$. If the covariance matrix is constructed by postmultiplying the data by its transpose,

$$\mathbf{Z} = \mathbf{D}\mathbf{D}' \tag{3.18}$$

then the covariance matrix will be of size $r \times r$. Either definition can be used and both will produce compatible results in the final analysis.

These definitions automatically place a latent statistical bias on the analysis. Each data point is inherently weighted in proportion to its absolute value. Large data points are given more statistical importance than are smaller values. Such bias is desirable when all the measurements are made relative to an absolute standard and all data points bear the same absolute uncertainty.

In certain problems we may wish to give equal statistical weight to each column of data. This is accomplished by normalizing each column of data. Normalization is carried out by dividing every element in a given data column by the norm of the column (i.e., by the square root of the sum of squares of all elements in the column). Normalization is warranted when the experimental error is directly proportional to the magnitude of the measurement, when each column of data involves a different property having significantly different orders of magnitude, or when the reference standard is arbitrary, thus prejudicing the magnitudes of the data points. This procedure is known as *correlation* and the resulting correlation matrix is denoted by $\mathbf{Z}_N$.

A thorough discussion of covariance and correlation, as well as other interesting aspects of the statistical nature of factor analysis, can be found in the text by Rummel.[4] In the mathematical development it is not necessary to constantly make a distinction between the original data matrix and the normalized data matrix because such distinction is not critical to the overall understanding of the mathematical steps that follow. If normalized data are used in the analysis, a normalized data matrix will emerge from the computations upon completion of the analysis.

3.2.2   Effect of Transposing the Data Matrix

If the size of the data matrix is $r \times c$ (r rows and c columns), the size of the covariance matrix will be $c \times c$. However, if the data matrix is transposed (i.e., rows and columns are interchanged) prior to forming the covariance matrix, the covariance matrix will be of size $r \times r$. Carrying out the factor analysis on either of these matrices will yield the same conclusion concerning the dimensionality of the factor space.

If covariance is used, then exactly the same eigenvectors and eigenvalues will emerge from the factor analysis regardless of whether the data matrix or its transpose is involved. This is not true if the correlation matrix is used instead of covariance.

3.2.3   Data Preprocessing

Four common methods of data processing have been used in factor analysis studies: (1) covariance about the origin, (2) covariance about the mean, (3) correlation about the origin, and (4) correlation about the mean. The relationship among these different methods depends on a simple linear transformation:

$$\mathbf{D}^{\#} = \mathbf{DA} + \mathbf{B} \tag{3.19}$$

Here $\mathbf{D}^{\#}$ rather than $\mathbf{D}$ represents the processed data that are subjected to factor analysis. The four methods differ in the definitions of $\mathbf{A}$ and $\mathbf{B}$: $\mathbf{A}$ is a diagonal matrix that adjusts the overall magnitude of each data column, consisting of diagonal elements a_{jj} only; $\mathbf{B}$ is a matrix in which all the b_{ij} elements in any one column are identical. This matrix shifts the origin of the factor space.

1. For covariance about the origin, C_o:

$$a_{jj} = 1 \qquad b_{ij} = 0 \tag{3.20}$$

2. For covariance about the mean, C_m:

$$a_{jj} = 1 \qquad b_{ij} = -\overline{d}_{.j} \tag{3.21}$$

3. For correlation about the origin, R_o, also known as *autoscaling*:

$$a_{jj} = \left(\sum_{i=1}^{r} d_{ij}^2 \right)^{-1/2} \qquad b_{ij} = 0 \tag{3.22}$$

4. For correlation about the mean, R_m:

$$a_{jj} = \left(\sum_{i=1}^{r} (d_{ij} - \overline{d}_{.j})^2 \right)^{-1/2} \qquad b_{ij} = -\overline{d}_{.j} a_{jj} \tag{3.23}$$

Here $\overline{d}_{.j}$, is the average value of the experimental data points of the jth column of the raw data matrix, and r is the total number of points in a column of data.

Rozett and Petersen[10] discussed the advantages and disadvantages of these four methods. In chemistry, covariance about the origin is preferred because it preserves the origin of the factor space, the relative lengths of the factor axes, and the relative error—a most desirable situation. By using covariance or correlation about the mean, we lose information concerning the zero point of the experimental scale. The addition of more data points will shift the origin of the factor space. By using correlation (about the origin or about the mean), we lose information concerning the relative size and relative error associated with the various data columns.

These four preprocessing techniques scale the elements in the data according to their columns. Analogous expressions can be applied to scale the elements according to their rows. An example of this is the convention introduced by geologists who represent the samples as data columns and the properties as data rows. According to this convention,[11] sample (column) pretreatment is called Q-mode factor analysis, and property (row) pretreatment is called R-mode factor analysis. For the Q-mode each data point is divided by the norm of its associated column vector: that is,

$$d_{ij}^{\#}(\text{Q-mode}) = d_{ij}/\|d_{\cdot j}\| \tag{3.24}$$

For the R-mode each data point is divided by the norm of its associated row vector: that is,

$$d_{ij}^{\#}(\text{R-mode}) = d_{ij}/\|d_{i\cdot}\| \tag{3.25}$$

Q-mode analysis attempts to identify groups of samples (columns) that are similar to each other. R-mode analysis identifies groups of properties (rows) that are inter-related in some fashion. The reason for choosing Q- or R-mode correlations is to compensate for different units and different orders of magnitude between the columns or between the rows. Q-mode analysis is equivalent to (3.22). R-mode analysis is equivalent to transposing the data matrix and then applying (3.22).

Other types of preprocessing procedures have also been invoked, such as range scaling and autoscaling. *Range scaling* converts the measurement to a relative position between the maximum and minimum values in the respective data row or column so the scaled value lies between 0 and 1. This is done by the following conversion:

$$d_{ik}(\text{scaled}) = \frac{d_{ik} - d_{\cdot k}(\text{min})}{d_{\cdot k}(\text{max}) - d_{\cdot k}(\text{min})} \tag{3.26}$$

Unfortunately, this method is supersensitive to outliers. There are two versions of *autoscaling*. One version, called *autoscaling to unit length*, is identical to correlation about the origin, involving mean centering and standardizing to unit length. The other version, called *autoscaling to unit variance*, involves mean centering and standardizing to unit variance by dividing by the standard deviation, s_k. For example, for the elements in the kth column,

$$d_{ik}(\text{autoscaled}) = \frac{d_{ik} - \overline{d}_{\cdot k}}{s_k} \tag{3.27}$$

where

$$s_k = \left(\frac{1}{r-1}\sum_{i=1}^{r}(d_{ik} - \overline{d}_{\cdot k})^2\right)^{1/2} \tag{3.28}$$

In many analytical techniques the measured data are normalized to a constant sum, which "closes" the data. In mass spectrometry, for example, the fragment peaks are usually reported relative to the largest peak set to 100. According to

Johansson, Wold and Sjodin,[12] such closure introduces a bias that may greatly affect the final outcome. Unfortunately, there exists no method that can "open" such data. Normalization should be used with caution and should be based on prior information about the raw data structure and the objective of the investigation.

Other data preprocessing methods include algebraic transformations of each data point, such as logarithms, exponentials, squares, square roots, and reciprocals. Preferably, these transformations should be based on scientific principles. The success of factor analysis often depends on the judicious choice of the data preprocessing scheme, an aspect that cannot be overemphasized.

3.2.4   Vector Interpretation

An insight into the overall operational details of factor analysis can be obtained from a vector viewpoint. From this perspective, the columns of the data matrix are viewed as vectors. The elements of the covariance matrix are generated by taking dot products (scalar products) of every pair of columns in the data matrix [see (3.17)]. On the other hand, to form the correlation matrix, each column of data is normalized before taking the scalar products. Each element of the correlation matrix represents the cosine of the angle between the two respective data column vectors. The diagonal elements are unity since they are formed by taking dot products of the vectors onto themselves.

If n eigenvectors are needed to reproduce the data matrix, all column vectors will lie in n space, requiring n orthogonal reference axes. This is best understood by studying a specific example. For this purpose consider a normalized data matrix, consisting of four data columns ($d_{.1}$, $d_{.2}$, $d_{.3}$, and $d_{.4}$), generated from two factors. From this matrix (not shown here), the following correlation matrix is formed:

$$
Z_N = \begin{array}{c} \\ d_{.1} \\ d_{.2} \\ d_{.3} \\ d_{.4} \end{array}
\begin{array}{cccc}
\mathbf{d}_{.1} & \mathbf{d}_{.2} & \mathbf{d}_{.3} & \mathbf{d}_{.4} \\
\left[\begin{array}{cccc}
1.00000 & 0.06976 & -0.58779 & 0.80902 \\
0.06976 & 1.00000 & 0.76604 & 0.64279 \\
-0.58779 & 0.76604 & 1.00000 & 0.00000 \\
0.80902 & 0.64279 & 0.00000 & 1.00000
\end{array}\right]
\end{array}
\qquad (3.29)
$$

The elements of this matrix are the cosines of the angles between the data column vectors. From the numerical values given in (3.29), the graphical representation illustrated in Figure 3.2 can be constructed. All four data vectors lie in a common plane. The problem is two-dimensional (i.e., only two factors are involved).

Each vector axis in Figure 3.2 corresponds to a column designee of the normalized data matrix. Each row designee of the normalized data matrix is represented by a point in the two-dimensional plane. A typical row-designee point is portrayed in the figure. The value of a data point associated with a given row and column is obtained by first drawing a line through the row-designee point, perpendicular to the appropriate column vector, and then reading the distance along the vector from the origin to the intersection. These projections are the normalized data values since the

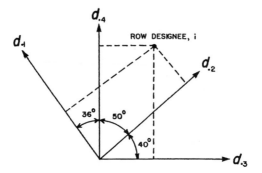

Figure 3.2   Vector relationships of data-column vectors used in forming the correlation of (3.29). Data points are obtained by the perpendicular projections of a row-designee point onto the respective column-designee vectors, $d_{.1}$, $d_{.2}$, $d_{.3}$, and $d_{.4}$.

vector axes represent normalized data columns. In a more general sense, the value of a point projected onto any axis is called the *projection*.

If one of the vectors in the figure did not lie in the plane, the space would be three dimensional. Three factors would be required to account for the data. Three axes would be required to locate the data points in the factor space. The rank of the data matrix would be three. The choice of reference axes is not totally arbitrary. If vector $d_{.1}$ projected out of the plane, for example, vector $d_{.1}$ and any two of the other three vectors could be used as reference axes. Vectors $d_{.2}$, $d_{.3}$, and $d_{.4}$ could not be used as reference axes since they lie in a common plane and do not span the three-dimensional space.

In many problems the factor space has more than three dimensions. It is impossible to sketch such multidimensional situations on two-dimensional graph paper. However, it is possible to extract all the necessary information with the aid of a computer. By using factor analysis the exact dimensions of the factor space can be determined. The eigenvectors that emerge from factor analysis span the factor space but do not coincide with the data vectors; they merely define the factor space in which all the experimental data points coexist.

When the correlation matrix of (3.29) is subjected to the decomposition (see Figure 3.1), two mutually orthogonal eigenvectors, c_1 and c_2, and their associated eigenvalues, λ_1 and λ_2, emerge:

$$\mathbf{c}_1 = \begin{bmatrix} c_{11} \\ c_{12} \\ c_{13} \\ c_{14} \end{bmatrix} = \begin{bmatrix} 0.5084 \\ 0.5084 \\ 0.0847 \\ 0.6899 \end{bmatrix} \qquad \mathbf{c}_2 = \begin{bmatrix} c_{21} \\ c_{22} \\ c_{23} \\ c_{24} \end{bmatrix} = \begin{bmatrix} 0.4909 \\ -0.4909 \\ -0.7144 \\ 0.0877 \end{bmatrix}$$

$$\lambda_1 = 2.070 \qquad\qquad \lambda_2 = 1.930$$

The c_{jk} coefficients that define the eigenvectors also measure the importance of each eigenvector on each data column. In general, it can be shown that each of the four data-column vectors can be expressed in terms of the basic eigenvectors in the following way:

$$\mathbf{d}_{\cdot k} = \sum_{j=1}^{n} \sqrt{\lambda_j} c_{jk} \mathbf{c}_j \qquad (3.30)$$

Accordingly, the four data-column vectors may be expressed in terms of the two eigenvectors:

$$\begin{aligned}
\mathbf{d}_{\cdot 1} &= 0.7314\mathbf{c}_1 + 0.6819\mathbf{c}_2 \\
\mathbf{d}_{\cdot 2} &= 0.7314\mathbf{c}_1 - 0.6819\mathbf{c}_2 \\
\mathbf{d}_{\cdot 3} &= 0.1218 c_1 - 0.9926 c_2 \\
\mathbf{d}_{\cdot 4} &= 0.9926\mathbf{c}_1 + 0.1218\mathbf{c}_2
\end{aligned} \qquad (3.31)$$

The validity of these equations can be verified by taking dot products between the vectors and comparing the results with the correlation matrix given in (3.29).

In a general sense, the $\sqrt{\lambda_j} c_{jk}$ coefficients are called *factor loadings* or simply *loadings*. The loadings are a measure of the relative importance of each eigenvector on each of the data-column vectors. They are similar to the "weightings" in regression analysis.

The geometrical relationship between the two eigenvectors and the four data-column vectors can be obtained by taking dot products between the eigenvectors and the data vectors as expressed in (3.31). Recall that the dot product equals the cosine of the angle between the two normalized vectors. The relationships so obtained are illustrated in Figure 3.3, from which the following observations can be made. The projections of each data vector onto the two eigenvectors represent the loadings of the eigenvectors on the respective data-column vector. These loadings are the coefficients in (3.31).

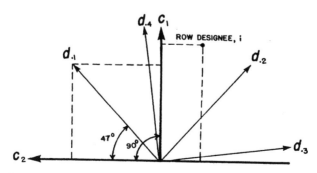

Figure 3.3 Geometrical location of eigenvectors $\mathbf{c}_1$ and $\mathbf{c}_2$ resulting from factor analysis. Dashed lines show the factor loadings of $\mathbf{c}_1$ and $\mathbf{c}_2$ on $\mathbf{d}_{\cdot 1}$, and the scores (projections) of row-designee i on factors $\mathbf{c}_1$ and $\mathbf{c}_2$.

Next, let us study the points that lie in the factor space. Recall that each row designee of the data matrix corresponds to a point in the factor space. Since there are r row designees, there are r points. For our example, all these points lie in the plane. The projection of a point onto a basis axis gives its score on the axis. Figure 3.3 shows the projections of one such point, i, projected on c_1 and on c_2. These projections are the values for elements r_{i1} and r_{i2}, respectively, in the row matrix. The row matrix, $\mathbf{R}$, is composed of all such projections onto the eigenvector axes.

Because there are only two basic axes, all the data information associated with this problem can be compressed and stored on two factors. For example, to locate a new row-designee point in the factor plane, only two measurements involving only two column designees are required. The projections of this point onto the two remaining column vectors yield predictions for the behavior of the new row designee.

The example given here is superficially small. In actual practice, the data matrix will consist of many more data columns. If the data contained 25 columns and the factor space were four dimensional, for example, then four measurements would be required to locate a row-designee point in the factor space. The projections of this single point onto the remaining 21 column-designee axes would yield 21 predicted values. This is interesting when we realize that these predictions are made in an abstract manner requiring no information about the true controlling factors. Considering the enormous number of chemicals that could comprise the rows, this feature of AFA offers useful possibilities to the chemist.

Equations (3.31) represent abstract solutions to the problem because the eigenvectors, c_1 and c_2, are devoid of any real physical or chemical meaning. For many chemical problems such solutions are sufficient and the factor analytical study is terminated at this point. However, we search for the real axes because there are many advantages to be gained if we succeed, as discussed in latter chapters. The real axes can often be found by using target transformation, described in Section 3.4. Target transformation allows us to test our ideas of what the real factors may be. Figure 3.4 illustrates the location of two hypothetical real factors, the dipole moment, μ, and the

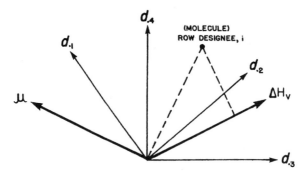

Figure 3.4   Two real factors (dipole moment, μ, and heat of vaporization, ΔH_v) responsible for four different data columns, $\mathbf{d}_{.1}$, $\mathbf{d}_{.2}$, $\mathbf{d}_{.3}$, and $\mathbf{d}_{.4}$.

heat of vaporization, ΔH_v. The projection of the row-designee points associated with various chemicals onto these two real axes yields predictions for the dipole moments and heats of vaporization. For the molecule shown in Figure 3.4, the dipole moment is predicted to be zero and the heat of vaporization is given by the distance along the ΔH_v axis. By identifying the real axes, we automatically arrive at a new method for predicting values associated with the real axes. This is another valuable feature of TFA.

3.3 FACTOR COMPRESSION AND DATA REPRODUCTION

3.3.1 Decomposition

The rank of the covariance or correlation matrix, and hence the dimensionality of the factor space, is determined mathematically by decomposing the covariance or correlation matrix into a set of eigenvectors. If the data were pure, exactly n eigenvectors would emerge. However, because of experimental error, the number of eigenvectors will equal either r, the number of rows, or c, the number of columns in the data matrix, whichever is smaller. For convenience we assume throughout this chapter that c is smaller than r and therefore that c eigenvectors emerge. Let $\mathbf{f}_1$, $\mathbf{f}_2, \ldots, \mathbf{f}_c$ be the basis set of unit vectors that defines the factor space [i.e., $\mathbf{f}_1' = (1, 0, 0, \ldots, 0)$, $\mathbf{f}_2' = (0, 1, 0, \ldots, 0)$, $\mathbf{f}_3' = (0, 0, 1, \ldots, 0)$, etc.]. These factor axes are orthonormal, so that

$$\mathbf{f}_j'\mathbf{f}_k = \delta_{jk} \tag{3.32}$$

where δ_{jk} is the Kronecker delta.

A data point can be viewed as a vector in factor space. Hence, instead of (3.1), a better representation is the following:

$$\mathbf{d}_{ik} = \sum_{j=1}^{c} r_{ij}\mathbf{f}_j c_{jk} \tag{3.33}$$

where $\mathbf{d}_{ik}$ is the data point vector.

The z_{ab} element of the covariance matrix is given by

$$z_{ab} = \sum_{i=1}^{r} \mathbf{d}_{ia}'\mathbf{d}_{ib} = \sum_{i=1}^{r} \left(\sum_{j=1}^{c} r_{ij}\mathbf{f}_j' c_{ja} \right) \left(\sum_{k=1}^{c} r_{ik}\mathbf{f}_k c_{kb} \right)$$

$$= \sum_{i=1}^{r} \sum_{j=1}^{c} r_{ij}^2 c_{ja} c_{jb} \tag{3.34}$$

The last step in (3.34) is true because $\mathbf{f}_j'\mathbf{f}_k = \delta_{jk}$. This equation shows that the entire covariance matrix can be decomposed into a sum of dyads times their corresponding eigenvalues:

$$\mathbf{Z} = \sum_{i=1}^{n} r_{i1}^2 \begin{bmatrix} c_{11} \\ c_{12} \\ \vdots \\ c_{1c} \end{bmatrix} [c_{11} \quad c_{12} \quad \cdots \quad c_{1c}]$$

$$+ \sum_{i=1}^{n} r_{i2}^2 \begin{bmatrix} c_{21} \\ c_{22} \\ \vdots \\ c_{2c} \end{bmatrix} [c_{21} \quad c_{22} \quad \cdots \quad c_{2c}]$$

$$+ \cdots + \sum_{i=1}^{n} r_{ic}^2 \begin{bmatrix} c_{c1} \\ c_{c2} \\ \vdots \\ c_{cc} \end{bmatrix} [c_{c1} \quad c_{c2} \quad \cdots \quad c_{cc}] \qquad (3.35)$$

In simpler notation (3.35) can be written

$$\mathbf{Z} = \lambda_1 \mathbf{c}_1 \mathbf{c}_1' + \lambda_2 \mathbf{c}_2 \mathbf{c}_2' + \cdots + \lambda_c \mathbf{c}_c \mathbf{c}_c' \qquad (3.36)$$

in which

$$\lambda_j = \sum_{i=1}^{r} r_{ij}^2 = \|\mathbf{r}_j\|^2 \qquad (3.37)$$

$$\mathbf{c}_j' = [c_{j1} \quad c_{j2} \quad \cdots \quad c_{jc}] \qquad (3.38)$$

In (3.37) λ_j is the eigenvalue associated with eigenvector $\mathbf{c}_j$. Note that the eigenvalue is the sum of squares of the projections of the row points onto the associated factor axis and hence represents the relative importance of the eigenvector.

Since there are c columns of data, c eigenvectors will emerge from the decomposition process (assuming that $c \leq r$). As shown in Chapter 4, only n of these (i.e., those associated with the largest eigenvalues) are required to account for the data. The remaining $(c - n)$ eigenvectors are the result of experimental error. If there were no errors in the data, exactly n eigenvectors would emerge. The number of important eigenvectors represents the dimensions of the factor space. Mathematically, this means that the true rank of the data matrix is equal to the dimensions of the factor space.

3.3.2 Principal Factor Analysis

There are a variety of mathematical ways to decompose the covariance matrix. Principal factor analysis (PFA), also known as principal component analysis (PCA), is by far the most widely used method. In PFA the eigenvectors are consecutively calculated so as to minimize the residual error in each step. Thus each successive eigenvector accounts for a maximum of the variation in the data. This section traces through the derivation in order to illustrate the philosophical and mathematical principles involved.

To locate the abstract eigenvectors, the following mathematical reasoning is involved. The principal eigenvectors constitute an optimized orthogonal coordinate system. Each successive eigenvector accounts for the maximum possible variance in the data. [A formal definition of variance is given by (4.67).] The eigenvector associated with the largest, most important eigenvalue is oriented in factor space so as to account in a least-squares sense for the greatest possible variance in the data. This primary vector passes through the greatest concentration of data points. The first eigenvector defines the best one-factor model for the data. Typically, the first factor accounts for a major fraction of the variance in the data.

This calculational procedure is continued stepwise until all c axes have been located. To find each succeeding axis, two conditions are imposed: (1) As much variance as possible is accounted for by each factor, and (2) the newest axis is mutually orthogonal to the set of axes already located. Each successive factor is responsible for a smaller fraction of the total variance in the data. The second principal axis is orthogonal to the first eigenvector and points in a direction that accounts for, as much as possible, the variation not accounted for by the first factor. The first two factors define a plane passing through the greatest concentration of data points. Points in the plane are specified by two coordinate axes in a two-factor model.

Eigenvectors associated with successively smaller eigenvalues account for less and less variance in the data. The smallest set of eigenvalues account for experimental error only. The cth eigenvector is situated in factor space so as to account for the last bit of experimental error. The complete set of c eigenvectors accounts exactly for every bit of data, including experimental error.

To keep track of the factors, we use parentheses to indicate the number of factors being considered. For example, $d_{ik}(m)$ is the reproduced data point in the ith row and the kth column calculated from the first m principal factors. Hence

$$d_{ik}(m) = \sum_{j=1}^{m} r_{ij}c_{jk} \tag{3.39}$$

where the sum is taken over the first m principal factors. The first principal component (factor) is obtained as follows. First, the residual error, $e_{ik}(1)$, is defined to be the difference between the experimental data point, d_{ik}, and the reproduced data point, $d_{ik}(1)$, based on one factor (the first principal factor):

$$e_{ik}(1) = d_{ik} - d_{ik}(1) \tag{3.40}$$

Inserting (3.39) into (3.40) gives

$$e_{ik}(1) = d_{ik} - r_{i1}c_{1k} \qquad (3.41)$$

To minimize the residual error, the method of least squares is applied. Accordingly. we take the derivative of the sum of squares of each residual error with respect to either the row factors or the column factors, depending on which type of designee we wish to focus attention. If we take the derivatives with respect to the row factors, we sum over c columns. On the other hand, if we take the derivatives with respect to the column factors, we sum over the r rows. The latter procedure yields

$$\sum_{i=1}^{r} \frac{de_{ik}^2(1)}{dc_{1k}} = 2c_{1k}\sum_{i=1}^{r} r_{i1}^2 - 2\sum_{i=1}^{r} r_{i1}d_{ik} \qquad (3.42)$$

According to the least-squares principle, this sum is set equal to zero, giving

$$\sum_{i=1}^{r} r_{i1}d_{ik} = c_{1k}\sum_{i=1}^{r} r_{i1}^2 \qquad (3.43)$$

Since k varies from 1 to c, there are c equations of this form, which in matrix notation can be expressed as

$$\mathbf{r}_1'\mathbf{D} = \mathbf{c}_1'\mathbf{r}_1'\mathbf{r}_1 \qquad (3.44)$$

Consistent with (3.37), λ_1 is defined by

$$\lambda_1 = \mathbf{r}_1'\mathbf{r}_1 = \sum_{i=1}^{r} r_{i1}^2 \qquad (3.45)$$

Inserting (3.45) into (3.44) and transposing give

$$\mathbf{D}'\mathbf{r}_1 = \lambda_1\mathbf{c}_1 \qquad (3.46)$$

According to (3.13), the complete data matrix can be written as

$$\mathbf{D} = \mathbf{r}_1\mathbf{c}_1' + \mathbf{r}_2\mathbf{c}_2' + \cdots + \mathbf{r}_c\mathbf{c}_c' \qquad (3.47)$$

Since the sum is taken over all c eigenvectors, this equation concerns $\mathbf{D}$, the complete data matrix, including experimental error, and not $\overline{\mathbf{D}}$, which involves summing over only n principal eigenvectors. Postmultiplying (3.47) by $\mathbf{c}_1$ and setting $\mathbf{c}_i'\mathbf{c}_j = \delta_{ij}$, so that the eigenvectors are orthonormal, we obtain

$$\mathbf{D}\mathbf{c}_1 = \mathbf{r}_1 \qquad (3.48)$$

Inserting (3.48) into (3.46) gives

$$\mathbf{D'Dc}_1 = \lambda_1 \mathbf{c}_1 \tag{3.49}$$

From the definition of the covariance matrix [see (3.17)], (3.49) leads to

$$\mathbf{Zc}_1 = \lambda_1 \mathbf{c}_1 \tag{3.50}$$

As described in Section 3.3.3, this expression is used to calculate the first principal eigenvector, $\mathbf{c}_1$, and its associated eigenvalue, λ_1.

The second principal component is obtained by considering the second residual error,

$$e_{ik}(2) = d_{ik} - d_{ik}(2) \tag{3.51}$$

which, from the definition given by (3.39), can be expressed as

$$e_{ik}(2) = e_{ik}(1) - r_{i2}c_{2k} \tag{3.52}$$

The error in the second principal component is minimized by applying the method of least squares to $e_{ik}(2)$ while keeping $e_{ik}(1)$ constant. This produces an expression analogous to (3.42), namely,

$$\sum_{i=1}^{r} \frac{de_{ik}^2(2)}{dc_{2k}} = 2c_{2k} \sum_{i=1}^{r} r_{i2}^2 - 2 \sum_{i=1}^{r} r_{i2}e_{ik}(1) \tag{3.53}$$

To minimize the error, this sum is set equal to zero, giving

$$\sum_{i=1}^{r} r_{i2}e_{ik}(1) = c_{2k} \sum_{i=1}^{r} r_{i2}^2 \tag{3.54}$$

There are c equations for this type, which in matrix notation take the form

$$\mathbf{r}_2'\mathbf{E}_1 = \mathbf{c}_2'\mathbf{r}_2'\mathbf{r}_2 \tag{3.55}$$

where $\mathbf{E}_1$ is an $r \times c$ error matrix composed of the first residual errors. Eigenvalue λ_2 is defined as

$$\lambda_2 = \mathbf{r}_2'\mathbf{r}_2 = \sum_{i=1}^{r} r_{i2}^2 \tag{3.56}$$

From (3.55) and (3.56),

$$\mathbf{E}_1'\mathbf{r}_2 = \lambda_2 \mathbf{c}_2 \tag{3.57}$$

Matrix E_1, however, can be written

$$E_1 = D - r_1 c_1' = r_1 c_1' + r_2 c_2' + \cdots + r_c c_c' \tag{3.58}$$

Postmultiplying (3.58) by c_2 and recalling that the eigenvectors are to be ortho-normal, we obtain

$$E_1 c_2 = r_2 \tag{3.59}$$

Inserting this equation into (3.57) gives

$$E_1' E_1 c_2 = \lambda_2 c_2 \tag{3.60}$$

From (3.44), (3.45), and (3.58),

$$E_1' E_1 = D'D - \lambda_1 c_1 c_1' \tag{3.61}$$

The first residual matrix is defined as

$$\mathscr{R}_1 = Z - \lambda_1 c_1 c_1' \tag{3.62}$$

Hence we conclude from (3.60), (3.61), and (3.62) that

$$\mathscr{R}_1 c_2 = \lambda_2 c_2 \tag{3.63}$$

This expression, as described in Section 3.3.3, can be used to calculate the numerical values of the second principal eigenvector, c_2, and its associated eigenvalue, λ_2.

To obtain the third eigenvector the method of least squares is applied to e_{ik} (3), yielding

$$\mathscr{R}_2 c_3 = \lambda_3 c_3 \tag{3.64}$$

where

$$\mathscr{R}_2 = Z - \lambda_1 c_1 c_1' - \lambda_2 c_2 c_2' \tag{3.65}$$

In this fashion the remaining eigenvectors are extracted in succession. In general, we find

$$\mathscr{R}_m c_{m+1} = \lambda_{m+1} c_{m+1} \tag{3.66}$$

where

$$\mathscr{R}_m = \mathbf{Z} - \sum_{j=1}^{m} \lambda_j \mathbf{c}_j \mathbf{c}_j' \tag{3.67}$$

3.3.3 Example of Decomposition Procedure

There are many different mathematical strategies that can be used to decompose a matrix into eigenvectors and eigenvalues. The strategies developed for computer computations differ considerably from those developed for hand computations. The iteration method described below is selected for pedagogical reasons.

Initially, numerical values for the elements of $\mathbf{c}_1$ are chosen at random. It is desirable, however, to normalize the eigenvectors at each stage so that we deal with an orthonormal set. According to (3.50), multiplication of this trial vector by $\mathbf{Z}$ yields $\lambda_1 \mathbf{c}_1$, where λ_1 is the normalization constant and $\mathbf{c}_1$ is the normalized eigenvector. The resulting values for $\mathbf{c}_1$ represent a new and better approximation for $\mathbf{c}_1$. Multiplication of the new approximate $\mathbf{c}_1$ by $\mathbf{Z}$ again gives $\lambda_1 \mathbf{c}_1$, yielding better values for the elements of $\mathbf{c}_1$. This iteration procedure is continued until the elements of $\mathbf{c}_1$ converge to constant values and (3.50) is obeyed.

Eigenvector $\mathbf{c}_2$ is obtained as follows. First, $\lambda_1 \mathbf{c}_1 \mathbf{c}_1'$ is computed and subtracted from the covariance matrix yielding the first residual matrix, $\mathscr{R}_1$ [see (3.62)]. Iteration based on (3.63) will yield $\mathbf{c}_2$ and λ_2. To start the iteration, arbitrary values for the elements of $\mathbf{c}_2$ are chosen. These values are multiplied by $\mathscr{R}_1$ as shown in (3.63), yielding a new normalized set of values for $\mathbf{c}_2$. The new values are then multiplied by $\mathscr{R}_1$ and the iteration is carried out until the elements of $\mathbf{c}_2$ converge to a constant set of values and (3.63) is satisfied.

To obtain $\mathbf{c}_3$, the second residual matrix, $\mathscr{R}_2$, is calculated by subtracting $\lambda_2 \mathbf{c}_2 \mathbf{c}_2'$ from the first residual matrix. Again, arbitrary values for $\mathbf{c}_3$ are chosen and (3.64) is applied iteratively until the equation is satisfied.

From the third, fourth, and so on residual matrices, the iteration method extracts $\mathbf{c}_4, \mathbf{c}_5, \ldots, \mathbf{c}_c$ and their associated eigenvalues. A detailed numerical example of the iteration procedure is presented in Chapter 5.

This method yields a unique set of mutually orthogonal eigenvectors that represent the coordinate axes of the data space. The first eigenvector that emerges from the iteration is associated with the largest eigenvalue and accounts for most of the variance of the data. The second eigenvector is orthogonal to the first and is also oriented in a direction that maximizes the projections of the data points onto this axis. In fact, each eigenvector that emerges from the iteration is orthogonal to all previous eigenvectors and is oriented in the direction that maximizes the sum of squares of all projections on the particular axis.

If there were no error in the data, exactly n eigenvectors would emerge. However, in chemistry, perfect data are impossible to obtain. As dictated by the mathematics, c eigenvectors (c being equal to the number of data columns) will always emerge from the decomposition. However, only the first n largest eigenvectors are required to account for the data within experimental error. The remaining $c - n$ eigenvectors merely account for experimental error and should be deleted from further consideration. Methods for deciding the value of n are discussed in Section 4.3.

3.3.4 Calculating the Column Matrix

The decomposition of the covariance matrix is tantamount to matrix *diagonalization*, as shown by the following observations. Equation (3.36) can be expressed as a product of matrices:

$$\mathbf{Z} = \mathbf{C}'\mathbf{\Lambda}\mathbf{C} \qquad (3.68)$$

where

$$\mathbf{C} = \begin{bmatrix} \mathbf{c}'_1 \\ \mathbf{c}'_2 \\ \vdots \\ \mathbf{c}'_c \end{bmatrix} = [\mathbf{c}_1 \quad \mathbf{c}_2 \quad \cdots \quad \mathbf{c}_c]' \qquad (3.69)$$

The eigenvectors, which result from the iteration, constitute the rows of the column matrix, $\mathbf{C}$. Note that (3.69) contains the complete set of c eigenvectors that account for all the data, including experimental error. Because the rows of $\mathbf{C}$ are orthonormal, its transpose equals its inverse:

$$\mathbf{C}' = \mathbf{C}^{-1} \qquad (3.70)$$

Hence (3.68) can be rearranged to

$$\mathbf{C}\mathbf{Z}\mathbf{C}^{-1} = \mathbf{\Lambda} \qquad (3.71)$$

The matrix on the right is a diagonal matrix containing eigenvalues as diagonal elements. All off-diagonal elements are zero. In this form matrix $\mathbf{C}$ is recognized as the *diagonalization matrix*. Because this matrix is composed of eigenvectors, it is commonly called the "eigenvector matrix."

3.3.5 Calculating the Row Matrix

Equation (3.2) can be rearranged to

$$\mathbf{R} = \mathbf{D}\mathbf{C}^{-1} = \mathbf{D}\mathbf{C}' \qquad (3.72)$$

Having obtained matrix $\mathbf{C}$ as described previously, we can calculate the complete set of numerical values for the elements of matrix $\mathbf{R}$ by carrying out the multiplication shown in (3.72). Each element of the row matrix represents the "projection" of a row designee onto the respective eigenvector.

Examination of the row matrix reveals that the columns of this matrix are mutually orthogonal. Proof of this is given below, where (3.72), (3.17), (3.70), and (3.71) have been used chronologically:

$$\mathbf{R'R} = (\mathbf{DC'})'(\mathbf{DC'})$$
$$= \mathbf{CD'DC'}$$
$$= \mathbf{CZC^{-1}}$$
$$= \mathbf{\Lambda} \tag{3.73}$$

From this we conclude that

$$\mathbf{r}'_j\mathbf{r}_j = \lambda_j \tag{3.74}$$
$$\mathbf{r}'_i\mathbf{r}_j = 0 \tag{3.75}$$

where $\mathbf{r}_j$, a column vector of row matrix $\mathbf{R}$, is associated with eigenvector λ_j. Combining (3.37) and (3.74), we see that

$$\mathbf{r}'_j\mathbf{r}_j = \lambda_j = \sum_{i=1}^{r} r_{ij}^2 \tag{3.76}$$

This is in accord with (3.45) and (3.56), as expected.

From (3.76) we see that the eigenvalue is the sum of squares of the projections of the row designees onto a factor axis. Since there are as many elements in $\mathbf{r}_j$ as there are rows in the data matrix, each element of $\mathbf{r}_j$ represents the projection or score of that row designee onto the factor axis. For this reason the row matrix is generally called the projection matrix or score matrix. Unlike $\mathbf{c}_j$, $\mathbf{r}_j$ is not normalized but can be normalized by dividing each element of $\mathbf{r}_j$ by the square root of eigenvalue λ_j.

3.3.6 Abstract Reproduction

Although there are c eigenvectors, only n eigenvectors are required to span the factor space. Consequently,

$$\mathbf{D} \cong \overline{\mathbf{D}} = \overline{\mathbf{RC}} = [\mathbf{r}_1 \quad \mathbf{r}_2 \quad \cdots \quad \mathbf{r}_n] \begin{bmatrix} \mathbf{c}'_1 \\ \mathbf{c}'_2 \\ \vdots \\ \mathbf{c}'_n \end{bmatrix} \tag{3.77}$$

There are n columns associated with the row matrix and n rows associated with the column matrix. The factor space is n dimensional.

By following the procedures described in the earlier sections, we can find a row matrix and a column matrix which, when multiplied together, reproduce the data matrix within experimental error. Unfortunately, no physical meaning can be attached to the resulting matrices since they represent mathematical solutions only. However, the abstract solutions can be used to classify row and column designees parametrically and to reproduce the data empirically. Since the abstract reproduction scheme makes use of the mathematical, abstract factors of the space, this procedure is called abstract factor analysis (AFA).

In chemistry, measurements invariably contain experimental error. Errors produce additional eigenvectors. Their retention in the factor scheme unnecessarily increases the dimensions of the factor space and yields predictions with an accuracy far beyond that which should be expected.

Because of experimental error, deciphering the dimension of the factor space is not an easy task. Various criteria have been developed for this purpose. These are discussed in considerable detail in Chapter 4. One simple criterion is based on comparing the original data matrix with that predicted from factor analysis using the abstract reproduction step. This method is described below.

To find the dimension of the factor space, we start with eigenvector $\mathbf{c}_1$, associated with the largest eigenvalue, λ_1. This eigenvector is most important and accounts for the maximum variance in the data. We now perform the following matrix multiplication:

$$\mathbf{D}_1 = \mathbf{r}_1 \mathbf{c}_1' \tag{3.78}$$

where $\mathbf{r}_1$ and $\mathbf{c}_1'$ are the respective vectors associated with λ_1. Matrix $\mathbf{D}_1$ calculated in this way is compared to the original data matrix. If the agreement between the calculated and experimental matrices is not within experimental error, we continue the analysis by employing the next most important eigenvector:

$$\mathbf{D}_2 = [\mathbf{r}_1 \quad \mathbf{r}_2] \begin{bmatrix} \mathbf{c}_1' \\ \mathbf{c}_2' \end{bmatrix} \tag{3.79}$$

If agreement is still not attained, we continue using the next important eigenvector and the next one, and so on, until we are able to reproduce the data satisfactorily. In this way we find

$$\mathbf{D}_n = [\mathbf{r}_1 \quad \mathbf{r}_2 \quad \cdots \quad \mathbf{r}_n] \begin{bmatrix} \mathbf{c}_1' \\ \mathbf{c}_2' \\ \vdots \\ \mathbf{c}_n' \end{bmatrix} = \overline{\mathbf{D}} \simeq \mathbf{D} \tag{3.80}$$

where $\mathbf{c}_n$ is the last eigenvector needed to reproduce the data. The factor space is n dimensional.

A word of caution is warranted at this time. If the true number of factors is greater than r or c (the number of rows or columns of the data matrix), the problem cannot be solved completely because there are more unknown factors than data. It is good practice to strive for an r or c, whichever is smaller, to be at least twice n. This may be accomplished by increasing the number of columns or rows in the data matrix.

In this manner we have found a mathematical procedure for expressing a data matrix as a product of two matrices in accord with (3.80). We have now completed the abstract reproduction loop shown in Figure 3.1 using the compressed factor space.

3.3.7 Singular Value Decomposition

Singular value decomposition (SVD) is another important mathematical technique for decomposing a matrix into eigenvectors. SVD is preferred to the method described in the previous sections because it is more robust and more accurate. It can distinguish between eigenvectors that differ minutely. For large matrices, involving thousands of points, the need for SVD becomes imperative. A discussion of the details of SVD goes beyond the scope of this text. A brief history of SVD is found in the study by Shrager.[13] Because of the increasing popularity of SVD, it is important for us to summarize the language and notation that have become standard with this technique. There are certain advantages of SVD notation that will become apparent in the latter sections of this text.

In SVD notation the data matrix is expressed as

$$\mathbf{D} = \mathbf{USV}' \tag{3.81}$$

Here $\mathbf{U}$ and $\mathbf{V}$ are orthonormal matrices such that $\mathbf{U}'\mathbf{U} = \mathbf{I}_r$ and $\mathbf{V}'\mathbf{V} = \mathbf{I}_c$, where $\mathbf{I}_r$ and $\mathbf{I}_c$ represent identity matrices. Matrix $\mathbf{S}$ is a diagonal matrix whose diagonal elements, called singular values, are equal to the square roots of the respective eigenvalues. Comparing (3.81) to $\mathbf{D} = \mathbf{RC}$, we see that

$$\mathbf{US} = \mathbf{R} \tag{3.82}$$

$$\mathbf{V}' = \mathbf{C} \tag{3.83}$$

$$\mathbf{S} = \mathbf{\Lambda}^{1/2} \tag{3.84}$$

The columns of $\mathbf{U}$ are the eigenvectors of $\mathbf{DD}'$ and thus span the row space of $\mathbf{D}$. The columns of $\mathbf{V}$ are the eigenvectors of $\mathbf{D}'\mathbf{D}$ and thus span the column space of $\mathbf{D}$. In fact,

$$\mathbf{U}'\mathbf{DD}'\mathbf{U} = \mathbf{S}^2 = \mathbf{V}'\mathbf{D}'\mathbf{DV} \tag{3.85}$$

The orthonormal properties of $\mathbf{U}$ and $\mathbf{V}$ greatly simplify the computations.[14] For example, the pseudoinverse of $\mathbf{D}$ or $\mathbf{D}'$ is simply

$$\mathbf{D}^+ = \mathbf{V}\mathbf{S}^{-1}\mathbf{U}' \qquad (\mathbf{D}')^+ = \mathbf{U}\mathbf{S}^{-1}\mathbf{V}' \tag{3.86}$$

Data matrix $\overline{\mathbf{D}}$ is reconstructed from the significant eigenvectors, $\overline{\mathbf{U}}$ and $\overline{\mathbf{V}}$, and their singular values, $\overline{\mathbf{S}}$, as follows:

$$\overline{\mathbf{D}} = \overline{\mathbf{U}}\overline{\mathbf{S}}\overline{\mathbf{V}}' \tag{3.87}$$

Matrices $\mathbf{D}^+\mathbf{D} = \mathbf{V}\mathbf{V}'$ and $\mathbf{D}\mathbf{D}^+ = \mathbf{U}\mathbf{U}'$ are projection matrices that span the factor space. Any vector, $\mathbf{y}$, of dimensionality c, or any vector, $\mathbf{x}$, of dimensionality r, when multiplied by the projection matrix with the same dimensionality will yield a predicted vector, $\hat{\mathbf{y}}$ or $\hat{\mathbf{x}}$, that lies inside the factor space. The remaining part of the test vector, not contained in this space, will be annihilated.

Matrices $\mathbf{I}_c - \mathbf{D}^+\mathbf{D} = \mathbf{I}_c - \mathbf{V}\mathbf{V}'$ and $\mathbf{I}_r - \mathbf{D}\mathbf{D}^+ = \mathbf{I}_r - \mathbf{U}\mathbf{U}'$ are projection matrices that span the space orthogonal to the factor space, called the null space. Multiplying test vectors by these matrices produces new vectors that are orthogonal to the factor space. These projections represent the residuals, $\delta\mathbf{x} = \mathbf{x} - \hat{\mathbf{x}}$ and $\delta\mathbf{y} = \mathbf{y} - \hat{\mathbf{y}}$, of the test vectors and are easily calculated as follows:

$$\delta\mathbf{x} = (\mathbf{I}_r - \overline{\mathbf{U}}\overline{\mathbf{U}}')\mathbf{x} \tag{3.88}$$

$$\delta\mathbf{y} = (\mathbf{I}_c - \overline{\mathbf{V}}\overline{\mathbf{V}}')\mathbf{y} \tag{3.89}$$

3.3.8   NIPALS Decomposition

Nonlinear iterative partial least squares (NIPALS) is an algorithm designed to extract eigenvalues and eigenvectors directly from the data without requiring premultiplication of the data matrix by its transpose.[15] Pairs of eigenvectors (rows and columns) are stripped out one at a time, from the largest to the smallest eigenvalue. Data reproduction can quickly be checked during the decomposition, and the process can be terminated at any desirable point without requiring extraction of all eigenvectors. In certain applications, this can reduce the analysis time considerably.

To demonstrate the underlying principles or NIPALS, using SVD notation, we rearrange (3.81) to read as follows:

$$\mathbf{D} = \Sigma\mathbf{u}_j s_j \mathbf{v}_j' \tag{3.90}$$

where $\mathbf{u}_j$ and $\mathbf{v}_j$ are, respectively, eigenvectors of $\mathbf{U}$ and $\mathbf{V}$, associated with singular value s_j, a scalar. The sum is taken over all vectors, real as well as null. Each term in the sum is a matrix having the same dimensions $(r \times c)$ as $\mathbf{D}$ and accounts for a portion of the data:

$$\mathbf{D} = \Sigma\mathbf{D}_j \tag{3.91}$$

where

$$\mathbf{D}_j = \mathbf{u}_j s_j \mathbf{v}_j' \tag{3.92}$$

The first step is to arbitrarily select some vector to represent $\mathbf{v}_1'$, such as $(1, 1, \ldots, 1)$. This vector is then normalized to unit length. Because the eigenvectors are orthonormal, multiplying $\mathbf{v}_1$ by $\mathbf{D}$, according to (3.90), should yield

$$\mathbf{D}\mathbf{v}_1 = \mathbf{u}_1 s_1 \tag{3.93}$$

This produces a vector, which when normalized to unit length, is an approximation to $\mathbf{u}_1$ with normalization constant s_1. According to (3.90), premultiplying $\mathbf{D}$ by $\mathbf{u}_1'$ gives

$$\mathbf{u}_1' \mathbf{D} = s_1 \mathbf{v}_1' \tag{3.94}$$

This also produces a vector that, when normalized to unit length, is a better approximation to $\mathbf{v}_1$. This sequence of multiplications is repeated over and over again, each time yielding better and better approximations for $\mathbf{u}_1$ and $\mathbf{v}_1$, until convergence is achieved and Eqs. (3.93) and (3.94) are obeyed.

The first residual matrix, $\mathbf{E}_1$, can then be calculated:

$$\mathbf{E}_1 = \mathbf{D} - \mathbf{D}_1 \tag{3.95}$$

where $\mathbf{D}_1$ is obtained from (3.92). The root mean square of the residual matrix is computed, compared to experimental error, and used as a criterion for continuing or stopping the decomposition.

To extract the second set of eigenvectors, the residual matrix is multiplied by some arbitrary $\mathbf{v}_2$, yielding an approximate $\mathbf{u}_2$, which in turn is used to generate a better approximation for $\mathbf{v}_2$. This is possible because, in general,

$$\mathbf{E}_{j-1}\mathbf{v}_j = \mathbf{u}_j s_j \tag{3.96}$$

$$\mathbf{u}_j' \mathbf{E}_{j-1} = s_j \mathbf{v}_j \tag{3.97}$$

where

$$\mathbf{E}_j = \mathbf{D} - \sum_{i=1}^{j} \mathbf{D}_i \tag{3.98}$$

After extracting the second set of eigenvectors, the third, fourth, and so forth sets are obtained by repeating the process with each successive residual matrix.

Because NIPALS involves a least-squares principle similar to that described in Section 3.3.2, this procedure extracts the eigenvalues in sequence from largest to smallest. Terminating the process at the proper factor level therefore preempts analysis of the null vectors, a task that could consume valuable computer time. The

NIPALS algorithm is particularly useful for specialized techniques such as partial least squares (PLS), which corrects the flaws in multiple regression analysis (MRA) and in principal component regression (PCR).

3.4   TRANSFORMATION

From the standpoint of a theoretical chemist, the analysis should not terminate here. The row and column factors in their abstract forms are not recognizable as physical or chemical parameters since the reference axes were generated to yield a purely mathematical solution. For scientific purposes we seek chemically recognizable factors. This can be accomplished by transforming the reference axes so that they become aligned with fundamental properties of the designees.

3.4.1   Physically Significant Parameters

Within certain limitations we can transform the axes and find many solutions that obey (3.2). Consider, for example, points that lie in a common plane. The positions of such points can be designated by the coordinates of the two distinct axes that lie in the plane. These axes may be rotated freely in the plane and any two distinct axes may be chosen to locate the points. In principle there are an infinite set of axes that can be used to define the plane and locate the data points. Similarly, in factor analysis the reference axes may be rotated as long as they are distinct and as long as they adequately span the factor space. In particular, we wish to transform the axes so that they are aligned with fundamental structural parameters of the row designees.

Transformation of the axes is accomplished by the following operation:

$$\hat{\mathbf{X}} = \overline{\mathbf{R}}\mathbf{T} \tag{3.99}$$

where $\mathbf{T}$ is the transformation matrix, of dimensions $n \times n$, and $\hat{\mathbf{X}}$ is the chemically recognizable row matrix in the new coordinate system. A least-squares method of obtaining the target transformation matrix is derived in Section 3.4.3.

The inverse of the transformation matrix is used to locate the column matrix in the new coordinate system. Equation (3.99) is first rearranged as follows:

$$\overline{\mathbf{R}} - \hat{\mathbf{X}}\mathbf{T}^{-1} \tag{3.100}$$

This equation is placed into (3.2), giving

$$\overline{\mathbf{D}} = \hat{\mathbf{X}}\mathbf{T}^{-1}\overline{\mathbf{C}}$$
$$= \hat{\mathbf{X}}\hat{\mathbf{Y}} \tag{3.101}$$

where $\hat{\mathbf{Y}}$, the column-factor matrix in the new coordinate system, is

$$\hat{\mathbf{Y}} = \mathbf{T}^{-1}\overline{\mathbf{C}} \tag{3.102}$$

In summary, by proper transformation of the coordinate axes, it is possible to find a row matrix that can be interpreted in chemical or physical terms. Because there are an infinite number of positions through which a set of axes may be rotated, there exists an infinite number of possible solutions resulting from factor analysis. Nevertheless, only certain orientations of the axes yield factors that correspond to recognizable parameters. It is advantageous to be able to test various physically significant parameters to determine whether or not they are real factors. The least-squares method called target transformation is most appropriate for this purpose.

3.4.2 Techniques

In the classical work of factor analysis, the terminology "rotation" rather than "transformation" is most commonly encountered. One employs abstract rotations when there exists little or no information about the true origin of the factors. The ultimate goal of abstract rotation is to extract meaningful factors that have the simplest factor structure. All rotation techniques attempt to locate a set of axes so that as many row points as possible lie close to the final factor axes, with only a small number of points remaining between the rotated axes.

There are many methods for obtaining rotation matrices. These methods are based on some intuitive criteria of the factor space, such as simple structure, parsimony, factorial invariance, partialing, casual exploration, and hypothetical structure.[4] These methods can be classified into one of two general categories: orthogonal rotations or oblique rotations. *Orthogonal rotations* preserve the angular dependence between the original set of eigenvectors that emerge from FA. Techniques such as quartimax and varimax belong to this class. *Oblique rotations* do not preserve the angles between the eigenvectors. Oblimax, quartimin, biquartimin, covarimin, binormamin, maxplane, and promax fall into this class.

Target transformation belongs to the category of oblique rotations. Since this is by far the most important technique for chemists, we study it in considerable detail. Those interested in learning details of the other methods listed above are advised to consult Rummel.[4] Before discussing target transformation, however, it is useful for us to examine briefly several of the popular methods of abstract rotations listed above.

Quartimax involves orthogonal rotation, preserving the angles between the axes. The basic principle of quartimax is best visualized by a two-dimensional example. As the orthogonal axes are rotated so that one axis approaches a data point, the projection (i.e., the loading) of the point on the axis increases. At the same time the point moves away from the other axis and its loading on that axis decreases. Quartimax searches for a set of orthogonal axes that group the points in clusters about each axis, with each point having either high or low loading on each axis.

According to Harman,[5] this can be achieved by rotating the axes to maximize the quartimax function, Q:

$$Q = \sum_{j=1}^{n} \sum_{k=1}^{c} \lambda_j^2 \tilde{c}_{jk}^4 \qquad (3.103)$$

Here λ_j is the jth eigenvalue and $\tilde{c}_{jk}$ is the loading of the kth data-column vector on the jth axis after the rotation has been completed. The sum is taken over all c data columns and over n eigenvectors, which are required to span the factor space. Unfortunately, quartimax tends to "overload" the first factor, producing one large general factor and many small subsidiary factors.

Varimax attempts to overcome the deficiency of the quartimax method. With varimax, the total variance, V, of the squared loadings is maximized:

$$V = \sum_{j=1}^{n} \left(\frac{1}{c} \sum_{k=1}^{c} (\lambda_j \tilde{c}_{jk}^2)^2 - \frac{1}{c^2} \left(\sum_{k=1}^{c} \lambda_j \tilde{c}_{jk}^2 \right)^2 \right) \qquad (3.104)$$

The varimax method of Kaiser[16] is currently the most popular of the orthogonal rotation schemes because of its ability to yield the same clusters regardless of the size of the data matrix.

Quartimin is essentially the same as quartimax except that the condition of orthogonality is removed. The eigenvector axes are rotated obliquely so that the loadings of a data point will be increased on one axis and decreased on all others. Thus the sum of the inner products of its loadings is reduced. According to Carroll,[17] this situation is achieved by minimizing the quartimin function, M:

$$M = \sum_{j<l=1}^{n} \sum_{k=1}^{c} \lambda_j \tilde{c}_{jk}^2 \lambda_l \tilde{c}_{lk}^2 \qquad (3.105)$$

Here j and l refer to the jth and lth oblique factors.

Oblimax involves oblique rotations where the number of low and high loadings on a given axis are increased by decreasing the loadings in the middle range. Saunders[18] showed that this can be accomplished by maximizing the kurtosis function, K:

$$K = \frac{\sum\limits_{j=1}^{n} \sum\limits_{k=1}^{c} \lambda_j^2 \tilde{c}_{jk}^4}{\left(\sum\limits_{j=1}^{n} \sum\limits_{k=1}^{c} \lambda_j \tilde{c}_{jk}^2 \right)^2} \qquad (3.106)$$

Covarimin[16] is an extension of varimax that permits oblique rotations. In this case the eigenvectors are rotated obliquely until the covarimin function, C, is minimized:

$$C = \sum_{j<l=1}^{n} \left(\frac{1}{c} \sum_{k=1}^{c} \lambda_j \tilde{c}_{jk}^2 \lambda_l \tilde{c}_{lk}^2 - \frac{1}{c^2} \sum_{k=1}^{c} \lambda_j \tilde{c}_{jk}^2 \sum_{k=1}^{c} \lambda_l \tilde{c}_{lk}^2 \right) \qquad (3.107)$$

In general, covarimin usually yields axes that are similar to those resulting from varimax.

These methods have been used extensively in the behavioral sciences but have not been fully explored in chemistry. Chemical investigators tend to favor the varimax method.

3.4.3 Target Transformation

Because target transformation plays an important role in chemistry, the underlying principles are discussed in this section in detail. Target transformation is unique because, in spite of the complexity of the data space, it allows us to search for the basic factors *individually*. This can be seen by examining (3.99), which concerns the mathematical operation involved in transforming the factor axes. This equation shows that $\hat{\mathbf{x}}_l$, the lth column of the newly transformed row matrix, is obtained by multiplying $\mathbf{t}_l$, the lth column of the transformation matrix, by the row matrix $\overline{\mathbf{R}}$:

$$\hat{\mathbf{x}}_l = \overline{\mathbf{R}}\mathbf{t}_l \tag{3.108}$$

where $\hat{\mathbf{x}}_l$ is called the *predicted vector* and $\mathbf{t}_l$ the associated transformation vector. A least-squares procedure is employed to generate a transformation vector that will yield a predicted vector, $\hat{\mathbf{x}}_l$, most closely matching $\mathbf{x}_l$, the *test vector* that is suspected to be a basic factor. The test vector is called the "target." The least-squares procedure minimizes the deviation between the test vector and the predicted vector, producing the best possible transformation vector for the individual target test being considered. The mathematical basis for obtaining the best $\mathbf{t}_l$ is described below.

Transformation vector $\mathbf{t}_l$ has components $t_{1l}, t_{2l}, \ldots, t_{nl}$. Each row of $\overline{\mathbf{R}}$ can be viewed as a row vector. The ith row of $\overline{\mathbf{R}}$ is a vector $\mathbf{r}_{i.}$ having components r_{i1}, $r_{i2}, \ldots, r_{in}$. Vector $\mathbf{r}_{i.}$ should not be confused with $\mathbf{r}_{.i}$, the ith column of the row-factor matrix. When $\mathbf{r}_{i.}$ is multiplied, dot-product-wise, by $\mathbf{t}_l$, the projection of the ith row entity on the new target-transformed coordinate axis is obtained:

$$\hat{x}_{il} = \mathbf{r}'_{i.}\mathbf{t}_l = r_{i1}t_{1l} + r_{i2}t_{2l} + \cdots + r_{in}t_{nl} \tag{3.109}$$

The sum in (3.109) is taken over all n principal factors.

Multiplying each row vector of the row matrix by $\mathbf{t}_l$ gives $\hat{x}_{1l}, \hat{x}_{2l}, \ldots, \hat{x}_{rl}$, which are the elements of $\hat{\mathbf{x}}_l$, the predicted vector. Each element of the predicted vector is then compared to the corresponding element of the test vector, $\mathbf{x}_l$, having components $x_{1l}, x_{2l}, \ldots, x_{rl}$. The difference between the value of $\hat{x}_{il}$ and the value of x_{il} is written as Δx_{il}:

$$\Delta x_{il} = \hat{x}_{il} - x_{il} = r_{i1}t_{1l} + r_{i2}t_{2l} + \cdots + r_{in}t_{nl} - x_{il} \tag{3.110}$$

To find the best $\mathbf{t}_l$, the deviation between the test vector and the predicted vector is minimized by setting the sum of the derivatives of the squares of the differences, given by (3.110), equal to zero. For example, the derivative of the square of the difference with respect to t_{1l} is

$$\frac{d(\Delta x_{il})^2}{dt_{1l}} = 2r_{i1}^2 t_{1l} + 2r_{i1}r_{i2}t_{2l} + \cdots + 2r_{i1}r_{in}t_{nl} - 2r_{i1}x_{il} \tag{3.111}$$

Similar expressions are obtained for each row designee. Summing over all the row designees and applying the least-squares criteria, we obtain

$$
\sum_{i=1}^{r} \frac{d(\Delta x_{il})^2}{dt_{1l}} = 0 = t_{1l} \sum_i r_{i1}^2 + t_{2l} \sum_i r_{i1} r_{i2} + \cdots
$$
$$
+ t_{nl} \sum_i r_{i1} r_{in} - \sum_i r_{i1} x_{il}
\tag{3.112}
$$

By repeating this calculation, the sums of squares of the deviations with respect to the remaining components of $\mathbf{t}_l$ (i.e., $t_{2l}, t_{3l}, \ldots, t_{nl}$) are obtained. This leads to the following set of simultaneous equations:

$$
\begin{aligned}
\Sigma r_{i1} x_{il} &= t_{1l} \Sigma r_{i1}^2 & + t_{2l} \Sigma r_{i1} r_{i2} + \cdots + t_{nl} \Sigma r_{i1} r_{in} \\
\Sigma r_{i2} x_{il} &= t_{1l} \Sigma r_{i1} r_{i2} + t_{2l} \Sigma r_{i2}^2 & + \cdots + t_{nl} \Sigma r_{i2} r_{in} \\
\vdots \qquad & \quad \vdots \qquad\quad \vdots \qquad\qquad \vdots \\
\Sigma r_{in} x_{il} &= t_{1l} \Sigma r_{i1} r_{in} + t_{2l} \Sigma r_{i2} r_{in} + \cdots + t_{nl} \Sigma r_{in}^2
\end{aligned}
\tag{3.113}
$$

In matrix notation (3.113) becomes

$$
\mathbf{a}_l = \mathbf{B} \mathbf{t}_l
\tag{3.114}
$$

where

$$
\mathbf{a}_l \equiv
\begin{bmatrix}
\Sigma r_{i1} x_{il} \\
\Sigma r_{i2} x_{il} \\
\vdots \\
\Sigma r_{in} x_{il}
\end{bmatrix}
\tag{3.115}
$$

$$
\mathbf{t}_l \equiv
\begin{bmatrix}
t_{1l} \\
t_{2l} \\
\vdots \\
t_{nl}
\end{bmatrix}
\tag{3.116}
$$

$$
\mathbf{B} \equiv
\begin{bmatrix}
\Sigma r_{i1}^2 & \Sigma r_{i1} r_{i2} & \cdots & \Sigma r_{i1} r_{in} \\
\Sigma r_{i1} r_{i2} & \Sigma r_{i2}^2 & \cdots & \Sigma r_{i2} r_{in} \\
\vdots & \vdots & & \vdots \\
\Sigma r_{i1} r_{in} & \Sigma r_{i2} r_{in} & \cdots & \Sigma r_{in}^2
\end{bmatrix}
\tag{3.117}
$$

Multiplying both sides of (3.114) by $\mathbf{B}^{-1}$ gives

$$
\mathbf{t}_l = \mathbf{B}^{-1} \mathbf{a}_l
\tag{3.118}
$$

Examination of (3.117) reveals

$$\mathbf{B} = \overline{\mathbf{R}}'\overline{\mathbf{R}} \tag{3.119}$$

and examination of (3.115) reveals

$$\mathbf{a}_l = \overline{\mathbf{R}}\mathbf{x}_l \tag{3.120}$$

where $\mathbf{x}_l$ is the test vector composed of the suspected parameters associated with the row designees. Thus

$$\mathbf{t}_l = (\overline{\mathbf{R}}'\overline{\mathbf{R}})^{-1}\overline{\mathbf{R}}'\mathbf{x}_l \tag{3.121}$$

Equation (3.121) is the central equation of the target factor analysis. The least-squares vector transformer, $\mathbf{t}_l$, a column of $\mathbf{T}$, is readily calculated by means of this equation without requiring any knowledge of the other contributing factors.

Because of (3.121), target transformation becomes a reality for the factor analyst. Numerical values for all quantities in this equation, except the test vector $\mathbf{x}_l$, are automatically calculated during the routine decomposition step. The test vector, of course, must be obtained from theory, empirical knowledge, or intuition. Deducing and formulating the test vector constitute the real chemical art involved in target factor analysis. To see whether or not a suspected factor is a true factor, one inserts the chosen test vector into (3.121) and the best transformation vector emerges.

Having obtained $\mathbf{t}_l$, we use (3.108) to obtain numerical values for the elements of $\hat{\mathbf{x}}_l$. We can then ascertain whether or not the following equation is obeyed within experimental error:

$$\hat{\mathbf{x}}_l \stackrel{?}{=} \mathbf{x}_l \tag{3.122}$$

If the suspected test vector $\mathbf{x}_l$ is a factor, each element of $\hat{\mathbf{x}}_l$ will equal the corresponding element of $\mathbf{x}_l$, within uncertainty limitations. If, on the contrary, it is not a true factor, the differences between the corresponding elements will be greater than expectations. Methods for judging the validity of a test vector are described in Section 4.6. The exciting feature of target testing is the fact that this is performed on an individual basis, focusing complete attention on a single factor, in spite of the multivariate nature of the data.

Before ending this discussion it is instructive for us to express (3.121) in several other ways. For example, the least-squares transformation vector can be written in a compact form as

$$\mathbf{t}_l = \overline{\mathbf{R}}^+\mathbf{x}_l \tag{3.123}$$

where $\overline{\mathbf{R}}^+$ is the pseudoinverse of $\overline{\mathbf{R}}$ (see Appendix A for a definition and discussion of the pseudoinverse). Another way of expressing $\mathbf{t}_l$, is obtained by inserting (3.73) into (3.121):

$$\mathbf{t}_l = \overline{\Lambda}^{-1}\overline{\mathbf{R}}'\mathbf{x}_l \tag{3.124}$$

Here $\overline{\Lambda}$ is a diagonal matrix composed of the primary eigenvalues only. This equation is convenient for programming purposes because it involves considerably less computations than (3.121).

Inserting (3.121), (3.122), and (3.123) into (3.108) leads to the following expressions for determining the predicted vector, bypassing the need to calculate the transformation vector:

$$\hat{\mathbf{x}}_l = \overline{\mathbf{R}}(\overline{\mathbf{R}'\mathbf{R}})^{-1}\overline{\mathbf{R}}'\mathbf{x}_l \tag{3.125}$$

$$\hat{\mathbf{x}}_l = \overline{\mathbf{R}\mathbf{R}}^+\mathbf{x}_l \tag{3.126}$$

$$\hat{\mathbf{x}}_l = \overline{\mathbf{R}\Lambda}^{-1}\overline{\mathbf{R}}'\mathbf{x}_l \tag{3.127}$$

$$\overline{\mathbf{x}}_l = \overline{\mathbf{U}\mathbf{U}}'\mathbf{x}_l \tag{3.128}$$

Each of these equations serves a special purpose and will be referred to, individually, as needed.

The discussion above focuses on testing column vectors, vectors that emulate the columns of the data matrix. Row vectors, vectors that emulate the rows of the data matrix, can also be target tested. In this case the least-squares transformation vector is

$$\mathbf{t}_i = \mathbf{y}_i\overline{\mathbf{C}}' \tag{3.129}$$

and the predicted vector is obtained by

$$\hat{\mathbf{y}} = \mathbf{y}_i\overline{\mathbf{C}}'\overline{\mathbf{C}} = \mathbf{y}_i\overline{\mathbf{V}\mathbf{V}}' \tag{3.130}$$

These expressions are analogous to (3.123) and (3.128) but have a simpler form because the rows of $\mathbf{C}$ are orthonormal whereas the columns of $\mathbf{R}$ are orthogonal but not normalized.

Uniqueness Tests. A special kind of target test called the *uniqueness test* should be carried out routinely for both row and column designees in every factor analysis (FA) study. The uniqueness test is designed to identify designees behaving atypically due to either chemically unique properties or gross errors in the data.

A uniqueness test vector contains an input value of 1 for the designee being tested for uniqueness and 0's for all other designees. A complete set of uniqueness vectors, $\mathbf{q}_k$, designed to test each column of the data matrix for uniqueness, is comprised of c vectors, each having r elements:

$$\mathbf{q}_1 = (1, 0, 0, \ldots, 0, 0, 0)'$$
$$\mathbf{q}_2 = (0, 1, 0, \ldots, 0, 0, 0)'$$
$$\vdots \qquad\qquad \vdots$$
$$\mathbf{q}_c = (0, 0, 0, \ldots, 0, 0, 1)'$$

A similar set of test vectors comprised of r vectors, each having c elements, is used to test for unique rows.

A designee is considered unique if the significance level of the F-test (see Section 4.6.3) is greater than 5 or 10%, or some other level deemed statistically significant for the problem at hand.

Uniqueness tests can pinpoint designees associated with unsuspected gross errors. An "outlier" point creates a separate, unique factor and may increase the factor size by as much as one unit. Thus uniqueness tests may indicate unreliable data. Unless a designee is considered to be chemically unique, uniqueness tests with high significance levels are adequate grounds for removing the associated row or column from the data matrix, and then redoing the factor analysis on the reduced matrix. Removal of the suspect designee may result in better reproduction and in some problems may reduce the rank of the data matrix by one unit.

Inspection of the intersections of unique rows and unique columns helps identify the particular offending data point. Predictions of these points can be made by target testing the deleted row and column while "free floating" these points in the factor space of the reduced data matrix (see Section 3.4.4).

The uniqueness test also serves as a preliminary classifier in that similar designees tend to have similar predicted values on the uniqueness tests. Hence the predicted uniqueness vector may indicate clusterings of the designees that have a factor in common. The use of the uniqueness test to identify clusters of similar solvents in gas–liquid chromatography (GLC) is described in Section 11.5.

Gemperline used the uniqueness test to separate unresolved chromatograms (see Section 6.2). Uniqueness vectors, in this case, represent idealistic elution profiles. When the retention time represented by a uniqueness vector corresponds to the retention time of a real component, the error in the fit reaches a local minimum.

Unity Test. Another important special test is the *unity test*. The unity test vector, $\mathbf{u}$, consists entirely of 1's. This vector tests for a constant factor common to all rows or to all columns:

$$\mathbf{u} = (1, 1, 1, \ldots, 1, 1, 1)' \tag{3.131}$$

Unity is expected to transform well if the data are measured relative to some arbitrary standard as done in nuclear magnetic resonance (NMR) chemical shift measurements. If the first principal factor is particularly important, unity will test well regardless of the physical situation. Data pretreatment, such as correlation or covariance about the origin, can be used to remove the constant factor from the problem, reducing the factor space by one unit. Results for unity tests from four chemical problems are shown in Tables 11.8, 11.9, 12.2, and 12.3.

3.4.4   Predicting Missing Points in Target Vectors

Free Floating. The least-squares method for transformation as developed in the previous section is completely general. It holds even if some of the values for a particular test vector are missing. In this case, however, appropriate terms must be removed from the summations in (3.113) through (3.117). Equations (3.73), (3.124) to (3.128) are not valid in this case. Vector t_l must be calculated by means of (3.118) after the deletions have been made. With these appropriate modifications, the predicted vector can be calculated with (3.108), and the results can be compared to the experimental vector.

This procedure, termed *free floating*, has a hidden advantage. Equation (3.108) automatically yields a predicted value for each designee, including those free floated in the test vector. Thus, the method can be used to predict missing points in valid test vectors.

Iterative Target Testing. Equations (3.121) and (3.124) through (3.129) can be adapted to handle incomplete targets by means of iterative target testing.[19] Arbitrary values, such as zero, are entered in the test vector for the missing elements. A new test vector is constructed from the original test vector by replacing these zeros with their predicted values, all other elements remaining unchanged. This process of replacing the missing points with their predictions is repeated until convergence is attained. Although the method is computationally different, the results of target iteration are exactly the same as those obtained by the free-floating method described previously.

Precaution. There is one important restriction on target testing. The number of test points in a test vector must be at least equal to n, the rank of the data matrix. The introduction of a test vector with insufficient test points will always yield a perfect fit of the test points. Such results, however, are totally meaningless. Extreme care must be exercised to include more test points than factors. For a detailed discussion of this point, refer to Section 4.6.

Illustration. Free floating can be illustrated with Figure 3.4, which shows the vector relationship between four columns (four properties) of a given data matrix. Using least-squares target transformation, we attempt to find an axis that lies in the factor plane and that has physical significance. If the dipole moment were a true factor, the projection of a row-designee (molecule) point on the dipole axis would correspond to the dipole moment of the molecule. To locate this axis, we need not know the dipole moments of all row-designee molecules. For a two-factor space a minimum of two dipole moments is required mathematically. However, it is advisable to use more than two test points.

In practice, a dipole moment vector is constructed by free floating all row-designee molecules whose dipole moments are not known. The transformation vector is calculated using only the partial set of data points, as described earlier. The computer output, which is the result of (3.108), yields dipole moments for all molecules, whether or not they were included in the test scheme. In this manner, dipole moments can be predicted. Although this was not the original intent of factor analysis, it does constitute an extremely useful and important fringe benefit.

3.4.5 Key Combination

Basic Factors. The ultimate objective of factor analysis is to obtain a basic set of factors that have real physicochemical meaning. Basic vectors can be tested individually and identified by applying the methods described in the previous sections. It is possible to find a sufficient number of acceptable test vectors but still not span the factor space because some of the test vectors lie in a common subspace. Because of the variety of ways of expressing real chemical factors and because of the multifaceted interrelationships among the factors, finding a "key set" of basic factors that adequately spans the data space is not a simple task. To find the key set, various combinations of acceptable basic vectors are assembled into matrix $\mathbf{X}_{\text{basic}}$ and the following calculation is carried out:

$$\mathbf{X}_{\text{basic}}\hat{\mathbf{Y}}_{\text{load}} = \hat{\mathbf{D}} \tag{3.132}$$

where

$$\hat{\mathbf{Y}}_{\text{load}} = \mathbf{T}^{-1}\overline{\mathbf{C}} \tag{3.133}$$

Each column of $\mathbf{X}_{\text{basic}}$ is an individually successful test vector that satisfactorily obeys (3.122). The transformation vectors associated with each of these successful tests are assembled into matrix $\mathbf{T}$:

$$\mathbf{T} = [\mathbf{t}_1 \quad \mathbf{t}_2 \quad \cdots \quad \mathbf{t}_n] \tag{3.134}$$

Matrix $\hat{\mathbf{Y}}_{\text{load}}$ is obtained by means of (3.133). Matrix $\hat{\mathbf{D}}$ is the reproduced data matrix that results from this target combination test.

Conversely, one can focus attention on testing basic vectors associated with the column-factor matrix, $\mathbf{Y}_{\text{basic}}$, by carrying out analogous computations:

$$\hat{\mathbf{X}}_{\text{load}}\mathbf{Y}_{\text{basic}} = \hat{\mathbf{D}} \tag{3.135}$$

where

$$\hat{\mathbf{X}}_{\text{load}} = \overline{\mathbf{R}}\mathbf{T}^{-1} \tag{3.136}$$

In either case, a key set of basic vectors is found when $\hat{\mathbf{D}}$, the combination reproduced data, adequately equals the original data matrix:

$$\hat{\mathbf{D}} \overset{?}{=} \mathbf{D} \tag{3.137}$$

Note that (3.132) involves row matrix $\mathbf{X}$ instead of $\hat{\mathbf{X}}$ and that (3.135) involves matrix $\mathbf{Y}$ instead of $\hat{\mathbf{Y}}$. If $\hat{\mathbf{X}}$ and $\hat{\mathbf{Y}}$ were employed instead of $\mathbf{X}$ and $\mathbf{Y}$, a unitary transformation would be performed, producing exactly the same data matrix as originally obtained from AFA based on the determined factor space. In other words,

$$\hat{X}\hat{Y} = \overline{R}TT^{-1}C = \overline{R}\,\overline{C} = \overline{D} \tag{3.138}$$

In effect, the model would be forced to fit the data. On the other hand, (3.132) and (3.135) are *not* unitary transformations and do *not* force the model to fit the data. Consequently, (3.132) and (3.135) are unique and severe tests of the model. The important distinction between these equations and (3.138) cannot be overstressed.

The mathematics involved in Eqs. (3.132) and (3.135) are equivalent to principal regression analysis (PRA). In essence, the combination step in factor analysis is identical to PRA.

For a specified combination of targets, assembled as a matrix, the MATLAB program labeled lfa.m (loading factor analysis) calculates the associated loading matrix and is included in Appendix C for the convenience of the reader.

Typical Factors. Since the columns of the data matrix lie in the factor space, a judicious choice of n data columns (or data rows) can be used to describe the n-dimensional factor space. Such a combination is called a key set of typical vectors.

To understand this, return to Figure 3.3, which illustrates the vector relationship between the four columns (four data vectors) of a given data matrix. This figure shows that the four data vectors lie in a common plane defined by two factors that emerge from the decomposition step. Since all four data-column vectors lie in a plane, any two data-column vectors can be used to locate a row-designee point. Predictions can then he made by reading the projections on the remaining data-column vectors. By means of the least-squares target transformation approach, the reference axes can be reassigned to coincide with an appropriate set of typical columns of the original data matrix. This is accomplished by using a column of the original data as a test vector. Obviously, each column will yield a successful test, since each column lies in the factor space. The test procedure, however, produces a transformation vector. A combination of such transformation vectors is needed to construct the transformation matrix.

In general, an arbitrary combination of data columns will not necessarily yield a transformation matrix that will be successful in reproducing the data. Only certain combinations, which span the total factor space, will work. The combination set that best accomplishes this task is called the *key combination set*. The lth column of the data matrix can be looked upon as a vector $\mathbf{d}_{.l}$, which represents test vector $\mathbf{x}_l$. The combination of typical vectors, $\mathbf{D}_{\text{key}}$, is a matrix constructed from only n data columns; that is,

$$X = D_{\text{key}} = [\mathbf{d}_{.a} \quad \mathbf{d}_{.b} \quad \cdots \quad \mathbf{d}_{.n}] \tag{3.139}$$

The transformation vector for each key data column is obtained from (3.124), which can be expressed as

$$\mathbf{t}_l = \overline{\Lambda}^{-1}\overline{R}'\mathbf{d}_{.l} \tag{3.140}$$

Using n of these transformation vectors in combination, a complete transformation matrix can be assembled as shown in (3.134). Various combinations are employed in an attempt to find the key combination set that best reproduces the data matrix. Hence we search for a set of n data columns that best satisfies

$$\|\mathbf{D}_{\text{key}} \hat{\mathbf{Y}}_{\text{load}} - \mathbf{D}\|^2 = \text{minimum} \qquad (3.141)$$

Of course, $\hat{\mathbf{Y}}_{\text{load}}$ is obtained by applying (3.133); $\mathbf{D}_{\text{key}}$, involving typical vectors, is analogous to $\mathbf{X}_{\text{basic}}$, involving basic vectors. The subscript "key" is inserted to signify the key combination set.

There are several advantages to using data columns as factors of the space. First, they are more easily visualized than the abstract factors produced by factor analysis. Second, their use precludes the need to identify the true controlling factors. Third, empirical predictions can be made quickly from the resulting equations. For many chemical problems, this is sufficient. Powerful uses of this technique are found in later chapters.

To further illustrate the principle involved in using the key combination set, let us return to the example portrayed in Figures 3.2 and 3.3. Here we are dealing with four data-column vectors that lie in a common plane. Because all four data-column vectors lie in the same plane, they are linearly dependent. Any two of the four vectors shown in Figure 3.2 will define the plane. For example, $\mathbf{d}_{.1}$ and $\mathbf{d}_{.3}$ may be chosen to be the bases. From the tips of vectors $\mathbf{d}_{.2}$ and $\mathbf{d}_{.4}$ lines parallel to the key set, $\mathbf{d}_{.1}$ and $\mathbf{d}_{.3}$ may be drawn as shown in Figure 3.5. The lengths of the intersections

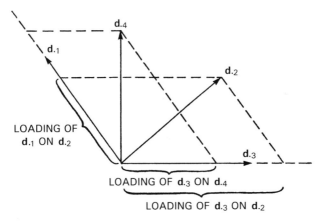

Figure 3.5 Factor loadings onto representative axes $\mathbf{d}_{.1}$ and $\mathbf{d}_{.3}$. Lines drawn parallel to $\mathbf{d}_{.1}$ and $\mathbf{d}_{.3}$ intersect $\mathbf{d}_{.1}$ and $\mathbf{d}_{.3}$ at points that correspond to the loadings.

on each key set axis represent the loadings of the typical vectors. From this diagram we can show that the four vectors can be represented as follows:

$$\mathbf{d}_{.1} = 1.0000\mathbf{d}_{.1} + 0.0000\mathbf{d}_{.3}$$
$$\mathbf{d}_{.2} = 0.7945\mathbf{d}_{.1} + 1.2330\mathbf{d}_{.3}$$
$$\mathbf{d}_{.3} = 0.0000\mathbf{d}_{.1} + 1.0000\mathbf{d}_{.3}$$
$$\mathbf{d}_{.4} = 1.2361\mathbf{d}_{.1} + 0.7265\mathbf{d}_{.3}$$

(3.142)

These linear equations have an important feature. The projection of a row-designee point on $\mathbf{d}_{.2}$ and $\mathbf{d}_{.4}$ can be calculated from the scores on $\mathbf{d}_{.1}$ and $\mathbf{d}_{.3}$, the axes. Thus a data set consisting of a large array of data columns can be reduced to a minimum set of representative data columns from which all the data can be generated. With these equations, we no longer need to compute the projections on the abstract eigenvectors resulting from factor analysis. Instead, we need only to store the data involving the chosen representative axes, $\mathbf{d}_{.1}$ and $\mathbf{d}_{.3}$.

Any individual point, d_{ik}, of the data matrix can be predicted from measurements of properties associated with key columns 1 and 3, since, according to (3.142),

$$d_{i2} = 0.7945d_{i1} + 1.2330d_{i3}$$
$$d_{i4} = 1.2361d_{i1} + 0.7265d_{i3}$$

(3.143)

Such equations have been used to predict the properties of new row designees, not included in the data matrix, from their key property measurements. An example of the practical use of this methodology for the prediction of proton NMR solvent shifts is described in Section 10.1.1. The use of typical sets for GLC studies are summarized in Section 11.6.

Iterative Key Set Factor Analysis. Using the combination step, factor analytical solutions can be expressed efficiently in terms of a minimum number of typical rows or columns, judiciously chosen from the data matrix itself. Abstract factor analysis can be used to determine the smallest number of typical vectors required to characterize the data. However, for a data matrix lying in n-dimensional factor space any arbitrary combination of n rows or n columns will not necessarily span the factor space. If there are p rows (or columns) and n factors, there are $p!/(p - n)!n!$ possible combinations. If p is 50 and n is 10, there are more than 10 billion combinations. Although a complete search of all combinations will ensure that the optimal set is found, the time required for such a search is impractical.

Techniques based on the principles of factor analysis have been designed to select the key set of typical rows and columns of a data matrix in a most efficient manner. One of these techniques is based on the premise that the most orthogonal set of typical vectors will best describe the factor space and thus will best represent the key set.

The strategy of key set factor analysis[20] (KSFA) is briefly outlined below. For convenience we focus attention on locating the key set of typical rows, but the arguments can just as well be applied to typical columns. Since only the vector directions are of interest and not their magnitudes, each row of the abstract row-factor matrix is normalized to unit length. After normalization, the ith row vector in n space is represented as

$$\mathbf{v}(i) = (v_{i1}, v_{i2}, \ldots, v_{in})' \tag{3.144}$$

where

$$v_{ik} = \frac{r_{ik}}{\left(\sum\limits_{j=1}^{n} r_{ij}^2 \right)^{1/2}} \tag{3.145}$$

The value of v_{ik} represents the relative importance of the kth eigenvector on the direction of the ith row. This normalization places the eigenvector representation of each row designee on the surface of an n-dimensional hypersphere in factor space defined by the primary eigenvectors.

The first eigenvector accounts for a maximum of the variation in the data and lies in a direction passing through the highest density of row vectors. For this reason the first key row should contain the least amount of the first eigenvector, indicated by the smallest projection on the first principal eigenvector. In other words, the first key row is the row with v_{i1} closest to zero.

The second key row is identified by finding the row most orthogonal to the first key row in two space defined by the first two eigenvectors associated with the two largest eigenvalues, accounting for a maximum variation in the data in two space. If $\mathbf{v}(i)$ is the first key row vector, then the jth row with $\mathbf{v}(j) = (v_{j1}, v_{j2})$ most orthogonal to $\mathbf{v}(i) = (v_{i1}, v_{i2})$ in two-dimensional space represents the second key row.

The third key row is obtained by considering the first three components of the normalized row vectors. It is the vector most orthogonal to the plane defined by the first two key row vectors in three space. Each successive key row is obtained by finding the row vector most orthogonal to the hyperplane defined by the previously determined key rows in the expanded factor space. In general, the mth key vector is obtained by using m components of the normalized row vectors associated with the m most important eigenvectors. The mth key row is that row that maximizes the determinant of the $m \times m$ matrix constructed from the $m - 1$ previously chosen key rows and the candidate row; namely,

$$\left| \det \begin{bmatrix} v_{i1} & v_{i2} & \cdots & v_{im} \\ v_{j1} & v_{j2} & \cdots & v_{jm} \\ \vdots & \vdots & \vdots & \vdots \\ v_{m1} & v_{m2} & \cdots & v_{mm} \end{bmatrix} \right| = \text{maximum} \tag{3.146}$$

This selection process is continued until n key rows are obtained so that

$$|\det[\text{KEY SET}]| = \left| \det \begin{bmatrix} \mathbf{v}(\text{KEY } 1)' \\ \vdots \\ \mathbf{v}(\text{KEY } i)' \\ \vdots \\ \mathbf{v}(\text{KEY } n)' \end{bmatrix} \right| = \text{maximum} \tag{3.147}$$

Pure noise points, generated by automatic recording devises, pose a serious problem with this procedure because the normalization procedure treats noise points as real data, giving them equal weight after normalization. In order to eliminate the possibility of including a noise row as a key factor, the noise rows must be identified by some screening process and removed from consideration. The importance of such screening cannot he overemphasized. Deleting all rows with a root-mean-square response less than five times the real error is recommended.

Because the selection process of KSFA is sequential, requiring truncation of the vectors for the determinant calculations, the procedure will not necessarily yield the most orthogonal set of data rows or columns especially for systems involving large numbers of factors. Iterative key set factor analysis[21] (IKSFA) represents a refinement of KSFA designed to correct for improper assignment caused by truncation. In IKSFA, the initial key set obtained from KSFA is treated as a first approximation to the true key set. Operating in n space, each key row (or column) is replaced in an efficient, iterative manner until the best key set in n space is obtained.

The first key row, v(KEY 1), in n space is replaced by the first row vector. If the determinant expressed by (3.147) is greater than the initial determinant, this row is a better representation, and the first key row is replaced by this row vector. Similarly, the second, third, fourth, and so forth row vectors are treated as replacement candidates, and v(KEY 1) is reassigned if the resulting determinant is larger. This process is repeated with the second key row, v(KEY 2), testing every row as a possible replacement for v(KEY 2), and replacing the key row whenever a larger determinant occurs. After every row has been tested, the process is repeated with the third key row, then the fourth key row, and so forth until all key rows have been examined and replaced when improvement in orthogonality occurs.

After this complete cycle, the iterative process is repeated, starting with the new key set, and replacing the new v(KEY 1), v(KEY 2), and so forth, whenever improvement occurs. The iteration is terminated when no replacements are required through a complete cycle.

An example of the iterative refinement is illustrated in Table 3.1, which concerns the mass spectra of 18 mixtures containing 10 components. The spectra consisted of 35 of the largest mass points between m/z 27 and 137. Each row in the table exhibits the new key set obtained after completion of the given iteration. The first row shows the key set of m/z designees obtained from KSFA. Replacement of m/z 83 by 111, as shown in the second row, yields a slightly larger determinant. A substantial increase in the determinant occurs at iterations 5 and 7, concerning replacement of KEY(5) and KEY(7). Convergence is confirmed by the second cycle, where no further improvement occurs. Comparison of the first and last rows illustrates the improvement achieved by IKSFA.

IKSFA can be used for a variety of applications. For example, it can be used for "preferred set selection." The object of preferred set selection is to identify the minimum set of measurements that best characterizes a multicomponent system.[22] The need for preferred set selection occurs most frequently in analytical spectroscopy, where it has been used for analytical calibration, wavelength selection, absorbance ratio monitoring, spectral isolation, modeling, and prediction. Care

TABLE 3.1 Iterative Refinement of the Key Set for Simulated Mass Spectra of 10 Component Mixtures[a]

| Cycle (i) | Iteration (i) | m/z^b | | | | | | | | | | DET(KEY SET) |
		KEY(1)	KEY(2)	KEY(3)	KEY(4)	KEY(5)	KEY(6)	KEY(7)	KEY(8)	KEY(9)	KEY(10)	
0[c]	0[c]	83	106	91	108	43	120	122	93	112	104	0.0893
1	1	**111**	106	91	108	43	120	122	93	112	104	0.0896
1	2	111	**44**	91	108	43	120	122	93	112	104	0.0956
1	3	111	44	91	108	43	120	122	93	112	104	0.0956
1	4	111	44	91	**79**	43	120	122	93	112	104	0.0973
1	5	111	44	91	79	**71**	120	122	93	112	104	0.1813
1	6	111	44	91	79	71	120	122	93	112	104	0.1813
1	7	111	44	91	79	71	120	**137**	93	112	104	0.2393
1	8	111	44	91	79	71	120	137	93	112	104	0.2393
1	9	111	44	91	79	71	120	137	93	112	104	0.2393
1	10	111	44	91	79	71	120	137	93	112	**121**	0.2409
2	1	111	44	91	79	71	120	137	93	112	121	0.2409
⋮	⋮	⋮	⋮	⋮	⋮	⋮	⋮	⋮	⋮	⋮	⋮	⋮
2	10	111	44	91	79	71	120	137	93	112	121	0.2409

[a]Reprinted with permission from K. J. Schostack and E. R. Malinowski, *Chemometrics Intell. Lab. Syst.*, **6**, 21 (1989).
[b]Key m/z replaced by refinement process are boldface.
[c]Initial key set selected by KSFA.

must be exercised when applying IKSFA because the method will always generate a solution. The validity of the results must ultimately be judged on the basis of chemical sense. For example, in certain cases the generation of negative concentrations or negative spectra may indicate that the method is not applicable.

3.5 CLASSICAL FACTOR ANALYSIS, COMMON FACTORS, AND COMMUNALITY

Classical factor analysis, as developed by the early pioneers,[23] is a search for the subset of factors that are common to the variables. This philosophy was adopted by the behavioral scientists because their data contain more factors than mathematically solvable. By focusing on the common factors, the problem is reduced to a practical size, yielding, albeit, only a part of the total solution.

Theoretically, the data are interpreted to be a linear combination of common factors and unique factors. *Common factors* are those factors that are shared between two or more variables of the data. *Unique factors* are those factors that are unique to a specific variable. Unique factors are composed of specific factors and error factors. *Specific factors* represent real, detectable sources of variability associated with a particular variable. *Error factors* are due to imperfections in the variable measurements and may be unique to the variable. The latter situation is illustrated by the following (transposed) loading matrix involving four variables:

$$
\begin{array}{ccccccccccccc}
\text{common} & & & & & \text{specific} & & & & & \text{error} & & \\
1 & 2 & 3 & 4 & 5 & 6 & 7 & 8 & 9 & 10 & 11 & 12 & 13 \\
\left| \begin{array}{ccc|cc}
c_{11} & c_{21} & c_{31} & c_{41} & c_{51} \\
c_{12} & c_{22} & c_{32} & & \\
c_{13} & c_{23} & c_{33} & & \\
c_{14} & c_{24} & c_{34} & &
\end{array} \right. & & & & &
\begin{array}{cccc}
& & & \\
c_{62} & & & \\
& c_{73} & c_{83} & \\
& & & c_{94}
\end{array} & & & &
\left. \begin{array}{cccc}
c_{10,1} & & & \\
& c_{11,1} & & \\
& & c_{12,1} & \\
& & & c_{13,1}
\end{array} \right|
\end{array}
$$

By combining the specific factors and error factors into unique factors the above matrix can be reduced conveniently to

$$
\begin{array}{ccccccc}
\text{common} & & & & \text{unique} & & \\
1 & 2 & 3 & 4 & 5 & 6 & 7 \\
\left| \begin{array}{ccc|c}
c_{11} & c_{21} & c_{31} & c_{41} \\
c_{12} & c_{22} & c_{32} & \\
c_{13} & c_{23} & c_{33} & \\
c_{14} & c_{24} & c_{34} &
\end{array} \right. & & & &
\left. \begin{array}{ccc}
& & \\
c_{52} & & \\
& c_{63} & \\
& & c_{74}
\end{array} \right|
\end{array}
$$

The basic equation of classical factor analysis has the following form

$$d_{ik} = \sum_j r_{ij}c_{jk}(\text{common}) + r_{iu}c_{uk}(\text{unique}) \tag{3.148}$$

Because the unique factors are uncorrelated, they do not contribute to the off-diagonal elements of the covariance matrix but do contribute to the diagonal terms.

The variance in the data, according to classical theory, is the sum of three terms—common variance, specific variance, and error variance:

$$\text{variance} = \overbrace{\text{common variance} + \underbrace{\text{specific variance} + \text{error variance}}_{\text{unique variance}}}^{\text{true variance}}$$

The *true variance* in the data is the sum of the common variance and the specific variance. The *unique variance* is the sum of the specific variance and the error variance.

In classical factor analysis, correlation about the origin is the usual practice because the variables, in most cases, have different units. Thus the correlation matrix will have unities along the diagonal, consisting of contributions from the common terms as well as from the uniqueness terms.

The total variance for variable k, a diagonal element of the correlation matrix, can be expressed as:

$$z_{kk} = h_k^2 + u_k^2 = 1 \quad \text{(for correlation)} \tag{3.149}$$

where h_k^2, called the *communality* of variable k, is due to n common variables,

$$h_k^2 = \sum_{j=1}^{n} c_{jk}^2 (\text{common}) \tag{3.150}$$

and u_k^2, called the *uniqueness* of variable k, is that part of the total variance not shared with any other variable,

$$u_k^2 = \sum_{j=1}^{n} c_{jk}^2 (\text{specific}) + c_{\cdot k}^2 (\text{error}) = c_{\cdot k}^2 (\text{unique}) \tag{3.151}$$

Equations (3.150) and (3.151) result from the fact that each variable (i.e., each column of $\mathbf{D}$) has been normalized, so that $\mathbf{Z} = \mathbf{D}'\mathbf{D} = (\mathbf{RC})'(\mathbf{RC}) = \mathbf{C}'\mathbf{R}'\mathbf{RC} = \mathbf{C}'\mathbf{C}$.

The common factor space is determined by replacing the diagonal elements of the correlation matrix by the communalities obtained by subtracting the uniqueness of each variable,

$$h_k^2 = z_{kk} - u_k^2 = 1 - u_k^2 \tag{3.152}$$

and then decomposing the reduced correlation matrix into eigenvalues.

The trouble that has plagued classical factor analysis since its inception is the fact that there exists no a priori way for determining the uniqueness of the variables. The results obtained are sensitive to the method used to estimate the communalities. For this reason, many scientists are reluctant to accept the classical method.

There are two general classes of methodologies for determining the communalities. In the first class the communalities are fixed in advance. This requires estimation of the communalities by invoking some principle such as the square multiple correlation coefficient followed by some iteration procedure to yield new estimates. In the second class the number of common factors is specified in advance.

There exists several statistical tests to help determine the number of common factors. The maximum-likelihood method is most popular in this class. Unfortunately, these methods are sensitive to the number of factors chosen. If the number of common factors is increased by one unit, an entirely different set of common factors is obtained. Even when the rank of the correlation matrix is fixed, the communalities may not be uniquely determined. This has promulgated a variety of techniques that invoke additional principles. Today, no single method has emerged as the "preferred" method of the classical analysts.

REFERENCES

1. E. R. Malinowski, *Diss. Abstr.*, **23** (8), Publ. No. 62-2027 (1963).
2. P. T. Funke, E. R. Malinowski, D. E. Martire, and L. Z. Pollara, *Sep. Sci.*, **1**, 661 (1966).
3. P. H. Weiner, E. R. Malinowski, and A. R. Levinstone, *J. Phys. Chem.*, **74**, 4537 (1970).
4. R. J. S. Rummel, *Applied Factor Analysis*, Northwestern University Press, Evanston, IL, 1970.
5. H. H. Harman, *Modern Factor Analysis*, 3rd rev. ed., University of Chicago Press, Chicago, 1967.
6. B. Fruchter, *Introduction to Factor Analysis*, Van Nostrand, New York, 1954.
7. A. L. Comrey, *A First Course in Factor Analysis*, Academic, Orlando, 1973.
8. D. N. Lawley and A. E. Maxwell, *Factor Analysis as a Statistical Method*, 2nd ed., Buttersworth, London, 1971.
9. P. Horst, *Factor Analysis of Data Matrices*, Holt, Rinehart and Winston, New York, 1965.
10. R. W. Rozett and E. M. Petersen, *Anal. Chem.*, **47**, 1301 (1975).
11. K. G. Joreskog, J. E. Klovan, and R. A. Reyment, *Geological Factor Analysis*, Elsevier, Amsterdam, 1976.
12. E. Johansson, S. Wold, and K. Sjodin, *Anal. Chem.*, **56**, 1685 (1984).
13. R. I. Shrager, *Chemometrics Intell. Lab. Syst.*, **1**, 59 (1986).
14. A. Lorber and B. R. Kowalski, *J. Chemometrics*, **2**, 93 (1988).
15. H. Wold, in P. Krishnaiah (Ed.), *Multivariate Analysis*, Academic, Orlando, 1966.
16. H. F. Kaiser, *Psychometrika*, **23**, 187 (1958).
17. J. D. Carroll, *Psychometrika*, **18**, 23 (1953).
18. D. R. Saunders, *Psychometrika*, **25**, 199 (1960).
19. T. H. Brayden, P. A. Poropatic, and J. L. Watanabe, *Anal. Chem.*, **60**, 1154 (1988).
20. E. R. Malinowski, *Anal. Chem. Acta*, **134**, 129 (1982).
21. K. J. Schostack and E. R. Malinowski, *Chemometrics Intell. Lab. Syst.*, **6**, 21 (1989).
22. D. L. Massart, A. Dijkstra, and L. Kaufman, *Evaluation and Optimization of Laboratory Methods and Analytical Procedures*, Elsevier, Amsterdam, 1978, Chap. 17.
23. L. L. Thurstone, *Multiple-Factor Analysis*, The University of Chicago Press, Chicago, 1947.

4

The simplest solution is the best solution.

EFFECTS OF EXPERIMENTAL ERROR ON TARGET FACTOR ANALYSIS

4.1   THEORY OF ERROR

In the absence of experimental error, factor analysis (FA) will yield the exact number of factors involved in a data matrix. Unfortunately, perfect data are not attainable, and the chemist is plagued with handling data that possess experimental uncertainty. Although in many instances we can make a reliable estimate of the error, we are often not sure of the uncertainty estimation. Experimental error blends into FA and tends to complicate the process at every decision-making step. For this reason it is important for us to study how errors weave into the factor analytical scheme. Only then can we make provisions to account for the perturbations produced by error. We learn in this chapter how special factor analytical techniques can be used to deduce not only the number of factors but also the experimental error, even when no such estimates are available.

4.1.1   Preliminary Considerations

Because of experimental error a "raw" data point, d_{ik}, is best represented as a sum of two terms:

$$d_{ik} = d_{ik}^* + e_{ik} \tag{4.1}$$

where d_{ik}^* represents a "pure" data point, free of experimental error, and e_{ik} is the experimental error associated with the data point.

Experimental error invariably produces a larger number of eigenvectors than is required by the pure factor space. Retention of all the resulting eigenvectors will lead to perfect reproduction of the raw data, including experimental error. Hence we need reliable criteria to help us choose the correct number of eigenvectors. Even when we do select the correct eigenvectors, it is impossible to remove all the error. In fact, we shall find that a part of the error mixes into the data reproduction scheme. In other words,

$$e_{ik} = e_{ik}^\dagger + e_{ik}^0 \tag{4.2}$$

where $e_{ik}^\dagger$ is that part of the error that mixes into the factor analytical scheme and e_{ik}^0, called the *residual error*, is that part that can be removed by deleting the unnecessary eigenvectors.

Inserting (4.2) into (4.1), we see that

$$d_{ik} = d_{ik}^\dagger + e_{ik}^0 \tag{4.3}$$

where

$$d_{ik}^\dagger = d_{ik}^* + e_{ik}^\dagger \tag{4.4}$$

Here $d^{\dagger}_{ik}$ is a reproduced data point using the proper number of eigenvectors. From (4.4) we see that the reproduced data point contains some error. From (4.3) we see that the residual error e^{0}_{ik} is simply the difference between the raw data and reproduced data point. Criteria for choosing the correct number of eigenvectors must be based on an understanding of how the experimental error enters and mixes into the factor analytical scheme. In this section we investigate this important aspect.

The raw data matrix is simply the sum of two matrices: a pure data matrix $\mathbf{D^*}$ and an error matrix $\mathbf{E}$:

$$\mathbf{D} = \mathbf{D^*} + \mathbf{E} \qquad (4.5)$$

All these matrices have the same size, $r \times c$, involving r rows and c columns. However, their dimensionalities are not the same. Although the pure data matrix is n dimensonal, the raw data matrix and the error matrix are either r or c dimensional, whichever is smaller. Throughout our discussion here we assume that c is smaller than r. Hence the error matrix is c dimensional.

Since the factor space of the pure data is n dimensional, a pure data point can be expressed as a linear sum of product terms:

$$d^*_{ik} = \sum_{j=1}^{n} r^*_{ij} c^*_{jk} \qquad (4.6)$$

This expression is identical to (3.1), except that an asterisk has been inserted as a superscript to distinguish the pure factors from factors that contain an amalgamation with error.

Although the error matrix is the same size as the data matrix, a larger number of eigenvector axes are required to span the error space than the pure data space. This is due to the fact that the error matrix consists of random values. Any orthogonal set of axes can be chosen to define the error space, as long as there is a sufficient number of axes. The number of axes required to span the error space will exactly equal the number of columns in the data matrix, assuming that $c < r$.

A set of n basis axes, called *primary axes*, is required to describe the raw data space within experimental error. However, all c axes are required to span the error space. In other words, there is a set of axes ($c - n$ in number), called *error axes* or *secondary axes*, that are associated solely with the remaining part of the error. Since the same n axes used to describe the raw data space may be used to describe part of the error space, the error associated with the raw data point, d_{ik}, can be represented as the following linear sum:

$$e_{ik} = \sum_{j=1}^{n} \sigma^{\dagger}_{ij} c_{jk} + \sum_{j=n+1}^{c} \sigma^{0}_{ij} c_{jk} \qquad (4.7)$$

Here c_{jk} is the kth component of the jth principal axis, σ_{ij} the projection of the ith row designee of the error matrix onto the jth primary axis, and σ^{0}_{ij} the corresponding projection onto the jth secondary axis. The sum is taken over all c eigenvector axes.

Placing (4.6) and (4.7) into (4.1), we find that

$$d_{ik} = \sum_{j=1}^{n} (r_{ij}^* c_{jk}^* + \sigma_{ij}^{\ddagger} c_{jk}) + \sum_{j=n+1}^{c} \sigma_{ij}^0 c_{jk} \tag{4.8}$$

Defining r_{ij} as follows:

$$r_{ij} = r_{ij}^* \frac{c_{jk}^*}{c_{jk}} + \sigma_{ij}^{\ddagger} \tag{4.9}$$

we obtain

$$d_{ik} = \sum_{j=1}^{n} r_{ij} c_{jk} + \sum_{j=n+1}^{c} \sigma_{ij}^0 c_{jk} \tag{4.10}$$

Based on this formulation, the complete factor analytical solution in matrix notation can be expressed as

$$\mathbf{D} = \mathbf{R}^{\#} \mathbf{C} + \mathbf{R}^0 \mathbf{C} \tag{4.11}$$

Here C is composed of the complete set of eigenvectors,

$$\mathbf{R}^{\#} \equiv \begin{bmatrix} r_{11} & \cdots & r_{1n} & 0 & \cdots & 0 \\ r_{21} & \cdots & r_{2n} & 0 & \cdots & 0 \\ \vdots & & \vdots & \vdots & & \vdots \\ r_{r1} & \cdots & r_{rn} & 0 & \cdots & 0 \end{bmatrix} \tag{4.12}$$

$$\mathbf{R}^0 \equiv \begin{bmatrix} 0 & \cdots & 0 & \sigma_{1,n+1}^0 & \cdots & \sigma_{1c}^0 \\ 0 & \cdots & 0 & \sigma_{2,n+1}^0 & \cdots & \sigma_{2c}^0 \\ \vdots & & \vdots & \vdots & & \vdots \\ 0 & \cdots & 0 & \sigma_{r,n+1}^0 & \cdots & \sigma_{rc}^0 \end{bmatrix} \tag{4.13}$$

If the zero elements, which are associated with the secondary eigenvalues, of $\mathbf{R}^{\#}$ are dropped from this $r \times c$ matrix, the usual $r \times n$ row-factor matrix, $\mathbf{R}^{\ddagger}$, is obtained. Similarly, upon deleting the secondary eigenvectors, C is reduced to $C^{\ddagger}$, the usual $n \times c$ column-factor matrix. The elements of $\mathbf{R}^0$ consist of nothing but error components and contain no useful information. Deletion of this matrix leads to factor compression, and a reproduced data matrix, $\mathbf{D}^{\ddagger}$, that differs slightly from the raw data matrix but is essentially the same within experimental error:

$$\begin{aligned} \mathbf{D}^{\ddagger} &= \mathbf{R}^{\#} \mathbf{C} \\ &= \mathbf{R}^{\ddagger} \mathbf{C}^{\ddagger} \end{aligned} \tag{4.14}$$

where $C^{\ddagger}$ contains only the n primary eigenvectors.

Using (3.36) and (3.37) we find that the covariance matrix can be decomposed into the following sum:

$$\mathbf{Z} = \sum_{j=1}^{n} \lambda_j^{\dagger} \mathbf{c}_j \mathbf{c}_j' + \sum_{j=n+1}^{c} \lambda_j^{0} \mathbf{c}_j \mathbf{c}_j' \tag{4.15}$$

where

$$\lambda_j^{\dagger} = \sum_{i=1}^{r} r_{ij}^2 \qquad \text{for } j = 1, \ldots, n \tag{4.16}$$

$$\lambda_j^{0} = \sum_{i=1}^{r} (\sigma_{ij}^0)^2 \qquad \text{for } j = n+1, \ldots, c \tag{4.17}$$

These equations show exactly how the error mixes into the factor analytical scheme. If there were no error, $\sigma_{ij}^{\dagger}$ and σ_{ij}^0 would equal zero, c_{jk} would equal c_{jk}^*, and (4.16) would reduce to

$$\lambda_j^{\dagger} = \sum_{i=1}^{r} r_{ij}^{*2} \tag{4.18}$$

and all the λ_j^0 would vanish. Instead of c eigenvalues, only n eigenvalues would result. The factor space would clearly be identified.

To see how the error mixes into the eigenvectors, we take advantage of the symmetry of the problem. The error could have been expressed in terms of axes associated with r_{ij} rather than c_{jk}. Instead of (4.7), we can write

$$e_{ik} = \sum_{j=1}^{n} r_{ij} \sigma_{jk}^{\dagger} + \sum_{j=n+1}^{c} r_{ij} \sigma_{jk}^{0} \tag{4.19}$$

where r_{ij} is the ith component of the jth axis, and $\sigma_{jk}^{\dagger}$ and σ_{jk}^0 are the projections of the kth column designee of the error matrix onto the jth axis of the primary and secondary axes, respectively.

We now make the following deductions, analogous to those employed to obtain (4.8):

$$\begin{aligned}
d_{ik} &= \sum_{j=1}^{n} (r_{ij}^* c_{jk}^* + r_{ij} \sigma_{jk}^{\dagger}) + \sum_{j=n+1}^{c} r_{ij} \sigma_{jk}^{0} \\
&= \sum_{j=1}^{n} r_{ij} c_{jk} + \sum_{j=n+1}^{c} r_{ij} \sigma_{jk}^{0}
\end{aligned} \tag{4.20}$$

where c_{jk} is defined as

$$c_{jk} \equiv c_{jk}^* \frac{r_{ij}^*}{r_{ij}} + \sigma_{jk}^\dagger \tag{4.21}$$

This equation shows precisely how the experimental error perturbs the components of the eigenvectors of factor analysis. If there were no error, σ_{jk}^0 and $\sigma_{jk}^\dagger$ would be zero, $r_{ij}^\dagger$ would equal r_{ij}^*, and c_{jk} would equal c_{jk}^*, as expected.

4.1.2   Primary and Secondary Factors

If the data matrix contained no error, the covariance matrix would be decomposed into a sum of n factors:

$$\mathbf{Z} = \sum_{j=1}^{n} \lambda_j \mathbf{c}_j \mathbf{c}_j' \tag{4.22}$$

However, because of experimental error, the decomposition will lead to a larger number of eigenvectors. In fact, since the covariance matrix is of size $c \times c$, where c is the number of columns in the data matrix, the decomposition will yield c factors. That is,

$$\mathbf{Z} = \sum_{j=1}^{c} \lambda_j \mathbf{c}_j \mathbf{c}_j' \tag{4.23}$$

As shown in (4.15), this sum can be separated into two groups. The first n terms in this sum are associated with the true factors but contain an admixture of error. The second set of terms consists of pure error. It is this second set of terms that should be omitted from further consideration.

The trace of the covariance matrix is invariant upon the similarity transformation expressed by (3.71). Hence the trace of the covariance matrix is related to the data points and the eigenvalues as follows:

$$\sum_{i=1}^{r} \sum_{k=1}^{c} d_{ik}^2 = \text{trace}(\mathbf{Z}) = \sum_{j=1}^{c} \lambda_j \tag{4.24}$$

Because of experimental error, the number of eigenvalues will always equal the number of columns in the data matrix, assuming, of course, that $c < r$. These eigenvalues can be grouped into two sets: a set consisting of the true eigenvalues but containing an admixture of error, and a set consisting of error only:

$$\sum_{j=1}^{c} \lambda_j = \sum_{j=1}^{n} \lambda_j^\dagger + \sum_{j=n+1}^{c} \lambda_j^0 \tag{4.25}$$

The first set, called the *primary eigenvalues*, consists of the first n members having the largest values. The second set, called the *secondary eigenvalues*, contains $c - n$ members having the smallest values. The secondary eigenvalues are composed solely of experimental error, and their removal from further consideration will, in fact, lead to data reproduction that will be more accurate than the original data (see Section 4.2). We must develop criteria to deduce how many of the smallest eigenvalues belong to this set.

Because the primary set of eigenvalues contain an admixture of error, their associated eigenvectors are not the true eigenvectors. Factor analysis of the improved data matrix, regenerated by the primary set of eigenvectors, will yield exactly the same set of primary eigenvalues and eigenvectors. The new secondary set will be extremely small because it is due to numerical roundoff. Consequently, further purification of the data by repeated factor analytical decomposition is not possible.

Using the primary eigenvectors, we can regenerate the data, which we label $d_{ik}^{\dagger}$. The regenerated data obey the following relationship analogous to (4.24):

$$\sum_{i=1}^{r} \sum_{k=1}^{c} d_{ik}^{\dagger 2} = \text{trace}[\mathbf{Z}^{\dagger}] = \sum_{j=1}^{n} \lambda_{j}^{\dagger} \tag{4.26}$$

By subtracting (4.26) from (4.24), we find that

$$\sum_{i=1}^{r} \sum_{k=1}^{c} (d_{ik}^{2} - d_{ik}^{\dagger\,2}) = \sum_{j=n+1}^{c} \lambda_{j}^{0}$$
$$= \sum_{i=1}^{r} \sum_{j=n+1}^{c} (\sigma_{ij}^{0})^{2} \tag{4.27}$$

where σ_{ij}^{0} is a row factor associated with the residual error. In other words, the sum of the differences between the squares of the raw data and the reproduced data equals the sum of the secondary eigenvalues. This sum is associated with the error removed by neglecting the secondary eigenvalues. Equation (4.27) is important because it shows the relationship among the raw data, the reproduced data, and the secondary eigenvalues.

Although the pure data matrix and error matrix are not mutually orthogonal, it is quite surprising that the reproduced data matrix and its associated residual error matrix are mutually orthogonal. To prove this, we first consider the perfect reproduction of the data matrix using all the eigenvectors, as shown in (4.11). This equation reproduces the data perfectly, including all the error. From (4.11) and (4.14), we conclude that

$$\mathbf{D} = \mathbf{D}^{\dagger} + \mathbf{E}^{0} \tag{4.28}$$

where

$$\mathbf{E}^{0} = \mathbf{R}^{0}\mathbf{C}$$

We note here that the residual error matrix, $\mathbf{E}^0$, is the difference between the raw data matrix and the abstract factor analysis (AFA)–reproduced data matrix. Furthermore, from (4.12) and (4.13), it is evident that

$$\mathbf{R}^{\#\prime}\mathbf{R}^0 = [\mathbf{0}] \qquad (4.29)$$

Using this equation, we find that

$$
\begin{aligned}
\mathbf{D}^{\ddagger\prime}\mathbf{E}^0 &= \{\mathbf{R}^{\#}\mathbf{C}\}'\{\mathbf{R}^0\mathbf{C}\} \\
&= \mathbf{C}'\mathbf{R}^{\#\prime}\mathbf{R}^0\mathbf{C} \\
&= \mathbf{C}'[\mathbf{0}]\mathbf{C} \\
&= [\mathbf{0}]
\end{aligned}
\qquad (4.30)
$$

Thus the AFA-regenerated data matrix and its associated residual error matrix are mutually orthogonal. This important fact will be used later to develop error criteria for determining the dimensions of the factor space.

4.1.3 Example

To illustrate how the experimental error perturbs the primary eigenvectors and produces secondary eigenvectors (which are composed solely of error), let us examine the simple one-factor data matrix[1,2] shown in Table 4.1. This matrix consists of two identical data columns labeled "pure data matrix." This matrix is obviously one dimensional since a plot of the points of the first column against the corresponding points of the second column yields a perfectly straight line with all

TABLE 4.1 Values Used to Construct an Artificial Raw Data Matrix and the Results of Factor Analyzing This Raw Data Matrix[a]

Pure Data Matrix, $\mathbf{D}^*$		Error Matrix, $\mathbf{E}$		Raw Data Matrix, $\mathbf{D} = \mathbf{D}^* + \mathbf{E}$		Reproduced Data Matrix Using One Factor, $\mathbf{D}^{\ddagger} = \mathbf{R}^{\ddagger}\mathbf{C}^{\ddagger}$	
1	1	0.2	0.0	1.2	1.0	1.0936	1.1052
2	2	−0.2	−0.2	1.8	1.8	1.7904	1.8095
3	3	−0.1	0.1	2.9	3.1	2.9846	3.0163
4	4	0.0	−0.1	4.0	3.9	3.9288	3.9705
5	5	−0.1	0.0	4.9	5.0	4.9671	4.9763
6	6	0.2	−0.2	6.2	5.8	5.9672	6.0305
7	7	0.2	−0.1	7.2	6.9	7.0118	7.0862
8	8	−0.2	0.1	7.8	8.1	7.9086	7.9920
9	9	−0.2	0.1	8.8	9.1	8.9033	8.9978
10	10	−0.1	0.2	9.9	10.2	9.9974	10.1036

[a]Reprinted with permission from E. R. Malinowski, *Anal. Chem.*, **49**, 606 (1977).

TABLE 4.2 Eigenvalues and Row-Factor Scores Resulting from Factor Analysis[a]

From Pure Data Matrix	From Raw Data Matrix	
$\lambda_1 = 770$	$\lambda_1^{\dagger} = 767.1514$	$\lambda_2^0 = 0.2886124$
r_{i1}^*	r_{i1}	σ_{i2}^0
1.41421	1.55487	0.14964
2.82843	2.54555	0.01344
4.24264	4.24333	−0.11901
5.65685	5.58569	0.10021
7.07107	7.00063	−0.03374
8.48528	8.48367	0.32765
9.89950	9.96895	0.26479
11.31371	11.24396	−0.15275
12.72792	12.65816	−0.14528
14.14214	14.21377	−0.13707

[a]Reprinted with permission from E. R. Malinowski, *Anal. Chem.*, **49**, 606 (1977).

points lying exactly on the one-dimensional line axis. When this data matrix is factor analyzed, via the covariance matrix, the results shown in Table 4.2 are obtained. Each row-factor score, r_{i1}^*, is the distance from the origin to the data point. The eigenvalue, 770, is simply the sum of the squares of the scores.

To see how experimental error perturbs these results, the following procedure is employed. First the "error matrix" shown in Table 4.1 is arbitrarily generated. This error matrix is then added to the pure data matrix to give the "raw data matrix," also shown in Table 4.1. The raw data matrix simulates real chemical data possessing experimental error. When the raw data matrix is factor analyzed, not one but two eigenvalues and their associated eigenvectors are produced. These are listed in Table 4.2.

Two eigenvectors emerge because the raw data points, unlike the pure data points, lie in a two-dimensional plane and not on a one-dimensional line. This is dramatically illustrated in Figure 4.1, where the scores of the first column of the raw data matrix are plotted against the corresponding scores of the second column. The direction of the primary axis is shifted slightly from its original angle of 45° so that $c_1 \neq c_1^*$. This occurs because part of the error cannot be removed by deleting the $c - n$ secondary eigenvectors. Each row-factor score, r_{i1}, is the distance along the primary axis to the point where the perpendicular projection of the data point intersects the axis. Note that the primary row factors of the raw data matrix are different from those of the pure data matrix. One can verify, after some tedious calculations, that these values obey (4.9).

The second eigenvector, c_2 that emerges consists of pure error. Each row factor, σ_{i2}^0, is the distance along the secondary axis (see Figure 4.1) from the origin to the place where the projection of the raw data point intersects the secondary axis. The

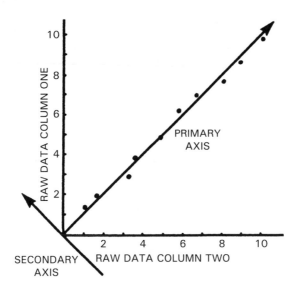

Figure 4.1   Geometrical relationships between the raw data points and the primary and secondary axes resulting from principal factor analysis of the raw data matrix. [Reprinted with permission from E. R. Malinowski, *Anal. Chem.*, **49**, 606 (1977).]

secondary eigenvalue, λ_2^0, has no physical meaning, since it is the sum of squares of the secondary scores, which contain nothing but error. By deleting this secondary axis, one obtains a reproduced raw data matrix that emulates the pure data matrix more accurately than does the original raw data matrix. This can be verified by comparing the reproduced and the raw data matrices with the pure matrix. The reproduced data matrix, using only the primary eigenvector, is given on the extreme right in Table 4.1.

4.2   AFA FOR DATA IMPROVEMENT

Abstract factor analysis reproduction has a built-in statistical feature. Because the secondary set of abstract eigenvalues are composed of pure error, their removal invariably will lead to data improvement. Like magic, this is accomplished without any a priori knowledge of the controlling factors. Although this is not the original intent of factor analysis, it does represent a valuable, unexpected benefit. In this section we attempt to determine quantitatively how much data improvement is possible using abstract factor analysis.

The raw experimental data matrix is, in reality, a sum of two matrices: a pure data matrix, having no error, and an error matrix. Our discussion in Section 4.1.2 shows that when we factor analyze the *pure* data matrix we obtain exactly n eigenvalues and n eigenvectors. If, on the other hand, we factor analyze the *raw* data, we obtain c eigenvalues and c eigenvectors. However, only n of these are associated with the true factors. The remaining $c - n$ eigenvalues and eigenvectors are composed of pure error.

Instead of factor analyzing the raw data matrix, let us examine what happens when we factor analyze the error matrix via its covariance matrix. The same eigenvectors used to describe the raw data can be used to describe the error space.[1,2] Of course, the resulting eigenvalues would be composed of pure error. Instead of (4.25), we would obtain the following analogous equation:

$$\sum_{j=1}^{c} \lambda_{je} = \sum_{j=1}^{n} \lambda_{je}^{\dagger} + \sum_{j=n+1}^{c} \lambda_{je}^{0} \tag{4.31}$$

where

$$\sum_{j=1}^{n} \lambda_{je}^{\dagger} = \sum_{i=1}^{r} \sum_{j=1}^{n} (\sigma_{ij}^{\dagger})^2 \tag{4.32}$$

$$\sum_{j=n+1}^{c} \lambda_{je}^{0} = \sum_{i=1}^{r} \sum_{j=n+1}^{c} (\sigma_{ij}^{0})^2 \tag{4.33}$$

In (4.31), the term on the left is equal to the sum of the squares of all the error points in the data matrix. This is true because the trace of the covariance matrix constructed from the error matrix is invariant upon the similarity transformation involved in the decomposition (diagonalization) process. Combining this fact with (4.31), (4.32), and (4.33), we find that

$$\sum_{i=1}^{r} \sum_{k=1}^{c} e_{ik}^2 = \sum_{i=1}^{r} \sum_{j=1}^{n} (\sigma_{ij}^{\dagger})^2 + \sum_{i=1}^{r} \sum_{j=n+1}^{c} (\sigma_{ij}^{0})^2 \tag{4.34}$$

The term on the left is the sum of the experimental error squared. It also represents the sum of the squares of the projections of the error points onto all c data-column axes. The first sum on the right concerns the projections of the error points onto the n primary eigenvector axes. This sum represents the error that mixes into the factor analytical process and cannot be removed. The second sum on the right concerns the projections onto the $c - n$ secondary axes, the axes that are removed from the analysis since their associated eigenvalues contain nothing but pure error. These three terms are related to the residual standard deviation (RSD) in the following way:

$$rc(\text{RSD})^2 = \sum_{i=1}^{r} \sum_{k=1}^{c} e_{ik}^2 \tag{4.35}$$

$$rn(\text{RSD})^2 = \sum_{i=1}^{r} \sum_{j=1}^{n} (\sigma_{ij}^{\dagger})^2 \tag{4.36}$$

$$r(c-n)(\text{RSD})^2 = \sum_{i=1}^{r} \sum_{j=n+1}^{c} (\sigma_{ij}^{0})^2 \tag{4.37}$$

Each of these expressions represents a different way of obtaining the residual standard deviation.

Placing (4.35), (4.36), and (4.37) into (4.34) and dividing through by rc, we conclude that

$$(\text{RSD})^2 = \frac{n}{c}(\text{RSD})^2 + \frac{c-n}{c}(\text{RSD})^2 \tag{4.38}$$

This important identity summarizes the theoretical arguments presented. The RSD can be interpreted[1,2] to be composed of two terms: *imbedded error* (IE) and *extracted error* (XE). In other words, the residual standard deviation, which is the real error (RE), can be expressed in a Pythagorean fashion as follows:

$$(\text{RE})^2 = (\text{IE})^2 + (\text{XE})^2 \tag{4.39}$$

where

$$\text{RE} = \text{RSD} \tag{4.40}$$

$$\text{IE} = \text{RSD}\sqrt{\frac{n}{c}} \tag{4.41}$$

$$\text{XE} = \text{RSD}\sqrt{\frac{c-n}{c}} \tag{4.42}$$

The imbedded error is due to the fact that only a fraction of the error from the data mixes into the factor analytical reproduction process. This error becomes imbedded into the factors and cannot be removed by repeated factor analysis. The extracted error is due to the fact that some error is extracted when the secondary eigenvectors are dropped from the scheme.

The IE is a measure of the difference between the pure data and the factor analysis reproduced data, whereas RE is a measure of the difference between the pure data and the raw experimental data. As a mnemonic we can symbolicially represent these statements in the form of a right triangle as shown in Figure 4.2.

Equation (4.41) shows that when n is less then c, IE is less than RE. Thus the error between the FA-reproduced data and the pure data is less than the original error

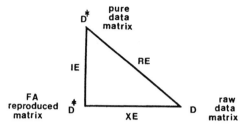

Figure 4.2 Mnemonic diagram of the Pythagorean relationship between the theoretical errors. [Reprinted with permission from E. R. Malinowski, *Anal. Chem.*, **49**, 606 (1977).]

between the raw data and the pure data. Hence by using the primary abstract eigenvectors of factor analysis, we can always improve the data (even when we are not able to identify the true underlying factors) simply by having more than n columns in our data matrix.

At this point it is informative to point out that the root-mean-square (RMS) error, defined by (4.46) in a later section is a measure of the difference between the FA-reproduced data and the raw experimental data. A quite interesting fact now becomes apparent: The RMS is identical to XE. This can be verified by placing (4.44) into (4.42) and comparing the result with (4.52).

Malinowski[1,2] reported the results of an error study using artificially constructed data matrices. His findings are summarized in Table 4.3. The first column gives n, the number of factors used to generate a pure data matrix. The second column shows the number of rows and columns, $r \times c$, in each data matrix. The third column gives the range of error in the error matrix that was added to the pure data matrix to yield the raw data matrix. The "true RMS of the errors" (i.e., the difference between the raw data and the pure data) of the various error matrices are listed in the fifth column.

TABLE 4.3   Summary of Results of Factor-Analyzing Model Data[a]

		Artificial Data and Artificial Error			Factor Analysis of Raw Data		
n	$r \times c$	Range of Pure Data	Range of Error	True RMS of Errors	RE(RSD)	IE	RMS of $(d_{ik}^{+} - d_{ik}^{*})$
1	10×2	1–10	−0.2–0.2	0.148	0.170	0.120	0.087
2	16×5	2–170	−0.08–0.08	0.041	0.041	0.026	0.026
2	16×5	2–170	−0.99–0.91	0.566	0.511	0.323	0.381
2	15×5	0–27	−1.9–2.0	1.040	0.962	0.608	0.726
3	10×6	0–32	−0.9–0.9	0.412	0.376	0.266	0.311
4	16×9	−581–955	−1.0–1.0	0.548	0.499	0.333	0.398
5	10×9	−137–180	−1.0–1.0	0.463	0.372	0.277	0.400

[a]Reprinted with permission from E. R. Malinowski, *Anal Chem.*, **49**, 606 (1977).

When these artificially generated raw data matrices were subjected to AFA, using the proper number of primary eigenvalues, the RE and IE values listed in the table were obtained. RE and IE were calculated from the abstract eigenvectors produced by the factor analysis by means of (4.44) and (4.41.). Theoretically, the calculated RE should be identical to the true RMS of the errors. Comparing values in the fifth column with the corresponding values in the sixth column of Table 4.3, we see that this is true (realizing that only the first digit of an error has significance).

The last column in Table 4.3 contains the RMS of the difference between the FA-reproduced data and the pure data. These values should be identical to the IE values calculated by the abstract eigenvalues. From Table 4.3 we see that this is generally so within the first digit. Perfect agreement should not be expected because of the statistically small number of data points used in the model analyses.

Using an excessive number of eigenvectors in the reproduction process, thus minimizing the RMS error, is a practice that must be avoided because this tends to minimize the extracted error. Extra factors simply reproduce unwanted experimental error. Conversely, an insufficient number of eigenvectors will not account for the true variables at play. The RMS error will be too large. We must therefore search for various criteria to judge the exact number of factors. This problem is attacked in the next section.

4.3   DEDUCING THE NUMBER OF FACTORS

Deducing the exact size of the true factor space is a difficult problem because of experimental error. The various techniques developed to solve this problem can be divided into two general categories: (1) methods based on a knowledge of the experimental error and (2) approximate methods requiring no knowledge of the experimental error. Obviously, methods in the first category are preferred when the error is known. Often such information is lacking and methods in the second category must be employed because they offer a solution, albeit of a more dubious nature.

4.3.1   Methods Based on Experimental Error

Various criteria have been developed for determining the size of the true factor space when the experimental error is known. Many of these criteria are described in several excellent texts[3-6] written for the social scientist rather than the physical scientist. To compare one criterion with another we apply the criterion, whenever possible, to the same data matrix chosen from the chemical literature. The infrared absorbance matrix of Bulmer and Shurvell[7] serves as a benchmark. This matrix consists of 1800 data points concerning the absorbances of 9 solutions of acetic acid in carbon tetrachloride measured at 200 different wavelengths in the carbonyl region. The standard deviation in absorbance was estimated to vary between 0.0005 and 0.0015 absorbance unit.

The rank of this absorbance matrix, according to Beer's law, is equal to the number of absorbing components. Since acetic acid may form dimers, trimers, and so on, it is not necessarily equal to unity.

When the absorbance matrix was subjected to AFA, using covariance about the origin, nine characteristic eigenvectors resulted.[7] The corresponding eigenvalues are listed in Table 4.4 in the order of decreasing value. The question we attempt to answer in this section is: Which of these eigenvectors belong to the primary set and which belong to the secondary set? In other words, what is the true rank of the example data matrix?

Residual Standard Deviation. As a guide to selecting the primary eigenvalues, many factor analysts use the residual standard deviation method. The RSD is defined by the following equation:

$$r(c - n)(\text{RSD})^2 = \sum_{i=1}^{r} \sum_{j=n+1}^{c} (\sigma_{ij}^0)^2 \tag{4.43}$$

The RSD in this expression is identical to the real error (RE) defined by (4.40). Using (4.17), we can rearrange (4.43) to read as follows:

$$\text{RSD} = \left(\frac{\sum_{j=n+1}^{c} \lambda_j^0}{r(c - n)} \right)^{1/2} \tag{4.44}$$

This equation is based on "covariance about the origin." If we use "correlation about the origin" instead, we would obtain the following expression:

$$\text{RSD} = \left(\frac{\sum_{i=1}^{r} \sum_{k=1}^{c} d_{ik}^2 \sum_{j=n+1}^{c} \lambda_j^0}{rc(c - n)} \right)^{1/2} \tag{4.45}$$

These equations provide us with an excellent criterion for deducing which eigenvalues belong to the primary and secondary sets. To begin the deductive process we first calculate the RSD on the basis of only one factor. In this case the largest, hence most important, eigenvalue represents the primary eigenvalue. All the remaining eigenvalues belong to the secondary set and are included in the summation as indicated in (4.44). We then compare the RSD calculated from (4.44) to the estimated experimental error. If the RSD is greater than the estimated error, we have not chosen a sufficient number of factors. If the RSD is approximately equal to the estimated error, we have chosen the proper number of factors. The factor space would be one-dimensional.

TABLE 4.4 Illustrative Example—Factor Analysis of the Molar Absorbances of the Carbonyl Region of Solutions Containing Acetic Acid in CCl_4[a]

n	Eigenvalue (λ)	Standard Error in λ	Real Error, RE[b]	Imbedded Error, IE	Indicator Function, IND × 10^5	Chi-Squared (χ^2)			Significance Level, %SL[d]
						Calculated	Expected[c]	3σ Misfit	
1	19.193396	0.00438	0.015461	0.005154	24.16	363,995.8	1592	983	0.0
2	0.368079	0.00295	0.003208	0.001512	6.55	10,479.8	1386	298	0.0
3	0.009065	0.00470	0.002110	0.001406	5.86	5,504.9	1182	152	6.5
4	0.004414	0.00372	0.000962	0.000642	3.85	780.9	980	0	24.5
5	0.000294	0.00393	0.000889	0.000662	5.56	561.5	780	0	33.5
6	0.000260	0.00475	0.000787	0.000643	8.74	329.0	582	0	43.6
7	0.000141	0.00315	0.000760	0.000670	19.0	194.6	386	0	50.0
8	0.000132	0.00356	0.000704	0.000663	70.4	81.5	192	0	59.7
9	0.000099	0.00351	—	—	—	0.0	0	0	—

[a]Reprinted with permission from J. T. Bulmer and H. F. Shurvell, *J. Phys. Chem.*, **77**, 256 (1973) and from E. R. Malinowski, *Anal. Chem.*, **49**, 612 (1977).
[b]Also known as the residual standard deviation (RSD).
[c]χ^2 (expected) = $(r - n)(c - n)$, where $r = 200$ and $c = 9$.
[d]From E. R. Malinowski, *J. Chemometrics*, **3**, 49 (1988).

On the other hand, if the RSD is greater than the estimated error, we then investigate the situation in which the two largest eigenvalues constitute the primary set. Once again we calculate the RSD by means of (4.44) and again compare it to the estimated error. If the RSD exceeds the estimated error, we repeat the process assuming that the three largest eigenvalues belong to the primary set. We continue in this manner, each time transferring the next largest eigenvalue from the secondary set to the primary set until we find that the RSD approximately equals the estimated error. When this occurs, we will have found which eigenvalues (and their associated eigenvectors) belong to the true primary and secondary sets.

As an example of this procedure, let us study the results of the illustrative example presented in Table 4.4. The RSD (RE) values were calculated from the eigenvectors by means of (4.44). Recalling that the error in the absorbances was estimated to be between 0.0005 and 0.0015, we conclude that there must be four recognizable factors because the largest RSD that does not exceed 0.0015 is 0.000962, corresponding to $n = 4$. For $n = 3$ the RSD (0.002110) is outside the acceptable range.

Because the residual standard deviation is an extremely useful concept in factor analysis, it is important for us to understand its full significance. If the errors are perfectly random, the error points will have a spherically symmetric distribution in the error space. Because of this distribution, the average of the sum of the squares of the projections of the data points onto any secondary eigenvector axis should be approximately the same. Hence the RSD should be the same, regardless of how many error axes are employed in the calculation. In reality, the error distribution will not be perfectly symmetric. In fact, the principal component feature of FA actually searches for axes that exaggerate the effects of this asymmetry. RSD values calculated from principal factor analysis (PFA) will, however, be approximately equal to the estimated error when the proper number of factors is employed. Principal factor analysis was used in obtaining the RSD values listed in Table 4.4.

The MATLAB program labeled sfa.m (significance factor analysis) computes RE as a function of the factor level and is included in Appendix C for the perusal of the reader.

Root-Mean-Square Error. The RMS error is defined by the equation

$$rc(\text{RMS})^2 = \sum_{i=1}^{r} \sum_{k=1}^{c} (d_{ik} - d_{ik}^\dagger)^2 \tag{4.46}$$

The term in parentheses on the right-hand side represents the difference between the raw data and the data regenerated by FA using only the primary set of eigenvectors. The sum is taken over all the data points. As strange as it may appear at first glance, this sum, involving the square of the difference, exactly equals the sum of the differences of the squares:

$$\sum_{i=1}^{r} \sum_{k=1}^{c} (d_{ik} - d_{ik}^\dagger)^2 = \sum_{i=1}^{r} \sum_{k=1}^{c} (d_{ik}^2 - d_{ik}^{\dagger 2}) \tag{4.47}$$

Since this equality is by no means obvious, proof is given here. First, recall that

$$e_{ik}^0 = d_{ik} - d_{ik}^\dagger \tag{4.48}$$

Now make the following sequence of observations:

$$d_{ik}^2 = (d_{ik}^\dagger + e_{ik}^0)^2$$
$$= d_{ik}^{\dagger 2} + 2d_{ik}^\dagger e_{ik}^0 + e_{ik}^{02}$$
$$\Sigma\Sigma d_{ik}^2 - \Sigma\Sigma d_{ik}^{\dagger 2} = 2\Sigma\Sigma d_{ik}^\dagger e_{ik}^0 + \Sigma\Sigma e_{ik}^{02}$$

The double sums are taken over $i = 1$ to $i = r$ rows and $k = 1$ to $k = c$ columns. But from (4.30)

$$\Sigma\Sigma d_{ik}^\dagger e_{ik}^0 = 0 \tag{4.49}$$

Therefore,

$$\Sigma\Sigma d_{ik}^2 - \Sigma\Sigma d_{ik}^{\dagger 2} = \Sigma\Sigma e_{ik}^{02} \tag{4.50}$$

Upon recalling (4.44), we find that

$$\Sigma\Sigma d_{ik}^2 - \Sigma\Sigma d_{ik}^{\dagger 2} = \Sigma\Sigma(d_{ik} - d_{ik}^\dagger)^2 \tag{4.51}$$

which is what we wanted to prove.

By substituting (4.51) into (4.46) and recalling (4.27), we conclude that

$$\text{RMS} = \left(\frac{\sum_{j=n+1}^{c} \lambda_j^0}{rc} \right)^{1/2} \tag{4.52}$$

By comparing this result with (4.44), an analogous equation concerning the residual standard deviation, we see that

$$\text{RMS} = \left(\frac{c - n}{c} \right)^{1/2} (\text{RSD}) \tag{4.53}$$

Although RMS and RSD are closely related, they measure two entirely different errors. The RMS error calculation measures the difference between raw data and FA-regenerated data. The residual standard deviation measures the difference between raw data and pure data possessing no experimental error. For this reason caution must be exercised when applying and interpreting these errors. Since c is greater than n, from (4.53) we see that RMS is less than RSD. The use of the RMS error as a criterion for rank analysis can be misleading and is not recommended.

Average Error. Several investigators have used the *average error*, $\bar{e}$, as a criterion for fit. The average error is simply the average of the absolute values of the differences between the original and regenerated data. This average is compared to an estimated average error. Eigenvectors are systematically added to the scheme until the calculated average error approximately equals the estimated error.

This method, in reality, is analogous to the root-mean-square error criterion, because, for a statistically large number of data points, the average error is directly proportional to the root-mean-square error:

$$\bar{e} = \left(\frac{2}{\pi}\right)^{1/2} (\text{RMS}) \tag{4.54}$$

Chi Squared. Bartlett[8] proposed using a *chi-squared* (χ^2) criterion when the standard deviation varies from one data point to another and is not constant throughout the data matrix. This method takes into account the variability of the error from one data point to the next. Its disadvantage is that one must have a reasonably accurate error estimate for each data point. In this case χ_n^2 is defined as follows:

$$x_n^2 = \sum_{i=1}^{r} \sum_{k=1}^{c} \frac{(d_{ik} - d_{ik}^{\dagger})^2}{\sigma_{ik}^2} \tag{4.55}$$

where σ_{ik} is the standard deviation associated with the measurable d_{ik}, $d_{ik}^{\dagger}$ is the value of the corresponding point regenerated from FA using the n largest eigenvalues, and the sum is taken over all experimental points. For each set of eigenvectors, χ_n^2 is compared to its expectation value given by the product

$$\chi_n^2(\text{expected}) = (r - n)(c - n) \tag{4.56}$$

The procedure for using the chi-squared criterion is as follows. First χ_1^2 is calculated from the data regenerated with the eigenvector associated with the largest eigenvalue. This value of chi-squared is then compared to the expectation value, $(r - 1)(c - 1)$. If χ_1^2 is greater than the expectation, the factor space is one dimensional or greater. Next, χ_2^2 is calculated using the two largest eigenvalues. If χ_2^2 is greater than the expectation value $(r - 2)(c - 2)$, the factor space is two-dimensional or greater. This procedure, using the largest eigenvalues, is continued until χ_n^2 is less than its corresponding expectation value $(r - n)(c - n)$. At this crossover point, the true n is estimated to be that which yields a χ_n^2 closest to its expectation value.

This behavior is illustrated by the benchmark (Table 4.4). In this case the standard deviations in absorbance were estimated for each data point, varying from 0.0005 to 0.0015. To calculate χ_n^2 this dispersion in the standard deviation was employed in (4.55). The crossover in χ_n^2 occurs between $n = 3$ (5504.9 > 1182) and $n = 4$ (780.9 < 980). Since 780.9 is closer to 980 than 5504.9 is to 1182, there are probably four factors. Another example of the chi-squared criterion is found in Section 9.1.3.

Standard Error in the Eigenvalue. Another method for deducing the factor space, based on a statistical criterion for the "vanishing" of an eigenvalue, was developed by Hugus and El-Awady.[9] They showed that the standard error in an eigenvalue is related to the standard deviations of the data points. In particular, they have shown that

$$
\sigma_m = \left(\sum_{j=1}^{c} \sum_{k=1}^{c} c_{mj}^2 c_{mk}^2 \sigma(\mathbf{Z})_{jk}^2 \right)^{1/2}
\tag{4.57}
$$

where σ_m is the standard error in the mth eigenvalue, c_{mj} and c_{mk} are the jth and kth components of the mth eigenvector [see (3.38)], and

$$
\sigma(\mathbf{Z})_{jk}^2 =
\begin{cases}
\sum_{i=1}^{r} (d_{ij}^2 \sigma_{ik}^2 + d_{ik}^2 \sigma_{ij}^2) & \text{for } j \neq k \\
\sum_{i=1}^{r} 4 d_{ij}^2 \sigma_{ij}^2 & \text{for } j = k
\end{cases}
$$

where σ_{ij} is the error in d_{ij}.

This statistical criterion allows the inclusion of individual error, which may vary from one data point to the next. The dimensionality of the factor space is taken to be the number of eigenvalues that have values larger than their respective standard error. Applying this criterion to the benchmark data (Table 4.4). we conclude that there are four eigenvectors, because the standard error in each eigenvalue is less than its eigenvalue for $n = 1$ to $n = 4$ but greater than its corresponding eigenvalue for $n = 5$ to $n = 9$. Another example of the use of this criterion is found in Section 9.1.3.

Distribution of Misfit. Another method for determining the dimensionality of the factor space involves studying the number of misfits between the observed and regenerated data as a function of the number of eigenvectors employed. A regenerated data point is classified as a misfit if its deviation from the observed value is three or more times greater than the standard deviation, σ, estimated from experimental information.

The benchmark data (Table 4.4) contained 1800 data points. With 3 eigenvectors, 152 regenerated data points had misfits greater than 3σ. With 4 factors no misfits were greater than 3σ. Hence the factor space is four dimensional. Another example of the 3σ misfit criterion is given in Table 9.1 in Section 9.1.3. A weak point of this method is the arbitrariness in deciding how many misfits can be tolerated.

Kankare[10] suggested that factor analysis be used to smooth the data by neglecting all data points with misfits greater than 3σ and substituting for them their FA-regenerated values. The purpose of the smoothing is to remove excessive errors introduced by those points whose deviation can only be considered accidental. In this case care must be exercised not to choose too small a value for n. After smoothing, the adjusted data is factor analyzed, hopefully yielding more reliable results. This process is somewhat risky because one might actually force the adjusted points to conform to a smaller factor space than that which truly exists, particularly so for points that are truly unique.

Residual Covariance Matrix. Ritter and co-workers[11] developed a method for choosing the number of factors, based on examining the residual errors in the covariance matrix. First, the data matrix is subjected to the "covariance about the mean" preprocessing procedure. Ritter and co-workers showed that the standard deviation, $e(z_{ij})$, in the covariance element z_{ij} can be calculated by means of the expression

$$e(z_{ij}) = \sigma(d_{ij})\left(\sum_{k=1}^{r}(d_{kj}^2 + d_{ki}^2)\right)^{1/2} \tag{4.58}$$

where $\sigma(d_{ij})$ is the standard deviation of the preprocessed data point d_{ij}, the sum being taken over all r elements in the respective data columns. The number of factors is the smallest value of n for which all elements in the residual covariance matrix approximately equal their corresponding standard deviations, that is, when

$$\mathscr{R}_{\text{minimum } n} \simeq [e(z_{ij})] \tag{4.59}$$

Reduced Error Matrix. A method for determining the rank of a matrix was devised by Wallace and Katz.[12] This method depends on the construction of a companion matrix **S**, whose elements are the estimated errors of the data matrix **D**. The method involves reducing the data matrix by a series of elementary operations to an equivalent matrix whose diagonal elements are maximized and whose elements below the principal diagonal are all zero. The error matrix **S** is continually transformed into an equivalent reduced error matrix during the reduction of **D**. This its done by a series of elementary operations based on the theory of propagation of errors. The rank of the data matrix is equal to the number of diagonal elements in the reduced data matrix, which are statistically nonzero. A diagonal element is arbitrarily considered to be nonzero if its absolute value is greater than three times the absolute value of the corresponding principal diagonal element of the reduced error matrix.

Interpretability of Target Tests. A promising method for finding the size of the factor space, without relying on a knowledge of the error in the data matrix, involves interpreting the target test vectors. Employing too few eigenvectors may lead to a poor fit between a predicted vector and its corresponding test vector, whereas employing too many eigenvectors will tend to reproduce experimental error and the agreement will be overly good. Unfortunately, to apply this method we need to know how much error can be tolerated in the test vector.

Another method for determining n from target testing involves comparing the predicted values with known values for a free-floated test point as a function of the number of factors employed. The true factor space is identified when the predicted values agree, within experimental error, with the known values. At the present time, the role of error in the target test is too poorly understood to make this approach a viable criterion.

These target methods are included in this section because they do require a knowledge of the error in the target.

Conclusion. The methods described in this section for determining the real factor space of a data matrix depend on an accurate estimate of the error. Each method may lead to a different conclusion. This problem was investigated by Duewer et al.,[13] who concluded that the determination of the true rank of data having uncertainty is not a trivial task and that none of the various rank determination criteria is clearly superior or completely satisfactory when used alone. The various criteria, examined together, afford a better guide than reliance on a single rule.

Duewer and co-workers based their conclusions on a manufactured set of data based on the dimensions of a rectangular solid, involving the following parameters: length (L), width (W), and thickness (T); three wraparound lengths $2(L + W)$, $2(L + T)$, and $2(W + T)$; and the three diagonals $(L^2 + W^2)^{1/2}$, $(L^2 + T^2)^{1/2}$, and $(W^2 + T^2)^{1/2}$. Random digits were generated for L, W, and T, and a true (pure) data matrix, consisting of the nine variables above, was formed from 25 sets of L, W, and T. To test the effects of analytical uncertainty, errors were added to the data and studies were conducted on both the true data matrix and the perturbed data matrix. Various error criteria (such as χ^2 and $\bar{e}$) for judging the rank of the perturbed data matrix were compared to the results obtained from the true data matrix. Data pretreatment methods such as C_o, C_m, R_o, and R_m (see Section 3.2.3) were also investigated.

The effects of adding artificial error to chemical data matrices have also been investigated by others.[14,15] Because these studies are quite detailed, interested readers are referred to the original studies.

4.3.2   Empirical Methods

In the previous section we studied various methods for determining the factor space when accurate estimates of the errors are known. Often these methods cannot be used because such information is either not available or is highly suspect. A more difficult problem is to deduce the factor space without relying on an estimation of the error. In this section we explore various methods that have been proposed to solve this challenging problem. In Section 4.4 we learn how these methods can be used to deduce not only the size of the factor space but also the size of the error.

Imbedded Error Function. The *imbedded error function* can be used to determine the number of factors in a data matrix without relying on any estimate of the error.[1,16] By inserting (4.44) into (4.41), we see that

$$
\mathrm{IE} = \left(\frac{n \sum_{j=n+1}^{c} \lambda_j^0}{rc(c - n)} \right)^{1/2}
\tag{4.60}
$$

The imbedded error is a function of the secondary eigenvalues, the number of rows and columns in the data matrix, and the number of factors. Because this information is always available to us when we perform factor analysis, we can calculate IE as a function of n, as n goes from 1 to c. By examining the behavior of the IE function as n varies, we can often deduce the true number of factors. The IE function should decrease as we use more and more primary eigenvectors in data reproduction. However, once we have exhausted the primary set and we begin to include secondary eigenvectors in the reproduction, the IE should increase. This should occur because a secondary eigenvalue is simply the sum of the squares of the projections of the error points onto an error axis. If the errors are distributed uniformly, their projections onto each error axis should be approximately the same. In other words, $\lambda_j^0 \simeq \lambda_{j+1}^0 \simeq \lambda_c^0$ and (4.60) becomes

$$
\mathrm{IE} \simeq n^{1/2} k \qquad \text{for } n > \text{true } n
\tag{4.61}
$$

In (4.61), k is a constant:

$$
k = \left(\frac{\lambda_j^0}{rc} \right)^{1/2}
\tag{4.62}
$$

These equations apply only when we have used an excessive number of eigenvectors in the reproduction process. Equation (4.61) shows that the IE will actually increase once we begin to use more factors than the true number required. In practice, a steady increase in IE will rarely be observed because the principal component feature of FA exaggerates the nonuniformity in the error distribution. Hence the secondary eigenvalues will not be exactly equal. The minimum in the IE function will not be clearly defined if the errors are not fairly uniform throughout, if the errors are not truly random, if systematic errors exist, or if sporadic errors exist.

Three examples illustrating the behavior of the IE function,[16] taken from chemical problems, are given next.

Example 1. Results of factor analyzing the infrared absorbances of acetic acid in CCl_4 solutions are given in Table 4.4. Here the IE shows a progressive decrease on going from $n = 1$ to $n = 4$ but shows no further decrease on going from $n = 4$ to $n = 8$. This gives evidence that four factors are responsible for the absorbance matrix. This conclusion is consistent with the other approaches described in Section 4.3.1.

Example 2. Table 9.1 concerns the factor analysis of the spectrophotometric absorbances of 38 solutions of $[(en)_2Co(OH)_2Co(en)_2]^{4+}$ measured at nine different wavelengths. The fact that the IE reaches a minimum at $n = 3$ gives evidence that there are three species responsible for the absorbance.

Example 3. Table 4.5 shows the results of factor analyzing the gas–liquid-chromatographic (GLC) retention indices of 22 ethers on 18 chromatographic columns. We see here that no minimum appears in the IE function. This misbehavior could be caused by any one or combination of the four reasons discussed earlier. Unfortunately, in this case we cannot use the IE functions to determine the number of factors present.

Factor Indicator Function. Malinowski[1,16] discovered an empirical function, called the *factor indicator function*, which appears to be much more sensitive than the IE function in its ability to pick out the proper number of factors. The factor indicator function (IND) is defined as

$$IND = \frac{RE}{(c - n)^2} \tag{4.63}$$

TABLE 4.5   Results of Factor Analyzing the GLC Retention Indices of 22 Ethers on 18 Chromatographic Columns[a]

n	RE	IE	IND	%SL[b]	n	RE	IE	IND	%SL[b]
1	22.28	5.25	0.07708	0.0	10	1.40	1.04	0.02187	28.7
2	7.25	2.42	0.02831	0.0	11	1.25	0.98	0.02553	24.2
3	5.30	2.16	0.02354	0.5	12	1.07	0.87	0.03975	23.2
4	4.06	1.91	0.02070	3.8	13	0.94	0.80	0.03748	30.4
5	3.24	1.71	0.01915	5.4	14	0.73	0.65	0.04586	38.4
6	2.76	1.59	0.01914	8.1	15	0.69	0.63	0.07618	36.2
7	2.42	1.51	0.01997	18.2	16	0.61	0.58	0.15261	45.5
8	2.05	1.36	0.02045	20.8	17	0.59	0.57	0.59012	57.4
9	1.71	1.21	0.02114	21.6	—	—	—	—	—

[a]Reprinted with permission from E. R. Malinowski, *Anal. Chem.*, **49**, 612 (1977).
[b]From E. R. Malinowski, *J. Chemometrics*, **3**, 49 (1988).

This function is composed of exactly the same variables as IE, namely, λ_j^0, r, c, and n. The IND function, similar to the IE function, reaches a minimum when the correct number of factors are employed. The minimum, however, is much more pronounced and, more important, often occurs in situations in which the IE exhibits no minimum.

To see the behavior of the IND function, let us examine the same three examples studied in the previous subsection. The first example concerns the absorbances of acetic acid in CCl_4, Table 4.4. In this table we see that the IND function reaches a minimum at $n = 4$, in agreement with our conclusions based on IE. In the second example the IND function for the cobaltic solutions, shown in Table 9.1, reaches a minimum at $n = 3$. This is also in agreement with our findings based on the IE function. The third example concerns the GLC retention indices. In Table 4.5, a minimum in the IND is clearly present at $n = 6$, whereas no minimum appears in the IE function. Apparently, the IND function somehow compensates for the principal component exaggeration of nonuniformity of the error distribution. The IND function is not fully understood at the present time, and it should be used with caution.

The MATLAB program labeled sfa.m (significance factor analysis) computes IND as a function of the factor level and is included in Appendix C for the perusal of the reader.

Variance. The factor analytical solution can be expressed as a linear sum of outer products of row-designee vectors and column-designee vectors:

$$D = r_1 c_1' + r_2 c_2' + \cdots + r_j c_j' + \cdots + r_c c_c' \tag{4.64}$$

Here r_j and c_j' are the row-designee and column-designee vectors associated with eigenvalue λ_j. Each outer product yields an $r \times c$ matrix, D_j, which accounts for a portion of the raw data matrix, so that

$$D = D_1 + D_2 + \cdots + D_j + \cdots + D_c \tag{4.65}$$

where

$$D_j = r_j c_j' \tag{4.66}$$

In other words, each D_j, which is associated with a particular eigenvector, c_j, and a particular eigenvalue, $\lambda_j = r_j' r_j$ [see (3.74)], accounts for some of the variance in the raw data matrix. The *variance* is defined as follows:

$$\text{Variance} = \frac{\sum_{i=1}^{r} \sum_{k=1}^{c} d_{ik}^2(j)}{\sum_{i=1}^{r} \sum_{k=1}^{c} d_{ik}^2} \tag{4.67}$$

where $d_{ik}(j)$ is an element of $\mathbf{D}_j$ and d_{ik} is an element of $\mathbf{D}$, the raw data matrix. The sums are taken over all $r \times c$ elements of the matrices. Based on arguments given in Section 4.1.1, one can easily show that the variance can be calculated directly from the eigenvalues; that is,

$$\text{Variance} = \frac{\lambda_j}{\sum\limits_{j=1}^{c} \lambda_j} \tag{4.68}$$

Because the variance measures the importance of an eigenvector, it can be used as a criterion for accepting or rejecting an eigenvector. In practice, eigenvectors having large variances are considered to be primary eigenvectors, whereas eigenvectors having small variances are considered to be secondary eigenvectors. Unfortunately, classifying the variance as large or small presents a problem. It is at this critical point in the process that various investigators diverge. Often, the factor analyst gives no justification for the cutoff point used in the variance classification, thus casting doubt on the conclusion.

Cumulative Percent Variance. The cumulative percent variance is a measure of the percentage of the total variance in the data that is accounted for by abstract reproduction. It is defined as follows:

$$\text{Cumulative percent variance} = 100 \left(\frac{\Sigma\Sigma d_{ik}^{\ddagger 2}}{\Sigma\Sigma d_{ik}^2} \right) \tag{4.69}$$

Here $d_{ik}^{\ddagger}$ is the value of a data point reproduced by AFA and d_{ik} is the raw, experimental data point. The sums are taken over all data points. Inserting (4.24) and (4.26) into (4.69), we see that the cumulative percent variance can be expressed in terms of the eigenvalues:

$$\text{Cumulative percent variance} = 100 \left(\frac{\sum\limits_{j=1}^{n} \lambda_j^{\ddagger}}{\sum\limits_{j=1}^{c} \lambda_j} \right) \tag{4.70}$$

For this reason the method is often referred to as the *percent variance in the eigenvalue.*

The percent variance criterion accepts the set of largest eigenvalues required to account for the variance within a chosen specification. The problem then becomes one of estimating exactly how much variance in the data must be accounted for. Arbitrary specifications such as 90, 95, or 98% variance do not provide reliable estimates for judging the number of factors because the calculated cumulative percent variance is sensitive to data pretreatment. In general, the method can be deceptively misleading and is not recommended unless one can make an accurate estimate of the true variance in the data. In practice, this cannot be done without a knowledge of the error.

Scree Test. The *Scree test*, proposed by Cattell,[17] is based on the observation that the residual variance should level off before the factors begin to account for random error. The residual variance is defined as follows:

$$\text{Residual variance} = \frac{\sum_{i=1}^{r} \sum_{k=1}^{c} (d_{ik}^2 - d_{ik}^{\dagger 2})}{rc} \tag{4.71}$$

From (4.46) and (4.47) we see that the residual variance is equal to the square of the RMS error, which can be calculated from the secondary eigenvalues by means of (4.52). The residual percent variance is defined as

$$\text{Residual percent variance} = 100 \left(\frac{\sum_{i=1}^{r} \sum_{k=1}^{c} (d_{ik}^2 - d_{ik}^{\dagger 2})}{\sum_{i=1}^{r} \sum_{k=1}^{c} d_{ik}^2} \right) \tag{4.72}$$

In terms of the eigenvalues, this expression takes the form

$$\text{Residual percent variance} = 100 \left(\frac{\sum_{j=n+1}^{c} \lambda_j^0}{\sum_{j=1}^{c} \lambda_j} \right) \tag{4.73}$$

When the residual percent variance is plotted against the number of factors used in the reproduction, the curve should drop rapidly and level off at some point. The point where the curve begins to level off or where a discontinuity appears is used to reduce the factor space. This is illustrated in Figure 4.3, taken from the work of Rozett and Petersen,[18] who investigated the mass spectra data of the 22 isomers of $C_{10}H_{14}$. These investigators used covariance about the mean, C_m; covariance about the origin, C_o; correlation about the mean, R_m; and correlation about the origin, R_o (see Section 3.2.3). They found that leveling is sensitive to the technique employed. For R_m and C_m, information concerning the zero point of the experimental scale is lost. For R_o and R_m, information concerning the relative sizes of the data points is lost. Such information is retained only in the C_o technique, which is the procedure recommended by Rozett and Petersen. For C_o, the leveling occurs at $n = 3$.

Average Eigenvalue. The *average eigenvalue criterion* proposed by Kaiser[19] has gained wide popularity among factor analysts. This criterion is based on accepting all eigenvalues with values above the average eigenvalue and rejecting all those with

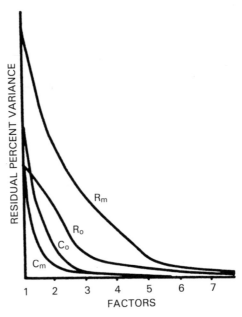

Figure 4.3 Residual percent variance as a function of the number of factors employed in data reproduction. See Section 3.2.3 for definitions of C_m, C_o, R_o, and R_m. [Reprinted with permission from R. W. Rozett and E. Petersen, *Anal. Chem.*, **47**, 1303 (1975).]

values below the average. If one uses correlation about the origin, so that the correlation matrix has unities for each diagonal element, the average eigenvalue will be unity. In this case the criterion accepts all eigenvalues above unity, the average value, and excludes all eigenvalues below unity. This procedure is more popularly known as the *eigenvalue-one criterion.*

Exner Function. Weiner and co-workers[20] suggested that the *Exner psi (ψ) function* be employed when a good measure of the experimental error is not available. This function is defined by

$$\psi = \left(\frac{\sum_{i=1}^{r} \sum_{k=1}^{c} (d_{ik} - d_{ik}^{\dagger})^2}{\sum_{i=1}^{r} \sum_{k=1}^{c} (d_{ik} - \bar{d})^2} \times \frac{rc}{rc - n} \right) \qquad (4.74)$$

Here $\bar{d}$ represents the grand mean of the experimental data. The ψ function can vary from zero to infinity, with the best fit approaching zero. A ψ equal to 1.0 is the upper limit of physical significance because this means that one has not done any better than simply guess that each point has the same value of the grand mean. Exner[21] proposed that 0.5 be considered the largest acceptable ψ value because this means that the fit is twice as good as guessing the grand mean for each point. Following Exner's reasoning, $\psi = 0.3$ is considered a fair correlation, $\psi = 0.2$ is considered a good correlation, and $\psi = 0.1$ is considered an excellent correlation. This method can only be expected to give a very crude estimate of the factor space.

Loading Distribution. Cattell[22] suggested that the factor loadings (i.e., the components of each eigenvector) be used to determine the factor space. Since each successive eigenvector involves more random error, the loadings should tend toward a normal distribution. The mean of the loadings of an eigenvector should be close to zero only if the eigenvector is an error eigenvector (i.e., a secondary eigenvector). Unfortunately, deciding which eigenvectors have mean loadings sufficiently small to be considered zero is arbitrary. Consequently, the method is subject to personal bias.

Submatrix Analysis. Cartwright[23] determined the number of chemical components (factors) by examining the eigenvalues of submatrices resulting from evolutionary processes (see Section 6.2). As an example, spectroscopic data are arranged in increasing (or decreasing) order of the evolutionary variable (pH, volume of titrant, time, etc.). Starting with only one spectrum, a series of submatrices are constructed by adding successive spectra to the previous submatrix. The eigenvalues of each submatrix are extracted, normalized to unit sum, and their logarithms are plotted as a function of the evolutionary variable (or sequence number). Visual inspection of the plot reveals the emergence of real factors above a threshold level. As seen in Figure 4.4 concerning phosphoric acid spectra from pH $= 0$ to pH $= 14$, four species (the acid and three conjugate bases) produce eigenvalues above the pool of error eigenvalues that conglomerate below the threshold. The method is restricted to those processes that follow the conditions required for evolving factor analysis (EFA).

Autocorrelation. Shrager and Hendler[24] identified the noise eigenvectors by use of an autocorrelation function. The method requires a large number of data rows, easily achievable with spectroscopic measurements. Columns of the abstract row-factor matrix that represent real signal will be smooth, whereas those representing noise will exhibit rapid fluctuations. A simple statistic for measuring smoothness is the autocorrelation coefficient of order 1:

$$\text{AUTO}_j = \sum_{i=1}^{r-1} \frac{r_{ij} r_{(i+1)j}}{\lambda_j} = \sum_{i=1}^{r-1} u_{ij} u_{(i+1)j} \tag{4.75}$$

In this expression division by the associated eigenvalue is required in order to normalize the vector to unit length. In singular value decomposition (SVD) notation, shown on the right, such division is not necessary because the required vectors are normalized. Autocorrelation coefficients greater than $+0.5$ are indicative of the smoothness of real signals. Values less than $+0.5$ are symptomatic of noise vectors.

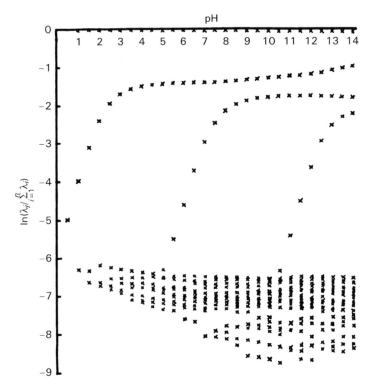

Figure 4.4   Logarithmic plot of eigenvalues of submatrices of phosphoric acid spectra as a function of pH. [From H. M. Cartwright, *J. Chemometrics*, **1**, 111(1987). Reprinted with permission of John Wiley & Sons, Ltd.]

The test can also be applied to the rows of the eigenvectors of the abstract column-factor matrix provided the matrix has a large number of columns. In this case the autocorrelation coefficient for the kth eigenvector is calculated as follows:

$$\text{AUTO}_k = \sum_{j=1}^{c-1} c_{jk}c_{(j+1)k} = \sum_{j=1}^{c-1} v_{kj}v_{k(j+1)} \qquad (4.76)$$

Division by λ_k is not necessary since the eigenvectors are already normalized. The term on the right is the SVD expression.

Frequency Distributions of Fourier Transformed Eigenvectors. Rossi and Warner[25] noticed that the Fourier transform of the eigenvectors have frequency distributions that can be used to distinguish the real eigenvectors from error eigenvectors. Transforms of the error vectors should contain more high-frequency components, produced by random noise in the original data. The success of the method depends on the frequency range chosen to generate the distribution plots, which are examined for plateaus or thresholds.

4.3.3 Statistical Methods

Cross Validation. Cross validation has been used to estimate the optimal number of principal components. Cross validation consists of deleting a row (or a set of rows) from the data matrix, carrying out the eigenanalysis of the reduced matrix, and target testing the deleted row(s) using various levels of the abstract factor space. The difference between the target points and the AFA predicted points are tabulated. This process is repeated until every row of the data matrix has been deleted only once. The errors in the target fit for each row are added to the tabulations. The "prediction sum of squares" of the errors, PRESS(j), are calculated for each of the j factor levels as follows:

$$\text{PRESS}(j) = \sum_{i=1}^{r} \sum_{k=1}^{c} (d_{ik} - \hat{d}_{ik}(j))^2 \qquad (4.77)$$

where $\hat{d}_{ik}(j)$ are the predicted elements of the deleted rows obtained with j factors.

Several criteria for determining the significant number of eigenvectors have been suggested by Wold[26] and co-workers.[27] One criterion[26] suggests that the optimal value for j is the level that yields the smallest PRESS(j). Another criterion[26] is based on the ratio, R, of successive prediction sum of squares:

$$R = \frac{\text{PRESS}(j+1)}{\text{PRESS}(j)} \qquad (4.78)$$

Starting with $j = 1$, if R is less than one, then the increased factor space yields better predictions; hence the procedure is repeated with $j = 2$, etc. until the ratio is greater than one, indicating that the added factor does not improve the predictions. Hence the number of factors is equivalent to that j level. A third criterion[27] is based on the assumption that the prediction errors have a normal distribution so an F test can be performed on the following F function:

$$F = \frac{\text{PRESS}(j-1) - \text{PRESS}(j)}{\text{PRESS}(j)}(c - j) \qquad (4.79)$$

Proceeding from $j = 1$ to $j = c$, a significant F value indicates that the jth factor contains real components.

Osten[28] examined four different criteria based on PRESS for determining the optimum factor level. The first method selected the level corresponding to the global minimum PRESS value. The second method chose the local minimum in PRESS corresponding to the smallest possible number of components. The third method selected the PRESS with a value below a preset (3 or 5%) threshold value, corresponding to the smallest number of factors. The fourth method was based on an F test designed to determine the significance of an additional factor:

$$F(q, rc - (n + 1)q = \frac{\mathrm{PRESS}(n) - \mathrm{PRESS}(n + 1)}{q} \\ \div \frac{\mathrm{PRESS}(n + 1)}{rc - (n + 1)q} \qquad (4.80)$$

where q is the number of coefficients calculated for each latent factor.

The first method, based on the global minimum in PRESS, which is the most popular method, gave the poorest results. Reasons for the disparities observed with cross-validation techniques are discussed in Osten's study.[28]

Eastman and Krzanowski[29] developed a cross-validation method based on predicting each element of the data matrix after deleting the corresponding row and column of the matrix. The method takes advantage of updating features of SVD that can be achieved with high-speed efficiency. After deleting each element of the data matrix, the PRESS(M) is calculated at the various factor levels by target testing the deleted row and column and then forming a composite prediction of the deleted point. To choose the optimum factor level, the following statistic, analogous to that used in regression analysis, was invoked:

$$W = \frac{\mathrm{PRESS}(M - 1) - \mathrm{PRESS}(M)}{D_M} \div \frac{\mathrm{PRESS}(M)}{D_R} \qquad (4.81)$$

where D_M is the number of degrees of freedom required to fit the Mth component and D_R is the number of degrees of freedom remaining after fitting the Mth component. The optimal value for M is the smallest M at which W is greater than unity.

Reduced Eigenvalue. Malinowski[30,31] studied the distribution of error eigenvalues resulting from AFA and concluded that the reduced eigenvalues associated with the error eigenvalues should be statistically equal. The reduced eigenvalue is defined as

$$\mathrm{REV}_j = \lambda_j / (r - j + 1)(c - j + 1) \qquad (4.82)$$

Because the reduced error eigenvalues are equally proportional to the standard deviation, an F test was invoked:

$$F(1, s - n) = \frac{\sum_{j=n+1}^{s} (r - j + 1)(c - j + 1)}{(r - n + 1)(c - n + 1)} \frac{\lambda_n}{\sum_{j=n+1}^{s} \lambda_j^0} \qquad (4.83)$$

Here s is equal to r or c, whichever is smaller. This is designed to test the null hypothesis

$$H_0: \qquad \mathrm{REV}_n = \mathrm{REV}^0_{\mathrm{pool}}$$

against the alternative hypothesis (a one-tail-test)

$$H_a: \qquad \mathrm{REV}_n > \mathrm{REV}^0_{\mathrm{pool}}$$

where $\mathrm{REV}^0_{\mathrm{pool}}$ is the weighted average of the pool of reduced error eigenvalues.

The Fisher variance ratio is the quotient of two variances obtained from two independent pools of samples with normal distributions. Because the eigenvectors are orthogonal, the condition of independence is satisfied. Although the error eigenvalues do not have a normal distribution, the reduced error eigenvalues should be normally distributed. The test statistic described above concerns the variance in the null vectors and not the variance in the residual points. Since real eigenvectors have contributions from structure as well as experimental error, their reduced values will be statistically greater than the pooled variance of the error eigenvalues.

The results of applying this test to infrared data are shown in the last column of Table 4.4. This column lists the percent significance level (%SL) corresponding to the F values obtained from (4.83). To determine the number of significant factors, the complete set of error eigenvalues must be identified. This is done by assigning the smallest eigenvalue to the pool of error eigenvalues and testing the next smallest eigenvalue for significance. If the significance level is less than some desired value (such as 5 or 10%), the null hypothesis is rejected and the alternative is accepted. Because 59.7 %SL is highly significant for $n = 8$, there is no reason to reject the null hypothesis. Therefore we add the eighth eigenvalue to the pool of error eigenvalues and then test the variance of the enlarged pool of error eigenvalues. We continue in this fashion, reading from the bottom of the table to the top, adding the next eigenvalue to the pool until we reach the desired (5 or 10%) significance level. In this case there are two significant factors at the 5% level and three factors at the 10% level.

The MATLAB program labeled sfa.m (significance factor analysis) computes REV and F as a function of the factor level and is included in Appendix C for the perusal of the reader. Please note that, in the output, $F(1, \mathrm{df})$ is converted into percent significance level (%SL).

Faber–Kowalski F Test. Faber and Kowalski[32] designed an F test based on the observations of Mandel.[33] Mandel argued that the secondary eigenvalues divided by the appropriate degrees of freedom, v_n, represent unbiased estimates of σ^2, the variance of the random error:

$$\sigma^2 = E[\lambda_n]/v_n \tag{4.84}$$

The expectation value, $E[\lambda_n]$, can be determined by taking the average of the largest eigenvalue resulting from decomposition of 1000 matrices of the same size composed solely of Gaussian errors with a standard deviation of one, $\sigma^2 = 1$. From this hypothesis, Faber and Kowalski developed a significance test based on

$$F_n(v_1, v_2) = \frac{v_2 \lambda_n}{v_1 \sum\limits_{j=n+1}^{c} \lambda_j} \tag{4.85}$$

where $v_1 = E[\lambda_1]$ and $v_2 = (r - n + 1)(c - n + 1) - v_1$.

This test is much more sensitive than the Malinowski F test. It can distinguish the primary from the secondary eigenvalues at the 1% significance level. Because the test can be applied from the top down, from the largest eigenvalue to the next largest eigenvalue, there is no need to compute all of the eigenvalues as required by the Malinowski F test.

Unfortunately the method is based on the premise that the uncertainties in the measurements are normally distributed. In many chemical experiments this may not be the case. In marked contrast, the Malinowski F test is based on the central limit theory. It tests for the equality of the reduced eigenvalues. Although the errors may not follow a normal distribution, their pooled samples, as expressed by the reduced eigenvalues, should follow a normal distribution, albeit, with a loss in the degrees of freedom and a loss in sensitivity.

Canonical Correlation. Whenever there are two or more matrices of the same size composed from the same intrinsic factors, the number of factors can be determined by canonical correlation.[34] The method is directly applicable to replicate data matrices, which, for simplicity, are assumed in the discussion that follows.

Let $D(1)$ and $D(2)$ represent replicate matrices that are subjected to singular value decomposition, each yielding two sets of eigenvector matrices $U(1)$ and $V(1)$, and $U(2)$ and $V(2)$, respectively. The first n vectors of these four matrices lie inside the real factor space, whereas the remaining eigenvectors are the result of noise. The first n vectors of $U(1)$ are highly correlated with the first n vectors of $U(2)$. Since noise vectors are uncorrelated, the $(n + 1)$th eigenvector of $U(1)$ is nearly orthogonal to the $(n + 1)$th eigenvector of $U(2)$. Hence canonical correlations between the first $n + 1$ columns of $U(1)$ and $U(2)$ should exhibit n large correlations and one small correlation. Similar behavior is expected between $V(1)$ and $V(2)$, which can be used to substantiate the conclusions.

Canonical correlations have been interpreted as the cosines of the angles between the subspaces, with values lying between 0 and 1. These correlations can be converted into statistical significant levels that are useful for determining n.

Multiple Sources of Error. The F tests described previously are based on the premise that the measurement errors are independent, normally distributed, and homoscedastic (i.e., constant variance). However, data gleaned from chemical instruments often contain a composite of undesirable artifacts that originate from several different sources, each with its own individual variance. For example, when an absorption spectrometer switches filters, a new uncertainty distribution is created. Changing sample cells or solvent introduces new perturbations. Pretreating the data (e.g., by dividing the absorptivities by concentrations, normalizing the data, smoothing, or taking logarithms) can also affect the error distributions. Factor analysis of pretreated data often leads to an excessive number of factors.

New uncertainty distributions will be introduced into the measurement whenever there is a sudden change in the experimental conditions. For example, in order to investigate the aggregation of methylene blue in water over a wide range of concentration from 2×10^{-6} to 1.6×10^{-2} M three different sample cells varying in path length were required to keep the absorbance measurements within the feasible range of the visible spectrometer.[35] A 400×33 absorptivity matrix was assembled from the absorbance measurements involving 18 samples using a 0.1000-cm cell, six samples using a 0.0188-cm cell, and nine samples using a 0.0041-cm cell. The absorptivities were calculated by dividing the absorbances by the respective path length and the respective concentration of methylene blue. This procedure introduced three different distributions of error into the resulting data matrix.

Both the IND function and the Malinowski F test gave evidence for 13 factors, far in excess of the expectations based on prior investigations. The RE at $n = 13$ was found to be 16, a value close to the standard deviation associated with the smallest of the three error distributions.

To determine the factor level corresponding to the next smallest error distribution, the eigenvalues belonging to the smallest null space should be deleted and the degrees of freedom should be reduced accordingly.[35] After this removal, the indicator function becomes

$$\text{IND}_n = \frac{1}{(c - n - m)^2} \sqrt{\frac{\sum_{k=n+1}^{c-m} \lambda_k}{r(c - n - m)}} \tag{4.86}$$

where m is the number of eigenvalues removed. The Malinowski F test becomes

$$F_n(1, c - n - m) = \frac{\lambda_n}{\sum_{j=n+1}^{c-m} \lambda_j} \frac{\sum_{j=n+1}^{c-m} (r - j + 1)(c - j + 1)}{(r - n + 1)(c - n + 1)} \tag{4.87}$$

The 13 largest eigenvalues were retained by setting $m = 20$. Application of (4.86) and (4.87), after this reduction, indicated 7 factors with an RE = 42. This error represents the standard deviation of the second error distribution. The 7 largest eigenvalues were retained by setting $m = 26$. After this reduction, (4.86) and (4.87) yielded 3 significant factors with an RE = 196, in agreement with expectations. The 3 factors represented the monomer, dimer, and trimer.

Another example, involving Cu(II) and ethylenediaminetetraacetate (EDTA) mixtures, is presented in the following subsection.

Ratio of Eigenvalues by Smoothed Ordinary Principal Component Analysis (RESO). The RESO method of Chen and co-workers[36] is based on examining the ratio of eigenvalues determined from smoothed and unsmoothed PCA spectra. Unlike the previous methods, RESO is limited to broadband spectra, such as those found in ultraviolet–visible (UV–VIS). The method is not applicable to sharply defined signals typical of mass spectra or high-resolution nuclear magnetic resonance (NMR).

RESO incorporates the concepts of smoothed principal component analysis (SPCA) introduced by Rice Ramsay and Silverman[37] in 1991 and simplified by Silverman[38] in 1996. SPCA is similar to PCA but imposes a penalty if the difference between the abstract spectral vectors is large. SPCA searches for a set of eigenvectors, $\mathbf{u}_i$, which maximizes the objective function $F(\mathbf{u}_i)$:

$$F(\mathbf{u}_i) = \mathbf{u}_i'\mathbf{DD}'\mathbf{u}_i/(\mathbf{u}_i' + k(\mathbf{u}_i'\mathbf{Q}'\mathbf{Qu}_i)) \qquad i = 1, 2, \ldots, c \qquad (4.88)$$

subject to

$$\mathbf{u}_i'\mathbf{u}_j + k(\mathbf{u}_i'\mathbf{Q}'\mathbf{Qu}_j)) = 0 \qquad i \neq 0 \qquad (4.89)$$

where

$$\mathbf{Q} = \begin{bmatrix} 1 & -2 & 1 & 0 & . & . & . & 0 \\ 0 & 1 & -2 & 1 & 0 & . & . & . \\ . & 0 & . & . & . & . & . & . \\ . & . & . & . & . & . & 0 & . \\ . & . & . & 0 & 1 & -2 & 1 & 0 \\ 0 & . & . & . & 0 & 1 & -2 & 1 \end{bmatrix}_{(r-2) \times r} \qquad (4.90)$$

The above maximizations are summarized by the following generalized eigenvalue–eigenvector problem:

$$\mathbf{DD}'\mathbf{u}_i = \lambda_i(\mathbf{I} + k\mathbf{Q}'\mathbf{Q})\mathbf{u}_i \qquad i = 1, 2, \ldots, c \qquad (4.91)$$

In these expressions, k is an arbitrary scalar that controls the degree of smoothness.

Eigenvectors, $\mathbf{u}_i$, and eigenvalues λ_i, can be gotten by rearranging (4.91) into a more convenient form. To accomplish this, the inverse of the term in the parenthesis is subjected to singular value decomposition as shown in (4.92).

$$(\mathbf{I} + k\mathbf{Q}'\mathbf{Q})^{-1} = \mathbf{U}_Q\mathbf{S}_Q\mathbf{V}_Q' \qquad (\mathbf{U}_Q = \mathbf{V}_Q) \qquad (4.92)$$

Inserting (4.92) into (4.91) and manipulating the result leads to an expression that can be used to calculate the smoothed data, $\mathbf{D}_{\text{SPCA}}$.

$$\mathbf{D}_{\text{SPCA}} = \mathbf{U}_Q \mathbf{S}_Q^{1/2} \mathbf{V}_Q' \mathbf{D} \tag{4.93}$$

Singular value decomposition of $\mathbf{D}_{\text{SPCA}}$ yields the desired quantities.

According to (4.88), eigenvectors that represent linear combinations of real spectra will be relatively unaffected by SPCA because they are smooth. Eigenvectors contaminated by noise will be smoothed by SPCA. Consequently, eigenvalues belonging to real eigenvectors will be weakly affected, whereas eigenvalues belonging to error eigenvectors will show a marked decrease in value. This decrease, expressed as the ratios of the eigenvalues, RESO, can be used to determine the number of chemical components:

$$\text{RESO} = \frac{\lambda_i^{\text{SPCA}}}{\lambda_i^{\text{PCA}}} \tag{4.94}$$

Figure 4.5 illustrates how SPCA smooths a real spectral eigenvector contaminated by noise. The pure noise eigenvectors are not shown in Figure 4.5.

Table 4.6 compares RESO to other techniques. The table concerns the analysis of the UV absorption spectra of 29 mixtures Cu(II) and EDTA, containing a constant amount of EDTA and an increasing amount of Cu(II). Three methods described earlier (IND, Malinowski F test, and Faber–Kowalski F test) clearly indicate seven factors. In contrast, RESO, based on $k = 1$ and $k = 10$, indicates four factors. This is evident by noting that the RESO values of the first four eigenvalues are close to unity, whereas, there is a marked decrease in RESO at $n = 5$ and beyond. The RE at $n = 4$ is 5×10^{-4}, which is closest to the expectation value. The RESO conclusion is further substantiated by the fact that window factor analysis (WFA) was only able to extract four concentration profiles.[39] Furthermore, when the absorption data was processed by equations (4.86) and (4.87), setting $m = 29 - 7 = 22$, in accord with the multiple sources of error theory, four factors were found.

4.4 DEDUCING EXPERIMENTAL ERROR

In Section 4.3.2 we learned how various criteria can be used to determine the number of factors. Once we know the true number of factors, we can calculate the real error (RE) in the data matrix without relying on any prior knowledge of the experimental error. This can be done because (4.44) and (4.40) lead to the expression

$$\text{RE} = \left(\frac{\sum\limits_{j=n+1}^{c} \lambda_j^0}{r(c - n)} \right)^{1/2} \tag{4.95}$$

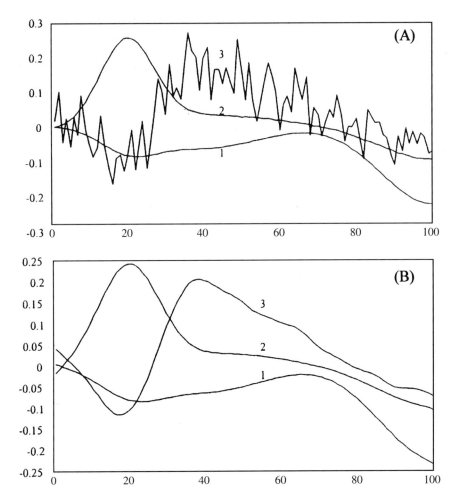

Figure 4.5   First three abstract spectral eigenvectors obtained from (A) PCA and from (B) SPCA. [Reprinted with permission from Z.-P Chen, Y.-Z. Liang, J.-H. Jiang, Y. Lang, J.-Y. Qian, and R.-Q. Yu, J. *Chemometrics*, **13**, 15 (1999). Copyright 1999 © John Wiley & Sons, Ltd.]

where the required quantities λ_j^0, r, c, and n are all known. These same quantities are involved in the IE and IND functions, which can be used to determine the number of factors [see (4.60) and (4.63)].

To see how these error functions can be used jointly, let us return to the three examples cited in Section 4.3.2. For the infrared absorbances of acetic acid (Example 1), the IE and IND functions lead us to surmise that $n = 4$ (see Table 4.4). From this we conclude that the RE is 0.000962, as shown in the table. This is in agreement with the estimate that the experimental error lies between 0.0005 and

TABLE 4.6   Comparison of RESO with Other Methods for Determining the Number of Chemical Components in a Mixture of Cu(II) and EDTA

n	RESO $(k = 1)$	RESO $(k = 10)$	RE $(\times 10^{-4})$	IND $(\times 10^{-7})$	Malinowski F Test	IND^a Eq. (4.86) $(\times 10^{-4})$	F Test^a Eq. (4.87)
1	0.99998	0.99991	139.59	178.05	0.0	8.375	0.0
2	0.99919	0.99656	17.32	23.761	0.0	1.588	0.0
3	0.97528	0.94538	10.20	15.081	0.0	1.556	5.1
4	0.99062	0.97334	5.28	8.455	0.0	1.398	4.3
5	0.79283	0.61446	4.23	7.341	1.7	2.535	26.1
6	0.81055	0.64809	3.67	6.943	6.1	9.315	45.9
7	0.77536	0.56964	3.19	6.586	6.5		
8	0.81503	9.67340	3.07	6.968	28.8		
9	0.83165	0.62871	2.97	7.424	30.5		
10	0.83544	0.65095	2.88	7.988	33.2		

^{a}Based on multiple sources of error theory using $m = 29 - 7 = 21$.

0.0015 absorbance unit. For the absorbances of the cobalt complexes (Example 2), we found $n = 3$. Applying (4.95), we find RE = 0.00104, in agreement with the known facts. For the GLC analysis (Example 3) we concluded that six factors were present (see Table 4.5). The RE corresponding to $n = 6$ is 2.76, in excellent agreement with the fact that the experimental error was estimated to be no greater than 3 units. Many other examples of the usage of the IE, IND, and RE functions can be found in the study by Malinowski.[16]

4.5   TESTING FACTOR ANALYZABILITY

To be truly factor analyzable, a data matrix must obey the linear-sum-of-products rule described by (1.1), must consist of a sufficient number of data columns, and must be free from excessive error. Malinowski[16] showed how the IE and IND functions can be used to decide whether a data matrix is factor analyzable. He studied a series of data matrices composed of random numbers. An example of the type of results obtained is shown in Table 4.7. The IE increases on going from $n = 1$ to $n = 2$ to $n = 3$, indicating, theoretically, that we are dealing with pure error space. Also, the IND increases from $n = 1$ to $n = 7$. Since this matrix is obviously not one dimensional, the minimum must occur at $n = 0$. If both the IE and IND functions behave in this fashion, the data are probably not factor analyzable. By studying the behavior of the IE and IND functions, we can weed out those data matrices that are not factor analyzable. This weeding-out process is an important adjunct to factor analytical studies and should be performed routinely.

TABLE 4.7   Results of Factor Analyzing a 10×8 Data Matrix Composed of Random Numbers Ranging from 4 to 99[a]

n	RE	IE	IND
1	27.00	9.54	0.55
2	24.66	12.33	0.69
3	21.88	13.40	0.88
4	18.51	13.09	1.16
5	14.42	11.40	1.60
6	11.02	9.54	2.75
7	6.47	6.05	6.47

[a]Reprinted with permission from E. R. Malinowski, *Anal. Chem.*, **49**, 612 (1977).

4.6   ERRORS IN TARGET TEST VECTORS

In this section we trace through the target testing process to develop error criteria that can be used to determine whether a test vector is a valid factor and whether the test vector will lead to data matrix improvement or degradation.

4.6.1   Theory

Arguments presented in Section 4.1 form the basis of the approach. There we learned how the experimental errors in the raw data matrix perturb the abstract factors. In fact, (4.9) and (4.21) show exactly how the errors contribute to the row and column factors, respectively. If the errors in the data matrix are reasonably small, the ratio c_{jk}^*/c_{jk} will be close to unity and (4.9) will reduce to the following:

$$r_{ij} = r_{ij}^* + \sigma_{jk}^{\ddagger} \tag{4.96}$$

Hence the abstract row-factor matrix, $\bar{\mathbf{R}}$, can be expressed as a sum of two matrices: $\mathbf{R}^*$, the pure row matrix, and $\mathbf{E}^{\ddagger}$, the associated error matrix:

$$\bar{\mathbf{R}} = \mathbf{R}^* + \mathbf{E}^{\ddagger} \tag{4.97}$$

Recall, from (3.121), that the elements of $\bar{\mathbf{R}}$ are used to yield a least-squares transformation vector, $\mathbf{t}_l$, for a given test vector, $\mathbf{x}_l$. The predicted vector, $\hat{\mathbf{x}}_l$, is calculated from these results by means of (3.108).

$$\hat{\mathbf{x}}_l = \bar{\mathbf{R}}\mathbf{t}_l \tag{3.108}, (4.98)$$

The *apparent error in the target test vector*, $\mathbf{e}_A$, is defined as the difference between the "predicted" vector, $\hat{\mathbf{x}}_l$, and the "raw" test vector, $\mathbf{x}_l$:

$$\mathbf{e}_A = \hat{\mathbf{x}}_l - \mathbf{x}_l \tag{4.99}$$

Placing (4.97) into (4.98) and then inserting the result into (4.99) gives

$$\mathbf{e}_A = \mathbf{E}^{\dagger}\mathbf{t}_l + \mathbf{R}^*\mathbf{t}_l - \mathbf{x}_l \tag{4.100}$$

If the test vector represents a true factor, and if both the test vector and the data matrix are pure (i.e., contain no experimental error), then

$$\mathbf{x}_l^* = \hat{\mathbf{x}}_l^* = \mathbf{R}^*\mathbf{t}_l^* \tag{4.101}$$

Here $\mathbf{x}_l^*$ is the pure test vector, $\hat{\mathbf{x}}_l^*$ is the pure predicted vector, and $\mathbf{t}_l^*$ is the transformation vector obtained using the pure data. Such situations almost never exist in chemistry. However, if the errors are small,

$$\hat{\mathbf{x}}_l^* \simeq \mathbf{R}^*\mathbf{t}_l \tag{4.102}$$

where $\mathbf{t}_l$ is the transformation vector obtained using raw data. Hence

$$\mathbf{e}_A = \mathbf{E}^{\dagger}\mathbf{t}_l + \mathbf{x}_l^* - \mathbf{x}_l \tag{4.103}$$

The real *error in the target test vector*, $\mathbf{e}_T$, is the difference between the pure test vector and the raw test vector:

$$\mathbf{e}_T = \mathbf{x}_l^* - \mathbf{x}_l \tag{4.104}$$

The *real error in the predicted vector*, $\mathbf{e}_P$, is defined as the difference between the predicted vector and the pure test vector:

$$\mathbf{e}_P = \hat{\mathbf{x}}_l - \mathbf{x}_l^* \tag{4.105}$$

From the equations above we see that

$$\mathbf{e}_A = \mathbf{e}_P + \mathbf{e}_T \tag{4.106}$$

$$\mathbf{e}_P = \mathbf{E}^{\dagger}\mathbf{t}_l \tag{4.107}$$

It is better to express the errors as root mean squares rather than as vectors. This is easily accomplished because the inner product of an error vector is related to the standard deviation. For example, the root mean square of the apparent error in the test vector (AET) is defined by

$$
\text{AET} = \left(\frac{\sum_{i=1}^{r}(\hat{x}_i - x_i)^2}{r} \right)^{1/2}
\tag{4.108}
$$

Here $\hat{x}_i$ and x_i are the ith elements of the predicted vector and the test vector, respectively. The sum, taken over all r elements of the vector, is equal to the dot product of error vector $\mathbf{e}_A$:

$$
\mathbf{e}_A'\mathbf{e}_A = \sum_{i=1}^{r}(\hat{x}_i - x_i)^2
\tag{4.109}
$$

Hence from (4.108) and (4.109), we obtain

$$
\mathbf{e}_A'\mathbf{e}_A = r(\text{AET})^2
\tag{4.110}
$$

The root mean square of the predicted vector (REP) is defined by

$$
\text{REP} = \left(\frac{\sum_{i=1}^{r}(\hat{x}_i - x_i^*)^2}{r} \right)^{1/2}
\tag{4.111}
$$

Here x_i^* is the ith element of the pure test vector. Realizing that

$$
\mathbf{e}_P'\mathbf{e}_P = \sum_{i=1}^{r}(\hat{x}_i - x_i^*)^2
\tag{4.112}
$$

and using (4.111) and (4.112), we conclude that

$$
\mathbf{e}_P'\mathbf{e}_P = r(\text{REP})^2
\tag{4.113}
$$

Similarly, in standard RMS form, the real error in the target vector (RET) is defined as

$$
\text{RET} = \left(\frac{\sum_{i=1}^{r}(x_i^* - x_i)^2}{r} \right)^{1/2}
\tag{4.114}
$$

Noting that

$$\mathbf{e}_P'\mathbf{e}_P = \sum_{i=1}^{r}(x_i^* - x_i)^2 \qquad (4.115)$$

and using (4.114) and (4.115), we find

$$\mathbf{e}_T'\mathbf{e}_T = r(\text{RET})^2 \qquad (4.116)$$

With this information we can convert (4.106), which shows the relationship between the error vectors, into a RMS relationship. To do so, we consider the inner product of both sides of 4.106:

$$\mathbf{e}_A'\mathbf{e}_A = \mathbf{e}_P'\mathbf{e}_P + \mathbf{e}_T'\mathbf{e}_T + \mathbf{e}_P'\mathbf{e}_T + \mathbf{e}_T'\mathbf{e}_P \qquad (4.117)$$

Since the elements of these vectors are random errors, being both positive and negative, the last two terms on the right should be relatively small in comparison with the first two terms, which are sums of squares. Hence

$$\mathbf{e}_A'\mathbf{e}_A = \mathbf{e}_P'\mathbf{e}_P + \mathbf{e}_T'\mathbf{e}_T \qquad (4.118)$$

Substituting (4.110), (4.113), and (4.116) into (4.118) yields

$$(\text{AET})^2 = (\text{REP})^2 + (\text{RET})^2 \qquad (4.119)$$

Figure 4.6 is a mnemonic representation of this Pythagorean relationship. The three different vectors ($\hat{\mathbf{x}}$, $\mathbf{x}^*$, and $\mathbf{x}$) concerning a single factor lie at the corners of the right triangle with $\mathbf{x}^*$ coincident with the right angle. RET and REP are the sides of the triangle and AET is the hypotenuse.

Equation (4.108) is a biased estimate of the error because it does not take into account the loss in degrees of freedom resulting from the use of n variables (factors) in the least-squares fitting procedure. Furthermore, in many situations, the test vector

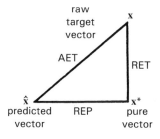

Figure 4.6   Mnemonic diagram of the Pythagorean relationship of the errors in the target test vector. [Reprinted with permission from E. R. Malinowski, *Anal. Chim. Acta*, **103**, 339 (1978).]

may be incomplete, having missing points, called "blanks." An unbiased estimate is obtained by reducing the denominator in (4.108) by n, the number of factors employed, and by b, the number of blanks in the test vector, and summing over the $r - b$ test points:

$$\text{AET} = \left(\frac{\sum\limits_{i=1}^{r-b} (\hat{x}_i - x_i)^2}{r - n - b} \right)^{1/2} \tag{4.120}$$

This expression is applicable for situations where the target is either complete or incomplete.

To obtain a numerical value for REP, we make the following observations. The inner product of $\mathbf{e}_P$, according to (4.107), can be written in terms of the error matrix $\mathbf{E}^\ddagger$ and the transformation vector $\mathbf{t}_l$:

$$\mathbf{e}_P' \mathbf{e}_P = (\mathbf{E}^\ddagger \mathbf{t}_l)'(\mathbf{E}^\ddagger \mathbf{t}_l) = \mathbf{t}_l' \mathbf{E}^{\ddagger\prime} \mathbf{E}^\ddagger \mathbf{t}_l \tag{4.121}$$

Since $\mathbf{E}^\ddagger$ is an $r \times n$ matrix, premultiplication by its own transpose will yield an $n \times n$ matrix with a trace equal to $rn(\text{RE})^2$, where RE is the real error in the data matrix [see (4.36) and (4.40)]. In comparison to the diagonal elements, the off-diagonal elements of this $n \times n$ matrix should be negligibly small because the elements of $\mathbf{E}^\ddagger$ are error points randomly scattered about zero. Because there are n elements along the diagonal, the average value of a diagonal element is $r(\text{RE})^2$. Hence as a reasonable approximation, we can write

$$\mathbf{e}_P' \mathbf{e}_P = r(\text{RE})^2 (\mathbf{t}_l' \mathbf{t}_l) \tag{4.122}$$

Inserting this into (4.113) gives

$$\text{REP} = (\text{RE}) \| \mathbf{t}_l \| \tag{4.123}$$

Equation (4.123) affords an easy method for calculating the real error in the predicted vector. A numerical value for RE is obtained from the eigenvalues during the abstract reproduction step, which also yields the size of the factor space. When a test vector is target transformed, numerical values for the transformation vector are generated. These values are then placed into (4.115) and a numerical value for REP is obtained.

From (4.120) and (4.123) numerical values for AET and REP can be calculated. Inserting these into (4.119), we can compute RET, the real error in the test vector. This error can be compared to our estimation of the error in the data used in constructing the test vector. If the calculated RET agrees with the estimated RET, the target vector is valid.

For any number of targets, the MATLAB program labeled tfa.m (target factor analysis) computes AET, REP, and RET for each target, based on a specified factor level, and is included in Appendix C for the convenience of the reader.

In the next two sections we develop empirical criteria for judging the validity of a test vector. Because of the approximations made, the derived error expressions will be valid only if the test vector is a true vector having sufficient accuracy. If the test vector is not a true representation of a valid factor, the resulting transformation vector will be incorrect, (4.102) will fail, and, consequently, all equations thereafter will be invalid. In the next section we develop empirical criteria to determine when this will occur.

4.6.2 SPOIL Function

In many situations, owing to lack of information concerning the experimental uncertainty, the error in the test vector cannot be estimated accurately. Even when error information is available, a direct comparison between the calculated and estimated RET does not give a clear indication of the errors that will be introduced into the reproduced data matrix by target combination. In addition to determining whether or not a test vector is acceptable as a true factor, we need some criterion to tell us whether a given test vector will improve or spoil the reproduced data matrix when that vector is used in the target factor analysis (TFA) combination step. The theory in Section 4.6.1 affords us with the necessary tools to develop such a criterion, called the SPOIL.[40]

Before defining the SPOIL function, let us examine (4.119) and Figure 4.6. If a given target vector were pure, having no error, RET would be zero and AET would equal REP. The predicted vector, obtained by the method of least squares, would be different from the pure test vector. In this situation the error in the predicted vector comes solely from the error in the data matrix. Thus

$$REP \simeq EDM \tag{4.124}$$

where EDM is the error contributed by the data matrix. This approximation holds even when $RET \neq 0$. Hence the apparent error in the target can be interpreted as a vector sum of two error sources: an error from the impure data matrix and an error from the impure target. Although somewhat oversimplified, this interpretation is conceptually satisfying.

If the real error in the target is less than the error contributed by the data matrix (i.e., if $RET < EDM \simeq REP$), the reproduced target will contain more error than the original target. When a good target is used in the combination step, it will tend to improve the regenerated data matrix. There is, therefore, a need to seek the most accurate data to describe the target.

On the other hand, if the real error in the target is greater than the error contributed by the data matrix (i.e., if $RET > EDM \simeq REP$), the reproduced target vector will be more accurate than the original impure target vector. Thus target testing can be used as a unique data improvement tool. This is true only if the target is a real factor and if the correct number of factors are employed in the transformation test. Unfortunately, however, such a target, when used in the combination step, tends to spoil the reproduced data matrix because it introduces additional error into the reproduction process.

The SPOIL function[40] is defined as

$$\text{SPOIL} = \frac{\text{RET}}{\text{EDM}} \simeq \frac{\text{RET}}{\text{REP}} \qquad (4.125)$$

If the SPOIL is less than 1.00, the combination-reproduced data matrix will be improved by the target vector. On the other hand, if the SPOIL is greater than 1.00, the reproduced data matrix will be spoiled by the target. In fact, the larger the SPOIL, the poorer will be the data matrix reproduction. Consequently, the errors in the target test points should be sufficiently small so that negligible error is introduced into the FA process. This means that we should strive for targets that yield the smallest possible SPOIL.

The SPOIL not only gives us a "sense" of how much error is in the target (relative to the data matrix) but it also provides us with a criterion for judging the overall validity of a suspected target. Since EDM is approximately equal to REP and since the equations used to evaluate RET and REP are not exact, we can only hope to develop rule-of-thumb criteria. By studying model sets of data, Malinowski[40] classified a target, on the basis of its SPOIL, as being acceptable, moderately acceptable, or not acceptable. A target is considered to be acceptable if its SPOIL lies between 0.0 and 3.0. If the SPOIL lies between 3.0 and 6.0, the target is moderately acceptable; although such a target is probably a true factor, the error is too large and its use in data matrix reproduction will lead to spoilage. A target is not acceptable if its SPOIL is greater than 6.0 because it will introduce excessive error into the reproduced data matrix.

For any number of targets, the MATLAB program labeled tfa.m (target factor analysis) computes the SPOIL for each target, based on a specified factor level, and is included in Appendix C for the convenience of the reader.

4.6.3 Statistical Methods for Target Testing

Several statistical tests based on variance ratio have been devised specifically for factor analytical target testing. Each method assumes that the associated error functions are normally distributed so Fisher F tests can be performed. Tabulated F values are compared to the ratio between the error variance associated with the fitted test vector and the error variance associated with the data matrix. The methods differ in the principles used to obtain the variances.

Lorber[41] developed an F test based on the theory of projection matrices and the distribution of quadratic forms. This method requires normalizing the columns of the data matrix as well as the target test vectors as a pretreatment process. AFA and TFA are then carried out in the usual manner. The variance in the target vector is estimated by

$$\text{Variance(target)} = \|\hat{\mathbf{x}} - \mathbf{x}\|^2/(r - c) \tag{4.126}$$

and the variance in the data matrix is estimated by

$$\text{Variance(data)} = \|\bar{\mathbf{D}} - \mathbf{D}\|^2/c(c - n) \tag{4.127}$$

The variance ratio is distributed as F with $(r - c)$ and $c(c - n)$ degrees of freedom.

$$F(r - c, c(c - n)) = \text{variance(target)}/\text{variance(data)} \tag{4.128}$$

This method is applicable only if $r > c$ and the test vector is complete (i.e., there are no missing points). It cannot be used to test row vectors (with column elements). The method also has a bias due to column-vector normalization, which weights small and large vectors equally. Although such weightings may be desirable in certain cases, methods free from such bias are preferable.

Malinowski[31] devised an F test that is free from the limitations of the Lorber test. He showed that the variance in a test vector can be expressed in terms of the reduced eigenvalues of the factor space. When this variance is compared to the variance of the pool of reduced eigenvalues, the following (one-sided) test statistic results:

$$F(r - n - b, s - n) = \frac{\displaystyle\sum_{j=n+1}^{s} (r - j + 1)(c - j + 1)}{(r - n + 1)(c - n + 1)} \times \frac{r\|\hat{\mathbf{x}} - \mathbf{x}\|^2}{(r - n - b) \displaystyle\sum_{j=n+1}^{s} \lambda_j^0 \|\mathbf{t}\|^2} \tag{4.129}$$

for column test vectors, and

$$F(c - n - b, s - n) = \frac{\displaystyle\sum_{j=n+1}^{s} (r - j + 1)(c - j + 1)}{(r - n + 1)(c - n + 1)} \times \frac{c\|\hat{\mathbf{y}} - \mathbf{y}\|^2}{(c - n - b) \displaystyle\sum_{j=n+1}^{s} \lambda_j^0 \|\mathbf{t}\|^2} \tag{4.130}$$

for row test vectors. Here s is either r or c, whichever is smaller.

The results of applying the Lorber and the Malinowski F tests to the mass spectra of four different mixtures of cyclohexane and cylohexene are shown in Table 4.8. At the 5% significance level the Lorber method correctly identifies the presence of these

TABLE 4.8   Comparison of F Tests Using Mass Spectra of Five Target Compounds for Mixtures of Cyclohexane and Cyclohexene

Target	Lorber Method[a]				Malinowski Method[b]			
	F	df_1	df_2	%SL	F	df_1	df_2	%SL
Cyclohexane	1.20	16	8	41.3	2.81	18	2	29.4
Cyclohexene	2.95	16	8	6.18	5.23	18	2	17.2
Bicyclo[3.1.0]hexane	9.26	16	8	0.21	16.07	18	2	6.0
Fluorocyclohexane	32.30	16	8	0.005	101.30	18	2	1.0
Bicyclopropyl	47.50	16	8	0.002	101.40	18	2	1.0
Cyclohexane[c]	—	—	—	—	1.52	10	2	46.0
n-Hexane[c]	—	—	—	—	3120.25	10	2	0.0

[a]From A. Lorber, *Anal. Chem.*, **56**, 1004 (1984). Copyright 1984 American Chemical Society.
[b]From E. R. Malinowski, *J. Chemometrics*, **3**, 49 (1988), John Wiley & Sons, Ltd.
[c]Spectra taken from a different library. The Lorber method is not applicable because spectral data points are missing in the target.

two constituents and rejects the others. The 10% level is required for the Malinowski method to achieve the same conclusions. The Lorber test[41] is more sensitive but is more restrictive in its applicability. The Malinowski method is less sensitive but has a wider range of applicability and can be applied to incomplete target tests as illustrated for the last two spectral vectors listed in Table 4.8.

It is important to note that the F tests described above can be expressed in terms of the apparent error in the target, AET, and the error emanating from the data matrix, EDM, namely:

$$F(a - n - b, s - n) = \left(\frac{AET}{EDM}\right)^2 \frac{\sum_{j=n+1}^{s}(r - j + 1)(c - j + 1)}{(r - n + 1)(c - n + 1)} \quad (4.131)$$

where a equals r for column test vectors and c for row test vectors.

When this test is applied to the target shown in Table 4.9, one finds $F(9, 5) = 4.12$ corresponding to a 6.67% significance level. This vector is significant at the 5% level but not significant at the 10% level.

For any number of targets assembled as a matrix, the MATLAB program labeled tfa.m (target factor analysis) computes $F(df_1, df_2)$ as well as df_1 and df_2 for each individual target, based on the specified factor level, and is included in Appendix C for the convenience of the reader.

4.6.4   Chemical Example

To illustrate the principles and usefulness of target error theory in chemistry, Malinowski[40] investigated the solvent dependence of ^{19}F magnetic resonance

TABLE 4.9   Target Testing ^{19}F Gas-Phase Shifts^a

Solute	Test Vector, x	Predicted Vector, $\hat{x}$	Difference, $\hat{x} - x$
CF_2Br_2	−2.27	−2.50	−0.23
$CFCl_3$	—	4.98	—
CF_2ClBr	4.64	4.52	−0.12
$CFCl_2CFCl_2$	—	71.62	—
sym-$C_6F_3Cl_3$	—	118.98	—
CF_2Cl_2	12.17	11.95	−0.22
cis-CFClCFCl	111.91	112.14	0.23
trans-CFClCFCl	—	125.62	—
C_6F_6	170.73	170.77	0.04
CF_3CCl_3	87.04	86.97	−0.07
CF_2CCl_2	95.67	95.98	0.31
CF_3CCF_3	59.29	59.41	0.12
C_4F_8	140.85	140.84	−0.01
CF_4	—	68.70	—
$C_6H_5CF_3$	—	69.94	—
$CF_3CHClBr$	—	82.66	—
α-C_6F_{14}	86.33	86.22	−0.11
β-C_6F_{14}	130.37	130.26	−0.11
γ-C_6F_{14}	126.73	126.44	−0.29
Theoretical	0.19	0.05	0.19
errors	RET	REP	AET

^{a}Reprinted with permission from E. R. Malinowski, *Anal. Chim. Acta*, **103**, 339 (1978).

chemical shifts. When the data matrix, which consisted of the fluorine shifts of 19 rigid, nonpolar solute molecules dissolved in 8 solvents, was subjected to AFA, 3 factors were found. This conclusion was based on the IE and IND functions, which yielded an RE equal to 0.035 ppm, in excellent agreement with the known uncertainty.

Based on theoretical arguments, the gas-phase shift of the solute was suspected to be one of the 3 basic factors. To see whether or not this was so, a gas-phase test vector, shown in Table 4.9, was constructed. This test vector contained only 12 points; 7 points had to be free-floated because their gas-phase shifts were not measured. The predicted vector yielded gas-phase shifts for all 19 solutes including the 7 free-floated solutes. The AET, shown at the bottom of the table, was calculated from the differences between the test and predicted points, via (4.120). The table also lists values for REP and RET, calculated by means of (4.123) and (4.119), respectively.

The SPOIL, calculated to be 3.8, indicated a moderately acceptable test vector. This implied that the experimental error in the vapor shifts was approximately four times the error in the solution shifts, in agreement with the known experimental estimates that the error in the vapor shifts was 0.14 ppm, whereas the error in the solution shifts was 0.035 ppm.

Because the REP is smaller than the RET, the predicted gas-phase shifts are more accurate than the measured shifts. We see here that, in certain situations, TFA can be used not only to identify a true factor but also to purify the target data.

4.7   ERRORS IN THE LOADINGS

In many studies the identities of the true factors may be known. Numerical values for the real row factors, which comprise the target vectors, may also be accurately known. Upon inserting these test vectors into the target combination procedure, one is faced with the task of estimating the reliability of the loadings (i.e., the column factors, $\hat{c}_{jk}$), which result from the combination transformation step. At the present time there exists three methods for estimating the errors in the loadings. One method is based on the theory of error for factor analysis and the other two methods are based on statistical procedures. The first method yields a root-mean-square error for each factor; the other methods yield the error and confidence limits for each individual loading. The methods are discussed in the next subsections.

4.7.1   Estimation Based on the Theory of Error

The RMS of the error in the factor loading, EFL, can be deduced from the theory of error[2,40] previously discussed. If the errors are reasonably small, $r_{ij}^*/r_{ij} \simeq 1$ and, according to (4.21), the abstract column factor can be expressed as

$$c_{jk} = c_{jk}^* + \sigma_{jk}^{\dagger} \tag{4.132}$$

which, in matrix form, can be written

$$\bar{\mathbf{C}} = \mathbf{C}^* + \mathbf{E}_c^{\dagger} \tag{4.133}$$

The abstract column matrix $\bar{\mathbf{C}}$ can be transformed into the real matrix $\hat{\mathbf{Y}}$ by means of (3.15). Hence

$$\hat{\mathbf{Y}} = \hat{\mathbf{Y}}^* + \mathbf{T}^{-1}\mathbf{E}_c^{\dagger} \tag{4.134}$$

where $\mathbf{T}^{-1}\mathbf{C}^* = \hat{\mathbf{Y}}^*$, assuming, of course, that $\mathbf{T}$ contains little or no error. Since $\hat{\mathbf{Y}}^*$ is the real matrix, free from error, it follows, from (4.134), that the error in the real loading matrix $\hat{\mathbf{E}}_y$ is simply

$$\hat{\mathbf{E}}_y = \hat{\mathbf{Y}} - \hat{\mathbf{Y}}^* = \mathbf{T}^{-1}\mathbf{E}_c^{\dagger} \tag{4.135}$$

In order to obtain the RMS errors we consider the product $\hat{\mathbf{E}}_y\hat{\mathbf{E}}_y'$, which from (4.135) yields

$$\hat{\mathbf{E}}_y\hat{\mathbf{E}}_y' = \mathbf{T}^{-1}\mathbf{E}_c^{\ddagger}\mathbf{E}_c^{\ddagger\prime}\{\mathbf{T}^{-1}\}' \tag{4.136}$$

According to the theory of error for AFA, $\mathbf{E}_c^{\ddagger}\mathbf{E}_c^{\ddagger\prime}$ is a diagonal matrix where the jth diagonal element equals $\sum_{k=1}^{k=c}(\sigma_{jk}^{\ddagger})^2 = c(\mathrm{RE})^2/\lambda_j^{\ddagger}$. Thus

$$\mathbf{E}_c^{\ddagger}\mathbf{E}_c^{\ddagger\prime} = c(\mathrm{RE})^2\mathbf{\Lambda}^{\ddagger-1} \tag{4.137}$$

where $\mathbf{\Lambda}^{\ddagger-1}$ is a diagonal matrix composed of the reciprocals of the primary eigenvalues. Putting (4.137) into (4.136) gives

$$\hat{\mathbf{E}}_y\hat{\mathbf{E}}_y' = c(\mathrm{RE})^2\hat{\mathbf{T}}\hat{\mathbf{T}}' \tag{4.138}$$

where

$$\hat{\mathbf{T}} = \mathbf{T}^{-1}\mathbf{\Lambda}^{\ddagger-1/2} \tag{4.139}$$

Since the jth diagonal element of $\hat{\mathbf{E}}_y\hat{\mathbf{E}}_y'$ equals $c(\mathrm{EFL})_j^2$, where $(\mathrm{EFL})_j$ is the RMS of the error in the jth factor loading,

$$(\mathrm{EFL})_j = \mathrm{RE}\|\hat{\mathbf{t}}_j\| \tag{4.140}$$

where $\hat{\mathbf{t}}_j$ is the jth row of $\hat{\mathbf{T}}$.

Equation 4.140 provides an easy method for determining the RMS error in each factor loading, because RE is obtained from AFA and $\hat{\mathbf{t}}_j$ is obtained from TFA. An example calculation involving the use of this equation is given in Section 5.2.3.

4.7.2   Jackknife Method

Weiner and co-workers[42] adopted the *jackknife method* of Mosteller and Tukey[43] to obtain not only the error in each loading factor but also the confidence limits. The basis of the jackknife method is best understood by means of a simple example involving integers. Consider the following five numbers: 3, 5, 7, 10, and 15. Any one of these numbers (say, 7) can be expressed as the difference between the weighted mean of all the numbers and the mean formed by excluding the number in question. For example,

$$7 = 5\left(\frac{3+5+7+10+15}{5}\right) - 4\left(\frac{3+5+10+15}{4}\right) \tag{4.141}$$

Although this calculation appears to be trivial for equally weighted integers, it is valuable when applied to other quantities, such as the weightings obtained from regression analysis or the loadings obtained from target factor analysis. In the latter case, instead of integers, we deal with the overall loading $\hat{y}_{jk}(\text{red}, i)$ obtained when the data matrix is reduced by removing the ith row. From these the effective loadings $\hat{y}_{jk}(\text{eff}, i)$ are calculated as follows:

$$\hat{y}_{jk}(\text{eff}, i) = r\hat{y}_{jk}(\text{all}) - (r - 1)\hat{y}_{jk}(\text{red}, i) \qquad (4.142)$$

where r corresponds to the number of rows in the complete data matrix. Values for $\hat{y}_{ik}(\text{eff}, i)$ are calculated by removing, systematically, each row of the data matrix, carrying out the combination target transformation each time, and applying (4.142).

Equation 4.142 is analogous to the integer calculation (4.141). The $\hat{y}_{jk}(\text{eff}, i)$ terms reflect the intrinsic effects of the data points that were dropped from the analysis. There are r values for each $\hat{y}_{jk}(\text{eff}, i)$. These values, according to Mosteller and Tukey,[43] are normally distributed even though the original $\hat{y}_{jk}(\text{all})$ values may not be normally distributed. Approximate confidence limits may be calculated by applying standard statistical methods to the effective loadings.

The average of the effective loadings, $\hat{y}_{jk}(\text{av})$, represents the "best" estimate of the true loadings:

$$\hat{y}_{jk}(\text{av}) = \frac{1}{r}\sum_{i=1}^{r}\hat{y}_{jk}(\text{eff}, i) \qquad (4.143)$$

The variance, s_c^2, of the effective loadings is calculated as

$$s_c^2 = \frac{\sum_{i=1}^{r}[\hat{y}_{jk}(\text{eff}, i) - \hat{y}_{jk}(\text{av})]^2}{r - 1} \qquad (4.144)$$

This permits an estimation of the confidence limits for each of the individual loadings. For example, based on the standard t distribution, the confidence interval for $\hat{y}_{jk}(\text{av})$ is $\pm ts_c/\sqrt{r}$, where t is Student's coefficient for a given confidence level.

4.7.3   Calculational Method

Although the jackknife method provides a detailed error analysis for each individual loading, the method is extremely time consuming because it requires repeated factor analysis of a series of reduced data sets where one observation (row or column) has been deleted each time. Recognizing this problem, Roscoe and Hopke[44] adopted the "calculational" method of Clifford,[45] which is based on the propagation of errors in multivariate analysis. The method relies on comparing the raw data to the reproduced data as summarized below.

For column test vectors, the error matrix, $\mathbf{E}$, is calculated as the difference between the raw data, $\mathbf{D}$, and the reproduced data, $\hat{\mathbf{D}}$. Each column of $\mathbf{E}$ is treated as an error vector, $\mathbf{e}_j$.

$$\mathbf{D} - \hat{\mathbf{D}} = \mathbf{E} = [\mathbf{e}_1 \quad \mathbf{e}_2 \quad \cdots \quad \mathbf{e}_j \quad \cdots \quad \mathbf{e}_c] \tag{4.145}$$

The fractional variance–covariance matrix of the loading errors, $\mathbf{V}$, is then calculated as

$$\mathbf{V} = \sigma_j^2 (\mathbf{X}'\mathbf{W}\mathbf{X})^{-1} \tag{4.146}$$

where

$$\sigma_j^2 = \mathbf{e}_j'\mathbf{W}\mathbf{e}_j / (r - n) \tag{4.147}$$

In these equations $\mathbf{W}$ represents a weight matrix. If weighted factor analysis has not been employed, $\mathbf{W}$ can be deleted from these expressions because its elements behave like the Kronecker delta. The standard deviations in the loadings, $\mathbf{s}_j$, are the square roots of the diagonal elements of $\mathbf{V}$,

$$\mathbf{s}_j = \begin{bmatrix} \sqrt{v_{11}} \\ \sqrt{v_{22}} \\ \vdots \\ \sqrt{v_{nn}} \end{bmatrix} \tag{4.148}$$

This is repeated for each column of (4.145). The standard deviation vectors thus obtained are then assembled as columns of the standard deviation matrix, $\mathbf{SD}$, having the same size $(n \times c)$ as the loading matrix, where each element represents the standard deviation of the corresponding element in the loading matrix:

$$\mathbf{SD} = [\mathbf{s}_1 \quad \mathbf{s}_2 \quad \cdots \quad \mathbf{s}_c] \tag{4.149}$$

For row test vectors, the rows of the error matrix are treated as error vectors.

$$\mathbf{D} - \hat{\mathbf{D}} = \mathbf{E} = [\mathbf{e}_1 \quad \mathbf{e}_2 \quad \cdots \quad \mathbf{e}_j \quad \cdots \quad \mathbf{e}_r]' \tag{4.150}$$

The fractional variance–covariance matrix of the loading errors, $\mathbf{V}$, is calculated as

$$\mathbf{V} = \sigma_j^2 (\mathbf{Y}'\mathbf{W}\mathbf{Y})^{-1} \tag{4.151}$$

where

$$\sigma_j^2 = \mathbf{e}_j'\mathbf{W}\mathbf{e}_j / (c - n) \tag{4.152}$$

Equation (4.148) is used to calculate the standard deviations in the loadings. These vectors are then assembled as rows of the standard deviation matrix.

The "calculational" method is incorporated in the MATLAB program listed as lfa.m in Appendix C.

4.7.4   Chemical Example

Weiner and co-workers[42] used the jackknife procedure to estimate the confidence limits of the distribution constants obtained by TFA of GLC retention volumes. Theoretically, when both the surface areas and the volumes of the liquid phases are inserted into the TFA combination scheme as test vectors, the resulting loadings are the distribution constants: K_A, the adsorption constant at the gas–liquid interface, and K_L, the bulk liquid distribution constant, respectively. Confidence limits obtained from the jackknife method were compared to those obtained by treating the same data with multiple linear regression analysis. In all cases, the distribution constants determined by both procedures overlapped with respect to their stated confidence limits. In all cases, the confidence limits calculated by the jackknife method were larger than those calculated using the standard error equation of regression analysis. This is expected since the jackknife procedure takes into account errors associated with both controlled and uncontrolled variables, whereas the error criteria used in linear regression take into consideration only those errors resulting from controlled variables. Of the two procedures, the jackknife method yields a truer picture of the confidence limits.

Roscoe and Hopke[44] used the Clifford calculational method[45] to obtain the loadings and the loading errors for the same GLC data. The values they obtained are shown in Table 4.10. In general, both the jackknife method and the calculational method yield, within their confidence regions, the same standard deviations. It is important to note that both methods tend to overestimate the errors in the loadings because with both methods the entire error in the analysis is assigned to the loading errors. The calculational method is preferred to the jackknife method because of the great reduction in computer time.

4.8   WEIGHTED FACTOR ANALYSIS

Weighted factor analysis is a form of data pretreatment (see Section 3.2.3) that enhances selected sections of the data matrix. Some form of weighting is required when the data rows (or columns) have different units. The discussion of weighted factor analysis has been relegated to the end of this chapter because the discussion makes use of error theory.

4.8.1   Rank Determination

The discussion of errors, up to this point, has been based on the assumption that the variance of the measurement errors is constant throughout the data. Hence the

TABLE 4.10 Factor Loadings and Their Associated Error for the Gas–Liquid Chromatography Solute Retention Study[a]

	K_A, Adsorption Constant ($\times 10^{-5}$)		
Solute	Calculation Method	Jackknife Method	Linear Regression
Carbon tetrachloride	5.18±0.09	5.11±0.11	5.27±0.05
Dichloromethane	4.42±0.15	4.32±0.17	4.50±0.04
Chloroform	12.4±0.6	12.06±0.11	13.56±0.10
Benzene	27.7±0.3	27.48±0.28	27.73±0.19
Toluene	34.2±0.1	34.19±0.10	34.35±0.25
n-Hexane	7.89±0.11	7.83±0.12	7.87±0.07
Cyclohexane	10.6±0.2	10.56±0.22	10.69±0.08
n-Heptane	31.0±0.3	30.94±0.25	30.99±0.15
2-Methylheptane	26.2±0.2	26.06±0.21	26.12±0.13
n-Octane	75.4±1.6	76.90±0.58	76.98±0.34

	K_L, Partition Constant		
Solute	Calculation Method	Jackknife Method	Linear Regression
Carbon tetrachloride	0.04±0.55	0.40±0.62	
Dichloromethane	0.12±0.90	0.83±1.33	
Chloroform	3.2±3.9	4.41±4.40	
Benzene	0.0±1.9	1.14±2.43	
Toluene	1.29±0.64	1.23±0.86	
n-Hexane	2.96±0.65	3.40±0.79	3.14±0.47
Cyclohexane	25.2±1.2	25.51±1.67	24.39±0.57
n-Heptane	19.8±1.5	20.37±1.17	20.19±1.05
2-Methylheptane	16.2±1.4	16.80±1.08	16.49±0.90
n-Octane	24.8±9.9	21.44±2.46	21.27±2.40

[a]Reprinted with permission from B. A. Roscoe and P. K. Hopke, *Anal. Chim. Acta*, **132**, 89 (1981), Elsevier Science Publishers.

principal component calculations were based on the covariance matrix constructed from the raw data. However, in many chemical processes the variance in the measurement errors is not constant throughout the experiment. For example, the noise present in absorption spectrometers is known to vary with respect to wavelength as well as line intensity. For this reason, decomposition of the covariance matrix can yield an incorrect rank estimation.

A generalized statistical weighting scheme for principal component analysis has been formulated by Cochran and Horne.[15] It is based on the assumption that the variance in each measurement error is a product of two terms: a_i, a contribution that varies from row to row, and b_j, a contribution that varies from column to column, so that

$$\mathrm{var}(e_{ij}) = a_i b_j \qquad (4.153)$$

Two nonsingular diagonal matrices, $\mathbf{A}$ and $\mathbf{B}$, are defined as

$$\mathbf{A} = \mathrm{diag}[a_1 \quad a_2 \quad \cdots \quad a_r]$$
$$\mathbf{B} = \mathrm{diag}[b_1 \quad b_2 \quad \cdots \quad b_r] \qquad (4.154)$$

Two nonsingular diagonal weighting matrices, $\mathbf{W}_r$ and $\mathbf{W}_c$, are defined as functions of (4.154):

$$\mathbf{W}_r = (a\mathbf{A})^{-1/2}$$
$$\mathbf{W}_c = (b\mathbf{B})^{-1/2} \qquad (4.155)$$

where a and b are arbitrary constants. The weighted data matrix, $\mathbf{D}_w$, is defined as

$$\mathbf{D}_w = \mathbf{W}_r \mathbf{D} \mathbf{W}_c \qquad (4.156)$$

Equations (4.155) were chosen so the eigenvectors of $\mathbf{D}_w$ are least perturbed by random measurement errors, approaching, as close as possible, the eigenvectors that would result from errorless data. Thus weighted principal component analysis (PCA) will yield a smaller variance estimate than unweighted PCA.

Eigenanalysis of $\mathbf{D}_w$, the weighted data, yields a weighted abstract row-factor matrix, $\mathbf{R}_w$, and a weighted abstract column-factor matrix, $\mathbf{C}_w$,

$$\mathbf{D}_w = \mathbf{R}_w \mathbf{C}_w \qquad (4.157)$$

which can be reduced to the proper rank by deleting the error vectors so that

$$\bar{\mathbf{D}}_w = \bar{\mathbf{R}}_w \bar{\mathbf{C}}_w \qquad (4.158)$$

The improved results obtained by weighted analysis requires additional experimentation to determine the elements of the weighting matrices, $\mathbf{W}_r$ and $\mathbf{W}_c$, a nontrivial task that can be time consuming and expensive. For certain types of data, there may be no other feasible alternative. Once the weighted data matrix is obtained, the dimensions of the factor space can be determined by the methods described in Section 4.3 using $\mathbf{D}_w$ in place of $\mathbf{D}$. The reproduced data matrix, $\bar{\mathbf{D}}$, based on n abstract factors is computed from the reproduced weighted matrix, $\bar{\mathbf{D}}_w$, in accord with Eq. 4.156, as follows:

$$\bar{\mathbf{D}} = \mathbf{W}_r^{-1} \bar{\mathbf{D}}_w \mathbf{W}_c^{-1} \qquad (4.159)$$

Unweighted PCA is a special case of weighted PCA, where the two weighting matrices are arbitrarily set equal to identity matrices. Use of the correlation matrix is a special case of weighted factor analysis, whereby one weighting matrix is the identity matrix and the other weighting matrix contains, as diagonal terms, norms of each of the respective data rows or data columns. Mean centering each data column (or data row) is also a form of weighted analysis in which importance is given to the distance from the respective mean. These techniques, unfortunately, are based on intuition rather than error theory. Consequently, the results may be biased. In some cases such bias is necessary or desirable. In other cases such bias may lead to deceptive conclusions. Sample correlation is used to avoid the problem of variable units. However, the new variables are not really standardized with respect to the population. This introduces the problem of interpreting *statistically* what actually has been computed. Detailed discussion of this important point goes beyond the scope of this text.

By increasing and decreasing selected elements of the weighting matrices, one can "focus" on segments of the data matrix. The rank, reliability, and interpretability of the results will depend strongly on the nature and degree of the weightings.

For those interested in learning more about the use of weightings in multivariate analysis of spectroscopic data, the chapter by Harris et al.[46] is highly recommended.

4.8.2 Target Tests

Target vectors should be weighted in accord with the weightings employed in the data pretreatment. For a column test vector, $\mathbf{x}$, the weighted test vector, $\mathbf{x}_w$, is calculated as follows:

$$\mathbf{x}_w = \mathbf{W}_r \mathbf{x} \tag{4.160}$$

For a row test vector, $\mathbf{y}'$, the weighted vector is

$$\mathbf{y}'_w = \mathbf{y}' \mathbf{W}_c \tag{4.161}$$

Using the weighted factor space as described in Section 4.8.1, the weighted test vectors can be evaluated by the statistical F tests described in Section 4.6.3.

Examples of the use of weighted factor analysis are found in the works by Hopke and co-workers concerning elemental analysis of airborne particles in an air pollution study[47] and lava flow for a geological investigation.[48]

4.9 GENERALIZED FORMULAS

For pedagogical reasons the discussions and development of the equations throughout this text are based on the presumption that the number of rows in the data matrix is greater than or equal to the number of columns ($r \geq c$). In practice this is easily achieved because the data matrix can always be transposed to fulfill this condition, if needed. One must, however, be extremely careful in applying these equations to situations where $r < c$.

For the general case, where s equals r or c, whichever is smaller, and l equals r or c, whichever is larger, previously developed equations take the following forms:

$$\text{RE} = \text{RSD} = \left(\frac{\sum_{j=n+1}^{s} \lambda_j^0}{l(s-n)} \right)^{1/2} \qquad \text{(4.40) (4.44) (4.95) (4.162)}$$

$$\text{IE} = \text{RSD}\sqrt{\frac{n}{s}} \qquad \text{(4.41) (4.163)}$$

$$\text{XE} = \text{RSD}\sqrt{\frac{s-n}{s}} \qquad \text{(4.42) (4.164)}$$

$$\text{IND} = \frac{\text{RE}}{(s-n)^2} \qquad \text{(4.63) (4.165)}$$

Precautions also must be taken when dealing with target vectors. For example, for row test vectors, where the elements of the vector emulate column designees, the theoretical equations developed previously take the following forms:

$$\text{AET} = \left(\frac{\sum_{k=1}^{c-b} (\hat{\mathbf{y}}_k - \mathbf{y}_k)^2}{c-n-b} \right)^{1/2} \qquad \text{(4.120) (4.166)}$$

$$\mathbf{e}_P' \mathbf{e}_P = \sum_{k=1}^{c} (\hat{\mathbf{y}}_k - \mathbf{y}_k^*)^2 \qquad \text{(4.112) (4.167)}$$

$$\mathbf{e}_P' \mathbf{e}_P = c(\text{REP})^2 \qquad \text{(4.113) (4.168)}$$

$$\mathbf{e}_p' \mathbf{e}_p = c(\text{RE})^2 (\mathbf{t}_l' \mathbf{t}_l) \qquad \text{(4.122) (4.169)}$$

Notice that (4.168) and (4.169) lead to (4.170), which is identical to (4.123).

$$\text{REP} = (\text{RE})\|\mathbf{t}_l\| \qquad \text{(4.123) (4.170)}$$

REFERENCES

1. E. R. Malinowski, in B. R. Kowalski (Ed.), *Chemometrics: Theory and Applications*, ACS Symp. Ser. 52, American Chemical Society, Washington, DC, 1977, Chap. 3.
2. E. R. Malinowski, *Anal. Chem.*, **49**, 606 (1977).

3. R. J. Rummel, *Applied Factor Analysis*, Northwestern University Press, Evanston, IL, 1970.

4. H. H. Harman, *Modern Factor Analysis*, 3rd rev. ed., University of Chicago Press, Chicago, 1967.

5. B. Fruchter, *Introduction to Factor Analysis*, Van Nostrand, New York, 1954.

6. A. L. Comrey, *A First Course in Factor Analysis*, Academic, Orlando, 1973.

7. J. T. Bulmer and H. F. Shurvell, *J. Phys. Chem.*, **77**, 256 (1973).

8. M. S. Bartlett, *Br. J. Psychol. Stat. Sect.*, **3**, 77 (1950).

9. Z. Z. Hugus, Jr., and A. A. El-Awady, *J. Phys. Chem.*, **75**, 2954 (1971).

10. J. J. Kankare, *Anal. Chem.*, **42**, 1322 (1970).

11. G. L. Ritter, S. R. Lowry, T. L. lsenhour, and C. L. Wilkins, *Anal. Chem.*, **48**, 591 (1976).

12. R. M. Wallace and S. M. Katz, *J. Phys. Chem.*, **68**, 3890 (1964).

13. D. L. Duewer, B. R. Kowalski, and J. L. Fasching, *Anal. Chem.*, **48**, 2002 (1976).

14. J. E. Davis, A. Shepard, N. Stanford, and L. B. Rogers, *Anal. Chem.*, **46**, 821 (1974).

15. R. N. Cochran and F. H. Horne, *Anal. Chem.*, **49**, 846 (1977).

16. E. R. Malinowski, *Anal. Chem.*, **49**, 612 (1977).

17. R. B. Cattell, *Multivariate Behav. Res.*, **1**, 245 (1966).

18. R. W. Rozett and E. Petersen, *Anal. Chem.*, **47**, 1303 (1975).

19. H. F. Kaiser, *Educ. Psychol. Meas.*, **20**, 141 (1960).

20. J. H. Kindsvater, P. H. Weiner, and T. J. Klingen, *Anal. Chem.*, **46**, 982 (1974).

21. O. Exner, *Collect. Czech. Chem. Commun.*, **31**, 3222 (1966).

22. R. B. Cattell, *Educ. Psychol. Meas.*, **18**, 791 (1958).

23. H. M. Cartwright, *J. Chemometrics*, **1**, 111 (1987).

24. R. I. Shrager and R. W. Hendler, *Anal. Chem.*, **54**, 1147 (1982).

25. T. Rossi and I. M. Warner, *Anal. Chem.*, **54**, 810 (1986).

26. S. Wold, *Technometrics*, **20**, 397 (1978).

27. R. N. Carey, S. Wold, and J. O. Westgard, *Anal. Chem.*, **47**, 1824 (1975).

28. D. W. Osten, *J. Chemometrics*, **2**, 39 (1988).

29. H. T. Eastman and W. J. Krzanowski, *Technometrics*, **24**, 73 (1982).

30. E. R. Malinowski, *J. Chemometrics*, **1**, 33 (1987).

31. E. R. Malinowski, *J. Chemometrics*, **3**, 49 (1988); **4**, 102 (1990).

32. N. M. Faber and B. R. Kowalski, *J. Chemometrics*, **11**, 53 (1997).

33. J. Mandel, *Technometrics*, **13**, 1 (1971).

34. X. M. Tu, D. S. Burdick, D. W. Millican, and L. B. McGown, *Anal. Chem.*, **61**, 2219 (1989).

35. E. R. Malinowski, *J. Chemometrics*, **13**, 69 (1999).

36. Z.-P. Chen, Y.-Z. Liang, J.-H. Jiang, Y. Lang, J.-Y. Qian, and R.-Q. Yu, *J. Chemometrics*, **13**, 15 (1999).

37. J. A. Rice Ramsay and B. W. Silverman, *J. R. Statist. Soc. B*, **53**, 233 (1991).

38. B. W. Silverman, *Ann. Statist.*, **24**, 1 (1996).

39. W. Den and E. R. Malinowski, *J. Chemometrics*, **7**, 89 (1993).

40. E. R. Malinowski, *Anal. Chim Acta*, **103**, 339 (1978).

41. A. Lorber, *Anal. Chem.*, **56**, 1004 (1984).

42. P. H. Weiner, H. L. Liao, and B. L. Karger, *Anal. Chem.*, **46**, 2182 (1974).

43. F. Mosteller and J. W. Tukey, in G. Lindzey and E. Aronson (Eds.), *The Handbook of Social Psychology*, 2nd ed., Vol. 2, Addison-Wesley, Reading, MA, 1986, p. 134.

44. B. A. Roscoe and P. K. Hopke, *Anal. Chim. Acta*, **132**, 89 (1981); **135**, 379 (1982).

45. A. A. Clifford, *Multivariate Error Analysis*, Applied Science, London, 1973.

46. J. M. Harris, S. D. Frans, P. E. Poston, and A. L. Wong, in H. L. C. Meuzelaar (Ed.), *Computer-Enhanced Analytical Spectroscopy*, Plenum, New York, 1989.

47. D. J. Alpert and P. K. Hopke, *Atmos. Environ.*, **15**, 675 (1981).

48. P. K. Hopke, D. J. Alpert, and B. A. Roscoe, *Comput. Chem.*, **7**, 149 (1983).

5

There is always a simpler solution.

NUMERICAL EXAMPLES OF TARGET FACTOR ANALYSIS

In order to Illustrate the mathematical concepts discussed in Chapters 3 and 4, this chapter presents the complete details of a factor analysis calculation using a model set of data. Reference is made to the pertinent equations of earlier chapters so that the calculations at each step can be followed systematically. An example that can be solved with a hand calculator within a very short period of time is studied so that those seriously interested in carrying out the calculations can do so without the aid of a computer. For this reason the calculations involve a 3×3 matrix. Since it is necessary to devise a problem with fewer factors than rows or columns of the data matrix, a two-factor space has been chosen. This also allows us to make two-dimensional plots that can be used to study the transformations of the eigenvectors.

5.1 MODEL SET OF PURE DATA

With the foregoing restrictions in mind, we arbitrarily construct three linear equations of the type

$$d_{ik} = x_{i1}y_{1k} + x_{i2}y_{2k} \tag{5.1}$$

Numerical values for the column factors α, β, and γ and for the row factors $a, b, \ldots, j$ were arbitrarily chosen. These values are shown in matrix form in (5.2). Matrix multiplication yields the data matrix on the right-hand side of (5.2):

$$
\begin{array}{c}
\text{Row} \\
\text{matrix} \\
\begin{array}{cc} x_{i1} & x_{i2} \end{array} \\
\begin{array}{c} a \\ b \\ c \\ d \\ e \\ f \\ g \\ h \\ i \\ j \end{array}
\begin{bmatrix}
0 & 4 \\
1 & -1 \\
2 & 0 \\
3 & 0 \\
4 & 3 \\
5 & -4 \\
6 & 5 \\
7 & 8 \\
8 & -2 \\
9 & -5
\end{bmatrix} \\
\mathbf{X}
\end{array}
\times
\begin{array}{c}
\text{Column matrix} \\
\begin{array}{ccc} \alpha & \beta & \gamma \end{array} \\
\begin{bmatrix}
2 & 5 & 2 \\
1 & 10 & -5
\end{bmatrix}
\begin{array}{c} y_{1k} \\ y_{2k} \end{array} \\
\mathbf{Y}
\end{array}
=
\begin{array}{c}
\text{Data matrix} \\
\begin{array}{ccc} \alpha & \beta & \gamma \end{array} \\
\begin{array}{c} a \\ b \\ c \\ d \\ e \\ f \\ g \\ h \\ i \\ j \end{array}
\begin{bmatrix}
4 & 40 & -20 \\
1 & -5 & 7 \\
4 & 10 & 4 \\
6 & 15 & 6 \\
11 & 50 & -7 \\
6 & -15 & 30 \\
17 & 80 & -13 \\
22 & 115 & -26 \\
14 & 20 & 26 \\
13 & -5 & 43
\end{bmatrix} \\
= \mathbf{D}
\end{array}
\tag{5.2}
$$

Factor analysis attempts to solve the reverse of what was just done in generating the data matrix. Starting with the data matrix, we want to obtain the row and column matrices.

5.1.1 Constructing Covariances and Correlation Matrices

The covariance matrix, which we label $\mathbf{Z}$, is calculated by taking the dot products of the column vectors of the data matrix shown in (5.2). In this case the normalization process is omitted. Carrying out the multiplication indicated by (3.17), we obtain

$$
\mathbf{Z} = \begin{bmatrix}
1,364 & 4,850 & 212 \\
4,850 & 24,725 & -5,230 \\
212 & -5,230 & 4,820
\end{bmatrix} \tag{5.3}
$$

Either the correlation matrix or the covariance matrix can be used in the factor analysis scheme. Both will yield results consistent with their own latent statistical criterion. The covariance matrix is used in the detailed example and thus the data matrix is generated directly. If we chose to use the correlation matrix, we would, of course, regenerate the normalized data matrix instead. The normalized data can be converted into the original data by dividing by the appropriate normalization constants. The normalization constant, N_k, is determined by taking the square root of the reciprocal of the sum of the squares of each element in the kth column of the data matrix:

$$N_k = \left(\frac{1}{\sum_i d_{ik}^2} \right)^{1/2} \tag{5.4}$$

In this way we obtain the normalized data matrix, $\mathbf{D}_N$:

$$
\mathbf{D}_N =
\begin{bmatrix}
0.10831 & 0.25538 & -0.28807 \\
0.02708 & -0.03180 & 0.10083 \\
0.10831 & 0.06360 & 0.05762 \\
0.16236 & 0.09539 & 0.08642 \\
0.29784 & 0.31798 & -0.10083 \\
0.16246 & -0.09539 & 0.43211 \\
0.46029 & 0.50877 & -0.18725 \\
0.59567 & 0.73135 & -0.37450 \\
0.37906 & 0.12719 & 0.37450 \\
0.35199 & -0.03180 & 0.61936
\end{bmatrix}
\tag{5.5}
$$

$$N_k \qquad 2.7076 \times 10^{-2} \qquad 6.3596 \times 10^{-3} \qquad 1.44037 \times 10^{-2}$$

The respective normalization constants are listed below each column in (5.5).

The correlation matrix $\mathbf{Z}_N$ is computed by taking dot products of the column vectors of the normalized data matrix in accord with (3.17). In this manner we obtain

$$
\mathbf{Z}_N =
\begin{bmatrix}
1.00000 & 0.83513 & 0.08268 \\
0.83513 & 1.00000 & -0.47908 \\
0.08268 & -0.47908 & 1.00000
\end{bmatrix}
\tag{5.6}
$$

5.1.2 Decomposition of the Covariance Matrix

The next step is to determine the eigenvectors that span the space. A theory of factor analysis states that the eigenvectors of the covariance matrix are the same vectors that span the space of the data matrix. To obtain the eigenvectors, we employ the method of iteration described in Sections 3.3.2 and 3.3.3.

To begin the iteration, we refer to (3.50).

$$\mathbf{Zc}_1 = \lambda_1 \mathbf{c}_1 \tag{3.50}, (5.7)$$

Here, c_1 is the first eigenvector and λ_1 is the corresponding eigenvalue. As a first approximation we arbitrarily set

$$\mathbf{c}_1' = (0.57735 \quad 0.57735 \quad 0.57735)$$

a normalized vector. In accord with (5.7), we multiply this vector by the covariance matrix:

$$\begin{bmatrix} 1,364 & 4,850 & 212 \\ 4,850 & 24,725 & -5,230 \\ 212 & -5,230 & 4,820 \end{bmatrix} \begin{bmatrix} 0.57735 \\ 0.57735 \\ 0.57735 \end{bmatrix} = \begin{bmatrix} 3,710.0 \\ 14,055.6 \\ -114.3 \end{bmatrix}$$

The resulting column vector on the right is then normalized by dividing each element by the square root of the sum of the squares of the elements:

$$\begin{bmatrix} 3,710.0 \\ 14,055.6 \\ -114.3 \end{bmatrix} = 14,537.4 \begin{bmatrix} 0.25520 \\ 0.96686 \\ -0.00786 \end{bmatrix}$$

The reciprocal of the normalization constant (14,537.4) is an approximation to λ_1. As a second approximation to $\mathbf{c}_1'$, we use the normalized vector

$$[0.25520 \quad 0.96686 \quad -0.00786]$$

resulting from the computation shown above. The new $\mathbf{c}_1$ is multiplied by $\mathbf{Z}$ and the resulting column vector is again normalized, giving a better approximation to $\mathbf{c}_1$ and λ_1. This process is repeated again and again, each time generating newer and better approximations to $\mathbf{c}_1$ and λ_1, until (5.7) is satisfied. Such iteration finally yields

$$\lambda_1 = 26,868.9$$
$$\mathbf{c}_1' = [0.180200 \quad 0.957463 \quad -0.225372]$$

To obtain the second eigenvector, we proceed by calculating the first-residual matrix, $\mathscr{R}_1$, as dictated by (3.62):

$$\mathscr{R} = \mathbf{Z} - \lambda_1 \mathbf{c}_1 \mathbf{c}_1' \qquad\qquad (3.62), (5.8)$$

The computations are as follows:

$$
\lambda_1 \mathbf{c}_1 \mathbf{c}_1' = 26,868.9
\begin{bmatrix}
0.180200 \\
0.957463 \\
-0.225372
\end{bmatrix}
[0.180200 \quad 0.957463 \quad -0.225372]
$$

$$
=
\begin{bmatrix}
872.49 & 4,635.82 & -1,091.21 \\
4,635.82 & 24,631.67 & -5,797.91 \\
-1,091.21 & -5,797.91 & 1,364.74
\end{bmatrix}
$$

Subtracting this matrix from the covariance matrix gives the first-residual matrix:

$$
\mathcal{R}_1 =
\begin{bmatrix}
491.51 & 214.18 & 1,303.21 \\
214.18 & 93.33 & 567.91 \\
1,303.21 & 567.91 & 3,455.26
\end{bmatrix}
$$

The second eigenvector is obtained by an iteration procedure analogous to the method used to obtain the first eigenvector, but involving (3.63):

$$
\mathcal{R}_1 \mathbf{c}_2 = \lambda_2 \mathbf{c}_2 \tag{3.63}, (5.9)
$$

To start the iteration, we arbitrarily choose values for $\mathbf{c}_2$, normalize $\mathbf{c}_2$, and continue to apply (5.9) until the normalized elements of $\mathbf{c}_2$ converge to constant values. In this way we find

$$
\lambda_2 = 4,040.2
$$
$$
\mathbf{c}_2' = [0.348777 \quad 0.151999 \quad 0.924790]
$$

The second-residual matrix is obtained by means of (3.65), which can be written

$$
\mathcal{R}_2 = \mathcal{R}_1 - \lambda_2 \mathbf{c}_2 \mathbf{c}_2' \tag{5.10}
$$

Carrying our this computation, we obtain

$$
\mathcal{R}_2 =
\begin{bmatrix}
0.0 & 0.0 & 0.1 \\
0.0 & 0.0 & 0.0 \\
0.1 & 0.0 & 0.0
\end{bmatrix}
$$

This residual is essentially zero. The small finite values are due to roundoff in the computations.

If correlation factor analysis had been used, the procedure would have been similar but the eigenvectors and eigenvalues would have been different because the computations would have involved $\mathbf{Z}_N$ instead of $\mathbf{Z}$.

5.1.3   Abstract Column and Row Matrices

Since two eigenvectors, c_1 and c_2, adequately account for the covariance matrix, we conclude that the data space is two dimensional. This is in accord with the known facts. The abstract column matrix is constructed from the eigenvectors as discussed previously [see (3.77)], so that

$$\bar{C} = \begin{bmatrix} c_1' \\ c_2' \end{bmatrix} = \begin{bmatrix} 0.180200 & 0.957463 & -0.225372 \\ 0.348777 & 0.151999 & 0.924790 \end{bmatrix} \tag{5.11}$$

According to (3.72), premultiplication of the inverse of the column matrix by the data matrix [given in (5.2)] yields the abstract row matrix. Because the eigenvectors are orthonormal, the inverse of the column matrix is simply its transpose. Carrying out this calculation, we find that

$$\bar{R} = [r_1 \quad r_2] = \begin{bmatrix} 43.5267 & -11.0207 \\ -6.1847 & 6.0624 \\ 9.3939 & 6.6143 \\ 14.0909 & 9.9214 \\ 51.4330 & 4.9630 \\ -20.0419 & 27.5564 \\ 82.5903 & 6.0669 \\ 119.9323 & 1.1084 \\ 15.8124 & 31.9674 \\ -12.1357 & 43.5401 \end{bmatrix} \tag{5.12}$$

Multiplying, vectorially, any row of R by any column of C regenerates the original data value. For example, for row-designee g and column-designee β, we find that

$$d_{g\beta} = (82.5903)(0.957463) + (6.0669)(0.151999) = 80.000$$

This number is precisely the same as that found in the data matrix shown in (5.2).

5.1.4   Target Transformation

Although we have been able to express the data matrix as a product of two matrices, one associated with row designees and one associated with column designees, the abstract row matrix (5.12) generated by factor analysis is quite different from the original row matrix (5.2). The reason for this difference lies in the fact that the factor analysis procedure selects a least-squares set of axes that span the factor space. At this point we have achieved the abstract reproduction shown in Figure 3.1.

The next step is to find a new set of axes that allows us to transform the abstract row and column matrices into the original matrices, which have chemical meaning. The relationship between these two sets of axes is best viewed graphically. In Figure 5.1 the results of the factor analysis are plotted using the two row-designee vectors c_1

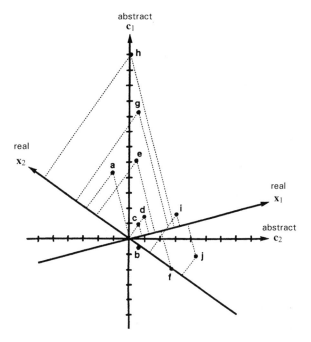

Figure 5.1 Geometrical relationship between the abstract factor axes and the real axes for a two-dimensional example problem.

and c_2 as the respective vertical and horizontal coordinate axes. To determine the score of a row designee on one of the new axes, we draw a line from the point perpendicular to the axis and read the relative distance along the axis. The scores (the projections) of the row-designee points onto the c_1 and c_2 axes give the values listed in (5.12). Upon careful inspection, we can find two other axes, labeled x_1 and x_2, which yield projections corresponding to the original model set of row factors listed in (5.2). Thus by proper transformation of the coordinate axes, we are able to find new axes, having physical significance, which express the scores in accord with those given in (5.2).

For two-dimensional problems, graphical transformation is easily accomplished. For higher-order dimensions, graphical techniques can display only two-dimensional projections, making it extremely difficult to find the original basic axes. Numerical methods, such as the least-squares method described in Section 3.4, play a vital role in locating the fundamental axes. The least-squares target transformation vector required to locate such a fundamental axis is readily obtained by carrying out the calculation expressed by (3.124):

$$\mathbf{t}_l = \bar{\mathbf{\Lambda}}^{-1}\bar{\mathbf{R}}'\mathbf{x}_l \qquad (3.124), (5.13)$$

Here the first matrix on the right is a diagonal matrix containing the reciprocals of the eigenvalues. The second matrix is the transpose of the abstract row matrix. The third term, the test vector $\mathbf{x}_l$ is a column of the original row matrix shown in (5.2), meant to simulate a real, basic factor.

In actual practice physically significant factors are tested one at a time by using (3.108):

$$\hat{\mathbf{x}}_l = \bar{\mathbf{R}}\mathbf{t}_l \qquad \text{(3.108), (5.14)}$$

If the suspected test vector $\mathbf{x}_l$ is a real factor, the regeneration, according to (3.108), will be successful. In other words, each element of $\hat{\mathbf{x}}_l$ will reasonably equal the corresponding element of $\mathbf{x}_l$.

For our problem

$$\bar{\Lambda} = \begin{bmatrix} 26{,}868.9 & 0.0 \\ 0.0 & 4{,}040.2 \end{bmatrix}^{-1} = \begin{bmatrix} 3.7218 \times 10^{-5} & 0.0 \\ 0.0 & 24.751 \times 10^{-5} \end{bmatrix} \qquad (5.15)$$

Inserting numerical values for $\bar{\Lambda}^{-1}$ and $\bar{\mathbf{R}}$ into (5.13), we find

$$\mathbf{t}_1 = \begin{bmatrix} 0.0016200 & -0.0002302 & \cdots & -0.0004517 \\ -0.0027277 & 0.0015005 & \cdots & 0.0107767 \end{bmatrix}\mathbf{x}_l \qquad (5.16)$$

5.1.5 Examples of Target Transformations

Next we give examples of two successful target transformations and one unsuccessful transformation. To test column x_{i1} of the row matrix given in (5.2), we let the elements of x_{i1}, constitute $\mathbf{x}_1$ and perform the multiplication expressed by (5.16), obtaining

$$\mathbf{t}_1 = \begin{bmatrix} 0.0563012 \\ 0.2223760 \end{bmatrix} \qquad (5.17)$$

We then carry out the calculation directed by (5.14), recalling that $\bar{\mathbf{R}}$ is the abstract row matrix given in (5.12). We obtain the results, labeled $\hat{\mathbf{x}}_1$, shown in Table 5.1. These results correspond exactly, except for errors due to computational roundoff, with column x_{i1} of (5.2), and therefore this test vector is a true factor of the space (as it indeed is, by design).

To test column x_{i2} of (5.2), we let the elements of x_{i2} constitute $\mathbf{x}_2$ and repeat the foregoing procedure, obtaining first

$$\mathbf{t}_2 = \begin{bmatrix} 0.0675947 \\ -0.0960138 \end{bmatrix} \qquad (5.18)$$

TABLE 5.1 Results of Target Tests

| Row Designee | Successful Tests | | | | Unsuccessful Test | |
	$\mathbf{x}_1$ Test	$\hat{\mathbf{x}}_1$ Predicted	$\mathbf{x}_2$ Test	$\hat{\mathbf{x}}_2$ Predicted	$\mathbf{x}_{unity}$ Test	$\hat{\mathbf{x}}_{unity}$ Predicted
a	0	-0.0001	4	4.0003	1	0.1376
b	1	0.9999	-1	-1.0001	1	0.1215
c	2	1.9997	0	-0.0001	1	0.3119
d	3	2.9996	0	-0.0001	1	0.4678
e	4	3.9994	3	3.0001	1	0.7270
f	5	4.9995	-4	-4.0005	1	0.6421
g	6	5.9991	5	5.0001	1	1.1077
h	7	6.9988	8	8.0004	1	1.3668
i	8	7.9990	-2	-2.0005	1	1.1787
j	9	8.9990	-5	-5.0008	1	1.2315

and then the values labeled $\hat{\mathbf{x}}_2$ shown in Table 5.1. Again, each value compares exactly, except for computational roundoff, with the corresponding value in (5.2). Hence we conclude that this vector is also a true factor of the space.

Only true test vectors will yield results compatible to the criteria discussed above. As an example of an unsuccessful transformation, consider the possibility that a factor of the space is the unity factor,

$$\mathbf{x}_{unity} = [1 \quad 1 \quad 1 \quad 1 \quad 1 \quad 1 \quad 1 \quad 1 \quad 1]'$$

which, in reality, is a test for a constant, a term independent of the row designee. Upon inserting this vector into (5.16), we find

$$\mathbf{t}_{unity} = \begin{bmatrix} 0.0111064 \\ 0.0313793 \end{bmatrix} \qquad (5.19)$$

Carrying out the calculation expressed in (5.14), we obtain the results shown in Table 5.1. Comparing $\mathbf{x}_{unity}$ with $\hat{\mathbf{x}}_{unity}$, we conclude that this test vector is not a factor of the space. Target transformation affords us with a unique opportunity to test vectors one at a time, without our having to specify all the vectors simultaneously as required in other methods, such as regression analysis.

5.1.6 Example of the Combination Step

The complete transformation matrix $\mathbf{T}$ is simply a combination of the transformation vectors:

$$\mathbf{T} = [t_1 \quad \cdots \quad t_n] \qquad (3.134), (5.20)$$

For our problem, using the two successful transformation vectors, (5.17) and (5.18), we find

$$\mathbf{T} = \begin{bmatrix} 0.0563012 & 0.0675947 \\ 0.2223760 & -0.0960138 \end{bmatrix} \tag{5.21}$$

According to (3.99), postmultiplying the abstract row matrix $\bar{\mathbf{R}}$ by $\mathbf{T}$ yields $\hat{\mathbf{X}}$, the row matrix in the new coordinate system:

$$\hat{\mathbf{X}} = \bar{\mathbf{R}}\mathbf{T} \tag{3.99), (5.22}$$

When this computation is performed, we obtain matrix $\hat{\mathbf{X}}$, which is identical to $\mathbf{X}$ in (5.2), except for computational error.

To obtain the column matrix $\hat{\mathbf{Y}}$ in the new coordinate system, we apply (3.102):

$$\hat{\mathbf{Y}} = \mathbf{T}^{-1}\bar{\mathbf{C}} \tag{3.102), (5.23}$$

Matrix $\bar{\mathbf{C}}$ is composed of the eigenvectors determined previously [see (5.11)]. Using standard mathematical methods, we calculate the inverse of the transformation matrix:

$$\mathbf{T}^{-1} = \begin{bmatrix} 4.6980 & 3.3075 \\ 10.8810 & -2.7545 \end{bmatrix} \tag{5.24}$$

Thus, using (5.23) and the values in (5.11) and (5.24), we find that

$$\hat{\mathbf{Y}} = \begin{bmatrix} 2.0002 & 5.0009 & 1.9999 \\ 1.0000 & 9.9995 & -4.9996 \end{bmatrix} \tag{5.25}$$

This matrix is identical to the original column matrix shown in (5.2).

In this manner we have completed the combination step illustrated in Figure 3.1, reproducing the experimental data with a set of physically significant linear parameters. The factor analysis is now essentially complete.

5.1.7   Relationship between Factor Axes and Data Space

We now show the relationship between the factor axes and the data space. Factor space is described by the minimum number of axes necessary to account for the data space. For our example problem the factor space can be described by the two columns of the abstract row matrix obtained from the factor analysis. As discussed earlier, these two columns are mutually orthogonal and can be normalized by dividing each column by the square root of its respective eigenvalue. The normalized row matrix $\bar{\mathbf{R}}_N$ [equivalent to $\bar{\mathbf{U}}$ in singular value decomposition (SVD) notation discussed in Section 3.3.7], for our problem is

$$
\bar{\mathbf{R}}_N = \begin{bmatrix}
0.265539 & -0.173384 \\
-0.037730 & 0.095377 \\
0.057308 & 0.104060 \\
0.085963 & 0.156089 \\
0.313772 & 0.078081 \\
-0.122268 & 0.433534 \\
0.503850 & 0.095448 \\
0.731659 & 0.017438 \\
0.096465 & 0.502930 \\
-0.074035 & 0.684999
\end{bmatrix} = \bar{\mathbf{U}}
\tag{5.26}
$$

In Figure 5.2 we have designated the vertical and horizontal axes as the two normalized axes $\mathbf{c}_1$ and $\mathbf{c}_2$ and have plotted points corresponding to the various row designees from the values given in (5.26). Upon examining the normalized data matrix of (5.5), we see that three axes can be drawn on this two-dimensional graph, each one being associated with a column of the normalized data matrix. The perpendicular projections, indicated by the dashed lines, from the points to the data axes (labeled $\mathbf{d}_\alpha$, $\mathbf{d}_\beta$, and $\mathbf{d}_\gamma$) intersect the axes at values corresponding to the "scores" of the row designees on the axes. These scores are identical to the data values listed in the normalized data matrix. Figure 5.2 is analogous to Figure 3.3 but shows considerably more detail. We can see clearly the geometrical relationships

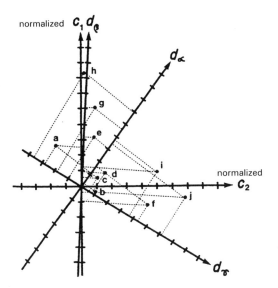

Figure 5.2 Geometrical relationship between the normalized abstract factor axes and the normalized data columns. The length of each axis is 1 unit.

between the data axes and the axes resulting from factor analysis. The data axes are not mutually perpendicular, nor do they coincide with the axes of the factor analysis. Any such correspondence would be a mathematical coincidence.

In Chapter 3 we learned that the correlation matrix is composed of dot products of the normalized data matrix. Each element of the correlation matrix represents the cosine of the angle between the respective data axes. Referring to (5.6), we conclude that the angle between $\mathbf{d}_\alpha$ and $\mathbf{d}_\beta$ should be $33°22'$, between $\mathbf{d}_\alpha$ and $\mathbf{d}_\gamma$ should be $85°16'$, and between $\mathbf{d}_\beta$ and $\mathbf{d}_\gamma$ should be $118°37'$. In Figure 5.2 we see that this is true.

5.1.8   Predictions Based on Typical Vectors

There is no need for the factor analyst to identify the true, basic factors if he or she is simply interested in predicting new data. For example. the analyst may be interested primarily in predicting data values for new row designees, $k, l, m, n, \ldots$, which were not included in the original data matrix. If d_β and d_γ for the new row designees were known, we can predict d_α values using a key set of typical vectors as described in Sections 2.7 and 3.4.5. Typical vectors are simply columns of the data matrix. Since column vectors $\mathbf{d}_\beta$ and $\mathbf{d}_\gamma$ lie in the factor space, as shown in Figure 5.2, they may be used to define the two-dimensional factor space instead of the principal eigenvectors $\mathbf{c}_1$ and $\mathbf{c}_2$. Hence $\mathbf{d}_\alpha$ can be expressed as a linear combination of $\mathbf{d}_\beta$ and $\mathbf{d}_\gamma$.

To achieve this mathematically, we employ (3.134) and (3.140), which, for two data columns, β and γ, leads to

$$\mathbf{T}_{\text{key}} = [\mathbf{t}_\beta \quad t_\gamma] = \mathbf{\Lambda}^{-1}\mathbf{R}'[\mathbf{d}_\beta \quad \mathbf{d}_\gamma] \tag{5.27}$$

Numerical values for the product $\mathbf{\Lambda}^{-1}\mathbf{R}'$ were evaluated as indicated in (5.16). Using these results and the two data columns given in (5.2), we find

$$\mathbf{T}_{\text{key}} = \begin{bmatrix} 0.0016200 & -0.0002302 & \cdots & 0.0004517 \\ -0.0027277 & 0.0015005 & \cdots & 0.0107767 \end{bmatrix} \begin{bmatrix} 40 & -20 \\ -5 & 7 \\ \vdots & \vdots \\ -5 & 43 \end{bmatrix}$$

$$= \begin{bmatrix} 0.9575 & -0.2254 \\ 0.1517 & 0.9248 \end{bmatrix} \tag{5.28}$$

According to (3.133), the transformed column-factor matrix is calculated as follows:

$$\hat{\mathbf{Y}} = \mathbf{T}_{\text{key}}^{-1}\bar{\mathbf{C}}$$

$$= \begin{bmatrix} 1.0056 & 0.2451 \\ -0.1649 & 1.0411 \end{bmatrix} \begin{bmatrix} 0.180200 & 0.9547463 & -0.225372 \\ 0.348777 & 0.151999 & 0.924790 \end{bmatrix}$$

$$= \begin{bmatrix} 0.2667 & 0.9973 & 0.0000 \\ 0.3334 & 0.0008 & 1.0000 \end{bmatrix} \tag{5.29}$$

From (3.139) and (3.132) we see that

$$\mathbf{D} = [\mathbf{d}_\alpha \quad \mathbf{d}_\beta \quad \mathbf{d}_\gamma] = \mathbf{D}_{key}\hat{\mathbf{Y}} = [\mathbf{d}_\beta \quad \mathbf{d}_\gamma]\hat{\mathbf{Y}} \tag{5.30}$$

Hence we find, using the first column of (5.29),

$$d_\alpha = 0.2667 d_\beta + 0.3334 d_\gamma \tag{5.31}$$

We can verify that this equation yields correct predictions for d_α by inserting values for d_β and d_γ given in (5.2) for any row designee. For example, for row designee e, (5.31) predicts that

$$d_\alpha = 0.2667(50) + 0.3334(-7) = 11.00 \tag{5.32}$$

which is in accord with the true value. Similarly, the d_α value for any new row designee can also be predicted from known values of d_β and d_γ, provided that the new role designee lies in the same factor space.

5.2   MODEL SET OF RAW DATA

The example studied in Section 5.1 is idealized because the data do not contain any error. Real data matrices inherently possess experimental uncertainties that tend to complicate the factor analysis. In this section we introduce artificial error into the data matrix and the target data. We then carry out the factor analysis and study some aspects of the theory of error described in Chapter 4.

5.2.1   Error in the Data Matrix

Our artificial error matrix, $\mathbf{E}$ is the following:

$$\mathbf{E} = \begin{bmatrix} -0.1 & 0.6 & 0 \\ 0.3 & 0.5 & 0.3 \\ -0.7 & 0.2 & -1.0 \\ 0.9 & 02 & -0.5 \\ 0 & -0.5 & 0.5 \\ 0.1 & -0.2 & 0.4 \\ -0.6 & 0.4 & -0.4 \\ 0.6 & -0.9 & 0.3 \\ -0.1 & -0.3 & 0.8 \\ 0 & 0.1 & 0 \end{bmatrix} \tag{5.33}$$

These errors were generated by randomly selecting numbers between -1.0 and $+1.0$, rounding the nearest tenth unit. The root mean square of these errors is

±0.477. When these errors are added, matrixwise, to the pure data matrix of (5.2), we obtain the raw data matrix:

$$
\mathbf{D}_{\text{raw}} = \begin{bmatrix}
3.9 & 40.6 & -20.0 \\
1.3 & -4.5 & 7.3 \\
3.3 & 10.2 & 3.0 \\
6.9 & 15.2 & 5.5 \\
11.0 & 49.5 & -6.5 \\
6.1 & -15.2 & 30.4 \\
16.4 & 80.4 & -13.4 \\
22.6 & 114.1 & -26.3 \\
13.9 & 19.7 & 26.8 \\
13.0 & -4.9 & 43.0
\end{bmatrix}
\tag{5.34}
$$

This matrix is a better representation of real chemical data because it contains some uncertainty, known to be approximately ±0.5.

From the raw data matrix, a somewhat different covariance (or correlation) matrix results. Carrying out the decomposition process as detailed in Section 5.1.2 leads to three eigenvectors instead of two as found for the pure data. The extra eigenvector is a secondary axis produced by the random error. The resulting eigenvalues, eigenvectors (grouped as the transposed, abstract column matrix), and the abstract row matrix are given in Table 5.2.

TABLE 5.2   Results of Factor Analyzing the Raw Data Matrix Using Covariance about the Origin

	Factor		
	1	2	3
Eigenvalue	26,760.421	4,090.5369	2.0717
Abstract column matrix (transposed)	0.180847	0.349120	0.919462
	0.956468	0.155291	−0.247089
	−0.229048	0.924121	−0.305838
Abstract row matrix	44.1189	−10.8161	−0.3292
	−5.7411	6.5011	0.0746
	9.6656	5.5084	−0.4036
	14.5264	9.8520	0.9064
	50.8233	5.5204	−0.1289
	−20.3982	27.8625	0.0670
	82.9352	5.8277	−0.6886
	119.2441	1.3044	0.6305
	15.2177	32.6784	−0.2836
	−12.1847	43.5148	0.0127
Real error (RE)	14.3049	0.4552	Undefined
Imbedded error (IE)	8.2589	0.3716	Undefined

From the eigenvalues we can calculate the real error (RE), assuming that only one or two eigenvalues belong to the primary set of eigenvectors. Referring to (4.44), we carry out the following computations:

1. Assuming that only the largest eigenvalue belongs to the primary set,

$$RE = \left(\frac{4{,}090.5369 + 2.0717}{10(3-1)}\right)^{1/2} = 14.3049 \qquad (5.35)$$

2. Assuming that the two largest eigenvalues belong to the primary set,

$$RE = \left(\frac{2.0717}{10(3-2)}\right)^{1/2} = 0.4552 \qquad (5.36)$$

The real error, 0.4552, calculated on the assumption that the two largest eigenvalues comprise the primary set, is in close agreement with the known root-mean-square error, 0.477. Hence the third factor belongs to the secondary set and contains nothing but error.

As discussed in Section 4.2, the removal of the secondary eigenvectors will lead to data improvement. Using (4.60), we can calculate the imbedded error (IE), the error that remains in the reproduced data matrix after deletion of the secondary eigenvectors. The results of such computations are shown in Table 5.2. By deleting the third eigenvector, we see that the error in the data is reduced from 0.455 to 0.372.

5.2.2 Error in the Target Data

In this section we study the effects of error on target transformation. First, we randomly select errors to be added to the two basic test vectors, x_{i1} and x_{i2}, utilized in Section 5.1.5. These errors, in a matrix form compatible with the data matrix, are

$$\begin{bmatrix} 0.1 & -0.1 \\ -0.2 & 0 \\ 0 & 0.2 \\ -0.1 & -0.2 \\ 0.2 & 0.1 \\ 0.1 & -0.1 \\ -0.1 & 0 \\ 0 & 0.1 \\ -0.2 & 0.1 \\ 0.1 & 0.1 \end{bmatrix}$$

The root mean squares (RMSs) of these two columns of errors are 0.130 and 0.118, respectively. When these two error vectors are added matrixwise to the row matrix of (5.2), we obtain the two impure test vectors shown in Table 5.3, labeled "Impure Factor 1" and "Impure Factor 2." These two test vectors simulate real chemical data that contain experimental error.

When we subject these two impure targets to individual least-squares target transformations using the two principal factors of Table 5.2, we obtain the predicted vectors listed in Table 5.3. The apparent error in the target (AET) for each of these tests is obtained by calculating the root mean square of the differences between the test and predicted points, in accord with (4.120). The real error in the predicted vector (REP) is calculated by multiplying the RE in Table 5.2 by the lengths of the respective transformation vectors listed in Table 5.3. For example, according to (4.123), for impure factor 1, we find that

$$REP = RE\|t\| = 0.4552[(0.05165)^2 + (0.22085)^2]^{1/2} = 0.1037 \qquad (5.37)$$

Using the values obtained for AET and REP, we then calculate the real error in the target (RET) by means of (4.119). For both impure test vectors, the RET values,

TABLE 5.3   Results of Testing Impure Targets on the Raw Data Matrix

Row Designee	Impure Factor 1		Impure Factor 2		Unity Test	
	Test	Predicted	Test	Predicted	Test	Predicted
a	0.1	0.088	3.9	4.027	1.0	0.154
b	0.8	1.113	−1.0	−1.000	1.0	0.139
c	2.0	1.759	0.2	0.146	1.0	0.280
d	2.9	2.991	−0.2	0.072	1.0	0.470
e	4.2	4.073	3.1	2.958	1.0	0.739
f	5.1	5.008	−4.1	−3.998	1.0	0.643
g	5.9	5.944	5.0	5.125	1.0	1.106
h	7.0	6.983	8.1	8.030	1.0	1.370
i	7.8	8.071	−1.9	−2.014	1.0	1.190
j	9.1	8.926	−4.9	4.900	1.0	1.223
Least squares	0.05615		0.06836		0.01114	
Transformation vector	0.22085		−0.09346		0.03123	
AET	0.192		0.140		0.580	
REP	0.104		0.053		0.015	
RET	0.161		0.130		0.580	
Known RMS error in the target	0.130		0.118		(Unknown)	
SPOIL	1.56		2.46		38.4	
$F(8, 1)$	1.51		3.10		664	
%SL	55.9		41.2		3.0	

0.161 and 0.130, are in excellent agreement with the known RMS errors in each vector, 0.130 and 0.118, respectively.

The SPOIL values, calculated by taking the ratio of the RET and REP values as defined by (4.125), are 1.56 and 2.46, respectively. Both of these values fall in the range of acceptability.

To apply the F statistic, we use (4.131), recognizing that $a = r = 10$, $s = c = 3$, $n = 2$, and $b = 0$ (no blanks), and that EDM = REP, where EDM is the error contributed by the data matrix. Applying (4.131) to impure factor 1, we find

$$F(8, 1) = \left(\frac{0.192}{0.104}\right)^2 \frac{(10 - 3 + 1)(3 - 3 + 1)}{(10 - 2 + 1)(3 - 2 + 1)} = 1.51 \qquad (5.38)$$

Similarly, for impure factor 2 we find $F(8, 1) = 3.10$. These two tests are highly significant because the significance levels associated with these two F values, 55.9 and 41.2%, respectively, are far in excess of the 5 or 10% level. According to the SPOIL and the F test, both of these vectors are recognized as true factors.

When the unity test vector is targeted, the results (given in Table 5.3) clearly show that this vector is not acceptable because the SPOIL, 38.4, is excessively large and the significance level, 3.0%, is less than 5%. These tests demonstrate how we can identify real factors and reject those that are false.

5.2.3   Error in the Factor Loadings

Using impure factors 1 and 2 in combination, we obtain, by the procedure described in Section 5.1.6, the following transformed column matrix:

$$\hat{\mathbf{Y}} = \begin{bmatrix} 2.0038 & 4.9155 & 2.0530 \\ 0.9996 & 9.9541 & -5.0368 \end{bmatrix} \qquad (5.39)$$

These loadings contain some error as a consequence of the errors in the data matrix. A comparison of the foregoing values with the pure values given in (5.2) leads to the following error matrix:

$$\bar{\mathbf{E}}_y = \hat{\mathbf{Y}} - \mathbf{Y}^* = \begin{bmatrix} 0.0038 & -0.0845 & 0.0530 \\ -0.0004 & -0.0459 & 0.0368 \end{bmatrix} \qquad (5.40)$$

Hence the RMS errors associated with the first and second factor loadings (EFL) are, respectively,

$$(EFL)_1 = \{[0.0038)^2 + (-0.0845)^2 + (0.0530)^2] \div 3\}^{1/2}$$
$$= \pm 0.058 \qquad (5.41a)$$
$$(EFL)_2 = \{[(-0.0004)^2 + (-0.0459)^2 + (0.0368)^2] \div 3\}^{1/2}$$
$$= \pm 0.034$$

$$(5.41b)$$

With real chemical data we cannot perform such RMS calculations because we never know the values of the pure loadings, $\hat{\mathbf{Y}}^*$. In order to estimate the RMS of the loading errors for each factor, we make use of (4.140):

$$(\text{EFL})_j = \text{RE}\|\hat{\mathbf{t}}_j\| \qquad \text{(4.140), (5.42)}$$

where $\hat{\mathbf{t}}_j$ is a row of $\hat{\mathbf{T}}$, is defined by (4.139):

$$\hat{\mathbf{T}} = \mathbf{T}^{-1}\bar{\mathbf{\Lambda}}^{1/2} \qquad \text{(4.139), (5.43)}$$

For the example problem,

$$
\hat{\mathbf{T}} = \begin{bmatrix} 0.05615 & 0.06836 \\ 0.22085 & -0.09346 \end{bmatrix}^{-1} \begin{bmatrix} (26{,}760.4)^{-1/2} & 0 \\ 0 & (4{,}090.5)^{-1/2} \end{bmatrix}
$$
$$
= \begin{bmatrix} 0.02808 & 0.05254 \\ 0.06636 & -0.04315 \end{bmatrix} \qquad \text{(5.44)}
$$

Hence

$$(\text{EFL})_1 = (0.4552)\{(0.02808)^2 + (0.05254)^2\}^{1/2} = \pm 0.027 \qquad \text{(5.45a)}$$

$$(\text{EFL})_2 = (0.4552)\{(0.06636)^2 + (-0.04315)^2\}^{1/2} = \pm 0.036 \qquad \text{(5.45b)}$$

Considering that these calculations are based on a statistically small number of data points, the results compare very favourably with those based on a knowledge of $\hat{\mathbf{Y}}^*$.

In the real world, problems are seldom as simple as this. In chemistry, data matrices usually require more than two dimensions. In spite of the complexity of the factor space, target factor analysis provides an efficient and realistic attack on such problems.

6

The simplest solution is the hardest to find.

EVOLUTIONARY METHODS

This chapter is an exposition of several important chemometric methods that are based on evolutionary principles of factor analysis. *Evolutionary factor analysis* is a general classification encompassing factor analytical techniques that take advantage of evolutionary processes that occur in chemistry. These methods make use of the fact that each chemical species has a single unique maximum in its evolutionary profile. Such processes can be affected by chromatography, reaction kinetics, titration, variations in pH, and so on. Evolutionary methods can be separated into several subdivisions: (1) modeling methods, (2) self-modeling methods, (3) rank annihilation, and (4) transmutation.

Modeling methods are those methods that require specification of the complete functional form of the evolutionary profile, whereas self-modeling methods are those methods that allow factor analysis to unravel the profiles of the individual components, subject to certain constraints or principles. Modeling methods tend to be more force fitting and are called "hard" models, whereas self-modeling methods tend to be more revealing but more ambiguous and are called "soft" models.

Rank annihilation and transmutation methods require a second set of data. The second set is the calibration set, containing the component(s) of interest measured under the same experimental conditions employed during the mixture measurements. The transmutation technique differs from all the other techniques because it does not require a two-way data matrix. Instead it operates on one-way data, typical of a single detector system.

Evolutionary processes have also been factor analyzed using three-way blocks of data. These techniques are described in Chapter 7.

6.1   MODELING METHODS

Kankare[1] was the first investigator to use models with adjustable parameters in factor analytical studies. Shrager[2−4] and Harris and co-workers[5−9] are the propounders of the modeling methodology. Models, based on well-established scientific knowledge and theory, are formulated to express the evolutionary profile, c_j, of component j as a function of instrumental and phenomenological parameters, g_{ijk} and p_{ijk}, respectively:

$$\mathbf{c}_j = f(g_{ijk}, p_{ijk}) \tag{6.1}$$

A unique function is required for every process. The beauty of the method is that values for all the parameters, as required for regression analysis, need not be known nor specified. Instead, the parameters are treated as variables in n space determined by abstract factor analysis (AFA).

As a starting point, values are arbitrarily assigned to each of these parameters and are assembled into a profile matrix, $\mathbf{C}_{\text{model}}$. The loading matrix, $\mathbf{L}$, is then computed by the pseudoinverse:

$$\mathbf{L} = \mathbf{D}\mathbf{C}_{\text{model}}^{+} \tag{6.2}$$

These matrices are multiplied to regenerate the data,

$$\hat{\mathbf{D}} = \mathbf{L}\mathbf{C}_{\text{model}} \tag{6.3}$$

The fit is evaluated by examining χ^2, the sum of squares of the differences between the raw data and the predicted data.

$$\chi^2 = \sum \sum (d_{ik} - \hat{d}_{ik})^2 \tag{6.4}$$

The parameters are varied by SIMPLEX optimization until χ^2 reaches a minimum.[5]

The fit can be improved by imposing a penalty if the loading, l_{ij}, has no physical meaning. For example, if, theoretically, the loadings cannot be negative as in the case of absorption spectroscopy, then the SIMPLEX optimization should be based on an error function such as

$$ERR = \chi^2 \left[1 + \frac{P_{neg} \sum\sum |l_{ij} < 0|}{\sum\sum l_{ij}} \right]$$ (6.5)

where P_{neg} is a constant between 1 and 100, specifying the penalty for negative loading.

Fluorescence Spectra. Knorr and Harris[5] analyzed the fluorescence emission spectra for a single-sample mixture by this technique. Time-resolved fluorescence emission spectra were recorded at nanosecond intervals after pulsed excitation of the sample. If the excited states of each component decay independently by first-order kinetics, then the concentration of excited states of component j at time $k\,\Delta t$ can be expressed as

$$c_{jk} = I_k \exp(k\,\Delta t/\tau_j)$$ (6.6)

where I_k is a convolution integral incorporating all contributions to temporal dispersion such as finite pulse and detector speed, Δt is the time interval, and τ_j is the lifetime of j. The method, of course, assumes that the response of the detector is linear.

Only one parameter, the fluorescence lifetime, for each component needs to be varied to achieve optimization. In comparison to other techniques this represents a tremendous reduction in effort and time to resolve complex mixtures.

Gas Chromatography/Mass Spectra (GC/MS). Modeling methods[6] have been applied to GC/MS of mixtures that exhibit overlapping chromatographs as well as spectra. In this case the rows of $\mathbf{C}_{model}$ represent the chromatographic profiles of the components. Theoretically, these profiles are expected to have a Gaussian distribution convoluted with a single-sided exponential to allow for tailing due to column dead space. The characteristic tailing decay time, τ, and the number of theoretical plates, N, can be determined from measurements with pure samples. Hence, the entire test matrix depends on one parameter, the retention time, t_{Rj}, for each component. Accordingly,

$$c_{jk} = g_{jk} \exp(k\,\Delta t/\tau)$$ (6.7)

where Δt is the time interval between spectral scans and

$$g_{jk} = \sigma_j^{-1}(2\pi)^{-1/2} \exp[(k\Delta t - t_{Rj})^2/2\sigma_j^2]$$ (6.8)

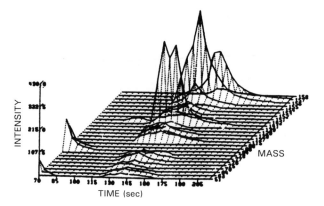

Figure 6.1 GC/MS data matrix of a ternary mixture, showing solvent peak on the left. [Reprinted with permission from F. J. Knorr, H. R. Thorsheim, and J. M. Harris, *Anal. Chem.*, **53**, 321 (1981). Copyright 1981 © American Chemical Society.]

where

$$\sigma^2 = t_{Rj}/N^{1/2} \tag{6.9}$$

Data and results obtained for a ternary mixture of 2-ethylnaphthalene, 1,3-dimethyl-naphthalene, and 2,6-dimethylquinoline are illustrated in Figures 6.1, 6.2, and 6.3.

Liquid Chromatography with Diode Array Ultraviolet (UV) Detection. The model approach can resolve overlapping liquid chromatographic peaks detected with a multiwavelength, photodiode-array, ultraviolet spectrometer without requiring any a

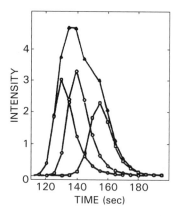

Figure 6.2 Chromatograms of the three components numerically isolated from the data illustrated in Figure 6.1. Upper curve (solid triangles) is the total ion current. [Reprinted with permission from F. J. Knorr, H. R. Thorsheim, and J. M. Harris, *Anal. Chem.*, **53**, 821 (1981). Copyright 1981 © American Chemical Society.]

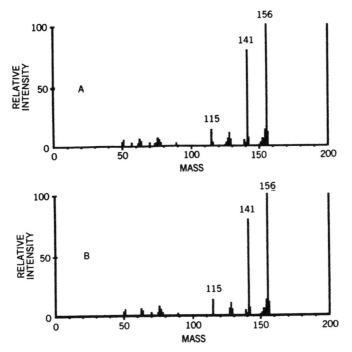

Figure 6.3 Mass spectrum (B) of one component, numerically isolated from the data in Figure 6.1, compared to the mass spectrum (A) of 1,3-dimethylnaphthalene. [Reprinted with permission from F. J. Knorr, H. R. Thorsheim, and J. M. Harris, *Anal. Chem.*, **53**, 321 (1981). Copyright 1981 © American Chemical Society.]

priori knowledge of the individual component spectra.[7] The elution profiles of the components are modeled with an exponential Gaussian function that allows for independent variation in retention time and band broadening.

$$c_{jk} = N_j \int_0^\infty \exp\left[-j\,\Delta t - t_{Rj} - t)^2/2\sigma_j^2\right] \exp[-j\,\Delta t/\tau]\,dt \qquad (6.10)$$

N_j is a normalization constant and the remaining terms are equivalent to those described for the GC/MS model. In this model, however, σ_j is treated as an independent parameter and is not modeled by (6.9). Thus for liquid chromatography/ultraviolet (LC/UV) there are $2n$ independent parameters, t_{Rj} and σ_j, for each of the n components.

The time constant, τ_j, accounts for tailing due to dead volume and is not expected to be strongly compound dependent. This variable can be replaced by a single global parameter, τ, characteristic of the instrument configuration and the flow rate, which are held fixed during operation. This global parameter can be ascertained by observations on chemically pure chromatographic peaks and need not be varied during SIMPLEX optimization.

Chromatograms composed of as many as eight overlapping components were successfully resolved by this method without recourse to the spectra of the pure components. In fact, the method predicts the spectra of the individual components.

Acid–Base Equilibria. Dissociation constants and ultraviolet/visible (UV/VIS) absorption spectra of monoprotic organic acid–base pairs have been determined by modeling factor analysis.[8] Multiwavelength pH–titration curves provide an appropriate data matrix. Modeling can be applied because the distribution curve of an acid and its conjugate base depends on pH and pK_a only. This becomes evident by examining theoretical expressions for the fraction of acid, α_{jk}, and fraction of base, β_{jk}, at a given pH:

$$\alpha_{jk} = [HA]/([HA] + [A^-]) = [1 + 10^{(pH-pK_a)}]^{-1} \tag{6.11}$$

$$\beta_{jk} = [A^-]/([HA] + [A^-]) = [1 + 10^{(pK_a-pH)}]^{-1} \tag{6.12}$$

To generate linearly independent vectors the evolutionary profile is expressed as the difference between these fractions:

$$c_{jk} = \beta_{jk} - \alpha_{jk} = \frac{10^{(pH-pk_a)} - 1}{1 - 10^{(pH-pK_a)}} \tag{6.13}$$

The redundant vectors arising from the sum of the fractions, $\alpha_{jk} + \beta_{jk} = 1$, are combined into a single constant row of $\mathbf{C}_{model}$, with the highest row number.

As a consequence of this procedure, the factors that constitute $\mathbf{L}$, the loading matrix, take the following form, in accord with Beer's law:

$$l_{ij} = (\epsilon_{A^-} - \epsilon_{HA})_{ij} b([HA] + [A^-])_j \tag{6.14}$$

Here ϵ is the molar absorptivity, b is the path length, and $([HA] + [A^-])_j$ is the total concentration of the acid–base pairs of species j, which is held invariant throughout the experiment. The first m columns of $\mathbf{L}$ concern the individual acid–base pairs. The last, $m + 1$, column contains the composite sum of all the absorbing species in the solution including contributions from components not dependent on pH. In this experiment, the size of the factor space is one unit larger than the number of acid–base pairs.

As in the previous studies, only one parameter per acid–base species is required for SIMPLEX optimization. The data matrix of a four-component mixture of the indicator dyes is illustrated in Figure 6.4. Analysis of these data gave the results shown in Table 6.1.

In a follow-up study Frans and Harris[9] recognized that the computation time could be greatly reduced by replacing (6.4) by an error fitting function that focuses on reproducing $\bar{\mathbf{C}}$, the abstract concentration matrix, rather than on $\mathbf{D}$; namely,

$$\chi^2 = \|\bar{\mathbf{C}}_{repro} - \bar{\mathbf{C}}\|^2 \tag{6.15}$$

TABLE 6.1 pK_a and UV/VIS Spectra from Model EFA of a Mixture of Dyes[a]

	pK_a Isolated	pK_a Mixture	Spectral Error Standard Deviation (AU)[b]	Relative Spectral Error (%)
Methyl orange	3.27	3.39	5.5×10^{-3}	2.5
Bromocresol green	4.77	4.80	3.8×10^{-3}	1.0
Chlorophenol red	6.07	6.05	3.4×10^{-3}	1.0
Phenol red	7.71	7.71	4.4×10^{-3}	0.9

[a] Reprinted with permission from S. D. Frans and J. M. Harris, *Anal. Chem.*, **56**, 466 (1984). Copyright 1984 American Chemical Society.
[b] Absorbance unit.

Because the elements of the abstract matrix are linear combinations of the true concentrations, the model concentration matrix may be transformed into the abstract factor space by premultiplication by a least-squares transformation matrix. This leads to

$$\bar{\mathbf{C}}_{repro} = \bar{\mathbf{C}}\mathbf{C}'_{model}(\mathbf{C}_{model}\mathbf{C}'_{model})^{-1}\mathbf{C}_{model} \qquad (6.16)$$

Shrager[4] proved that the differences that occur as a result of (6.4) and (6.15) are due to differences in weightings. The two methods become equivalent if (6.15) is weighted by the eigenvalues as shown in the following equation:

$$\chi^2 = \|\bar{\mathbf{D}} - \mathbf{D}\|^2 = \|\Lambda^{1/2}(\bar{\mathbf{C}}_{repro} - \bar{\mathbf{C}})\|^2 \qquad (6.17)$$

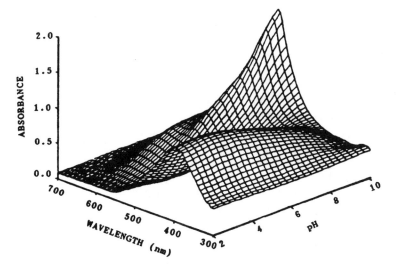

Figure 6.4 Absorbance matrix of a mixture of four dyes (see Table 6.1) generated by varying the pH from 2 to 10. [Reprinted with permission from S. D. Frans and J. M. Harris, *Anal. Chem*, **56**, 466 (1984). Copyright 1984 © American Chemical Society.]

6.2 SELF-MODELING METHODS

Self-modeling methods attempt to resolve mixed evolutionary profiles without presupposing a well-defined shape function. Instead, they impose known constraints concerning the general shape of the profile and spectra, such as (1) each component profile has only one maximum, (2) all points in the concentration profile must be greater than or equal to zero, within experimental uncertainty, and (3) all points in the component spectra must be greater than or equal to zero, within experimental uncertainty. Rule 3, however, is not valid for certain types of measurements such as circular dichroism and optical rotation. Other empirical rules, such as minimum overlap between spectra or between profiles, are usually imposed and the results are qualified by statements concerning upper or lower boundary estimations. These methods differ in their initial estimation of the concentration profile matrix and in their strategy to impose the constraints.

Iterative Target Transformation Factor Analysis (ITTFA). Gemperline[10] developed a self-modeling method for resolving overlapped peaks in high-performance liquid chromatography/UV (HPLC/UV). After generating an appropriate data matrix and subjecting the matrix to AFA to determine the number of spectroscopically visible components, Gemperline individually target tested the complete series of uniqueness vectors. These vectors contain unity for each digitized elution time, t_j; namely,

$$\mathbf{c}_{t_1,\text{test}} = (1, 0, 0, \ldots, 0, 0, 0)'$$
$$\mathbf{c}_{t_2,\text{test}} = (0, 1, 0, \ldots, 0, 0, 0)'$$
$$\vdots \qquad \qquad \vdots$$
$$\mathbf{c}_{t_c,\text{test}} = (0, 0, 0, \ldots, 0, 0, 1)'$$

(6.18)

The power of target factor analysis (TFA) lies in its ability to test a single target vector without requiring any information concerning the other components. Uniqueness test vectors represent idealistic elution profiles, affording an excellent starting point. When the retention time represented by $\mathbf{c}_{t_j,\text{test}}$ corresponds to the retention of a real component, the apparent error in the test (AET) (see Section 4.6.1) will reach a local minimum because the predicted profile is a better representation than the idealistic "uniqueness" profile. If more than n local minimum are found, then only the n vectors with the smallest minimum are selected.

Because negative regions are physically meaningless, these profiles are refined by setting to zero all points beyond the boundaries marked by the first negative regions encountered on the left and on the right of the peak maximum. New profiles are produced by target testing the refined profiles. The refinement is repeated until (1) the real error in the target (RET) is less than the real error in the predicted vector (REP) (see Section 4.6.1) or (2) the decrease in error of the truncated points, between successive iterations, becomes less than the predicted error (REP).

Gemperline[10] tested this technique with simulated mixtures of adenylic, cytidylic, guanylic, and uridylic acids. Random numbers were added to simulate experimental uncertainty. The results of the method were found to be in good agreement with the original profiles (see Figure 6.5).

A similar method was devised by Vandeginste et al.[11] during their investigation of HPLC/UV of mixtures of polynuclear aromatic hydrocarbons and mixtures of proteins. As an initial starting point these investigators subjected the abstract concentration matrix to varimax rotation [see (3.104) in Section 3.4.2], an orthogonal rotation that maximizes the total variance of the squared loadings. This aligns the abstract factors, as close as possible, along the unknown concentration profiles,

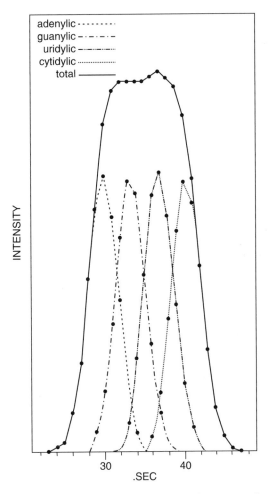

Figure 6.5 Elution profiles of components in acid mixtures predicted by Gemperline [Reprinted with permission from P. J. Gemperline *J. Chem. Inf. Comput. Sci.*, **24**, 206 (1984). Copyright 1984 © American Chemical Society.]

revealing the retention times (peak maxima) of the components. Uniqueness vectors, corresponding to these retention times, are then created and subjected to testing and refinement by iteration.

Each individual profile is refined by setting to zero any emerging point smaller than a given threshold and the minor peak of any doublet separated by one or more zeros. The iteration is stopped when any one of the following occurs: (1) refinement makes no further change in the target, (2) the correlation coefficient between the test and predicted vectors exceeds a prescribed value, or (3) the number of iterations reaches a prescribed maximum.

Evolving Factor Analysis (EFA). The model-free method of Gampp et al.[12] is based on repetitive eigenvalue analysis of a set of data matrices garnered from an evolutionary process. Eigenanalyses are performed on a series of matrices constructed by successively adding spectra to the previous matrix during the evolutionary process. When a new absorbing species begins to appear, an eigenvalue evolves from the pool of error eigenvalues, increasing in value in relation to its contribution to the enlarged data set. This procedure is called forward evolving factor analysis.

Backward evolving factor analysis is initiated by starting the eigenanalysis with the last two spectra and systematically adding spectra in the reverse order of collection. The resulting eigenvalues map out the disappearance of the components.

Eigenvalues from the forward and backward analyses are plotted on the same semilog graph as a function of the evolutionary variable. The region under the ith forward eigenvalue curve and the $(n + 1 - i)$th backward eigenvalue curve, common to both curves, maps out the concentration profile of the ith species. Typical results for the pH titration of Cu^{2+} by 3,7-diazanonane diamine using visible spectroscopic measurements are shown in the lower section of Figure 6.6. Because there are four components, the eigenvalue curves (forward–backward$'$) are connected as follows: 1–4$'$, 2–3$'$, 3–2$'$, and 4–1$'$. The curves connected accordingly compare very favorably with the results obtained directly from electron spin resonance shown in the upper section of the figure.

Gampp and co-workers[12] refined these profiles by the following iteration scheme. First, the concentration profiles are normalized to unit sum. Then the component spectra are calculated from the data matrix and the profile matrix by means of the pseudoinverse. All negative absorbances are set to zero. By applying the pseudo-inverse, new concentration profiles are obtained from the data matrix and the refined spectra. All negative concentrations are set to zero. Again the profiles are normalized to unit sum and the spectra are regenerated. This process of refining both the spectra and the concentrations is continued until convergence is achieved.

Gemperline and Hamilton[13] applied EFA to resolve seven ionic species of bismuth ion from ion chromatography. The data matrix portrayed in Figure 6.7 was generated by injecting a sample of bismuth perchlorate into a flowing stream of chloride ion in an ion chromatograph equipped with a UV/VIS diode array detector. After the midpoint of the sample plug (175 s) the components begin to elute in

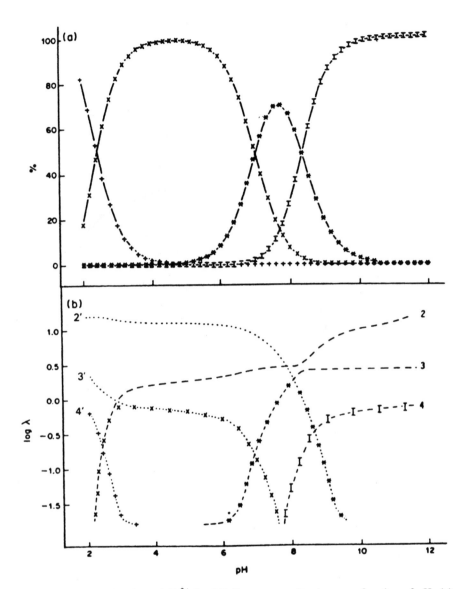

Figure 6.6 Complexation of Cu^{2+} by 3,7-diazanonane diamine as a function of pH. (*a*) Species distribution curve obtained from electron spin resonance data. (*b*) Logarithmic plot of the eigenvalues resulting from forward (unprimed) and backward (primed) EFA of the visible spectra. [Reprinted with permission from H. Gampp, M. Maeder, C. J. Meyer, and A. D. Zuberbuhler, *Talanta*, **32**, 1133 (1985). Copyright 1985 © Pergammon Press plc.]

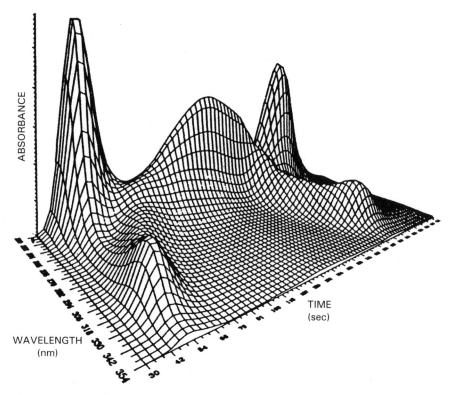

ABSORBANCE

WAVELENGTH
(nm)

TIME
(sec)

Figure 6.7 Absorption spectra of bismuth perchlorate in a flowing stream of hydrochloric acid. [Reprinted with permission from P. J. Gemperline and J. C. Hamilton, *J. Chemometrics*, **3**, 455 (1989). Copyright 1989 © John Wiley & Sons, Ltd.]

reverse order, producing a skewed, symmetric chromatographic profile. EFA was applied separately to the first half and to the second half. The resolved chromatograms are illustrated in Figure 6.8, and the UV spectra of the seven ions, obtained from the pseudoinverse computations, are illustrated in Figure 6.9.

Meader[14] has shown how EFA can be used to completely deconvolve an unresolved chromatogram. The method is similar to rank annihilation evolving factor analysis described in Section 6.3.2. The appearance and disappearance of the components are determined by forward and backward EFA of the chromatogram. The "windows" (i.e., the region of existence along the evolutionary axis) for each of the components are mapped out as accurately as possible, as shown in Figure 6.10.

These windows correspond to sections of the true concentration matrix, **C**. In fact, each column of the transposed matrix, shown in Figure 6.9, corresponds to a specific component. The unshaded portions correspond to the windows where the respective components exist. The shaded portions correspond to the regions where

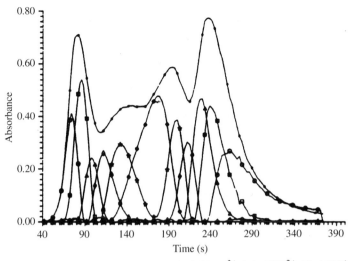

Figure 6.8 Chromatographic profiles of total Bi (*), Bi^{3+} (+), $BiCl^{2+}$ (*), $BiCl_2^+$ ($\diamond$), $BiCl_3$, ($\triangle$), $BiCl_4^-$ ($\square$), and $BiCl_5^{2-}$ plus $BiCl_6^{3-}$ (O) obtained by EFA of the data in Figure 6.7. [Reprinted with permission from P. J. Gemperline and J. C. Hamilton, *J. Chemometrics*, **3**, 455 (1989). Copyright 1989 John Wiley & Sons, Ltd.]

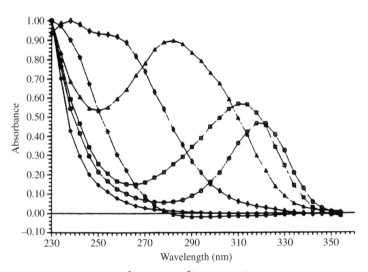

Figure 6.9 UV spectra of Bi^{3+} (+), $BiCl^{2+}$ (*), $BiCl_2^+$ ($\diamond$), $BiCl_3$ ($\triangle$), $BiCl_4^-$ ($\square$), and $BiCl_5^{2-}$ plus $BiCl_6^{3-}$ (O) obtained by EFA of the first half of the data in Figure 6.7. [Reprinted with permission from P. J. Gemperline and J. C. Hamilton, *J. Chemometrics*, **3**, 455 (1989). Copyright 1989 John Wiley & Sons, Ltd.]

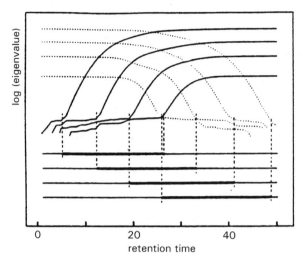

retention time

Figure 6.10 Concentration window for the ith component is defined by the rise of the ith eigenvalue in the forward EFA plot (—) and the $(n + 1 - i)$th eigenvalue in the backward EFA plot (...). [Reprinted with permission from M. Maeder, *Anal. Chem.*, **59**, 527 (1987). Copyright 1987 American Chemical Society.]

the component concentration is zero. This matrix is related to the abstract column-factor matrix by a simple transformation, which can be expressed as

$$\mathbf{C}' = \bar{\mathbf{C}}'_{abst}\mathbf{T} \tag{6.19}$$

where $\mathbf{T}$ is the appropriate transformation matrix. For a specific component (6.19) becomes

$$\mathbf{c}_j = \bar{\mathbf{C}}'_{abst}\mathbf{t}_j \tag{6.20}$$

where $\mathbf{t}_j$ is the appropriate transformation vector. These two equations are symbolically represented in Figure 6.11. When the component regions are deleted, the following equation, shown on the right in Figure 6.11, results:

$$\mathbf{c}_j^0 = \mathbf{C}_{abst}^0\mathbf{t}_j \tag{6.21}$$

where $\mathbf{c}_j^0$ is a zero vector. Equation (6.21) represents a homogeneous system of equations with the trivial solution $\mathbf{t}_j = 0$. The rank of $\mathbf{C}^0$ is $n - 1$ because component j has been deleted. For the nontrivial solution, one element of $\mathbf{t}_j$, can be chosen freely, then the rest of $\mathbf{t}_j$ can be calculated. Finally, the chromatographic profile, $\mathbf{c}_j$, of component j is determined by application of (6.20).

This procedure is repeated for all components, yielding the complete concentration matrix. A pseudoinverse computation produces the matrix of component

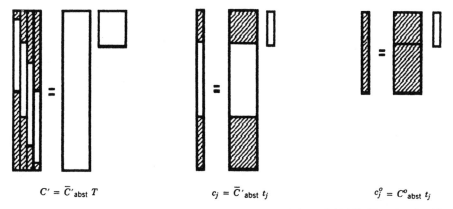

$$C' = \bar{C}'_{abst}\, T \qquad\qquad c_j = \bar{C}'_{abst}\, t_j \qquad\qquad c_j^o = C^o_{abst}\, t_j$$

Figure 6.11 Symbolic representation of Eqs. (6.19), (6.20), and (6.21). [Reprinted with permission from Maeder, *Anal. Chem.*, **59**, 527 (1987). Copyright 1987 American Chemical Society.]

spectra, which is used for quantification on a relative basis. Figure 6.12 shows the typical results. In spite of the fact that the central contour plot exhibits no signs of resolution, the profiles and spectra of the components are completely resolved. Obviously, determination of absolute concentrations requires calibration from external experiments.

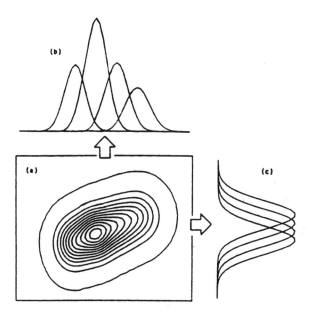

Figure 6.12 Chromatographic data represented as (*a*) a contour plot, (*b*) deconvolved into concentration profiles, and (*c*) absorption spectra. [Reprinted with permission from M. Maeder, *Anal. Chem.*, **59**, 527 (1987). Copyright 1987 American Chemical Society.]

Fixed-Size Window Movement. Peak purity is particularly important in the pharmaceutical industry. Unfortunately, low concentrations of impurities cannot be detected by EFA of HPLC/UV data. The fixed-size window movement (FSWM) method of Keller and Massart[15] overcomes this deficiency. The method is based on the EFA principle, using a fixed-size window instead of a progressively increasing window. For example, the first seven spectra, spectra 1 to 7, are assembled into a matrix and subjected to eigenanalysis; then spectra 2 to 8 are assembled and eigenanalyzed, and so forth. As in EFA, the logarithms of the eigenvalues are plotted as a function of the centers of the time windows. In addition to increased sensitivity, FSWM plots are easier to interpret.

Figure 6.13*a* is a simulated chromatogram of a two-component mixture where one of the components is an impurity amounting to 0.2%. The signal of the impurity cannot be detected by the naked eye. Figure 6.13*b* is a log plot of the concentrations, magnifying the impurity. Efforts to detect the minor component by classical EFA were unsuccessful. The results of applying FSWM, based on a seven-spectra window, are displayed in Figure 6.14. The first component begins to elute at time 12. The second component begins to elute at time 33. The two components begin to disappear at times 34 and 43. The shoulder of the first eigenvalue indicates a transfer of importance from the main constituent to the minor constituent. This is in accord with the postulate that the first component to elute is the first to disappear, and the last component to elute is the last to disappear.

If the resolution between the components in insufficient, this may not be true and classical EFA will lead to an incorrect conclusion. Figure 6.15 is a modification of Figure 6.14 where the resolution (R_s) has been changed from 1.0 to 0.2 and the impurity increased to 0.5%. Notice that the elution assumption is not valid in this situation. The absence of a shoulder in Figure 6.15*a* shows that the main component completely encompasses the signal of the minor constituent. In spite of this condition the profile of the minor component is completely resolved by FSWM as shown in Figure 6.16.

The fact that the baselines in the FSWM plots are flat is another important advantage that makes interpretation much easier than classical EFA.

6.3 RANK ANNIHILATION FACTOR ANALYSIS

Quantification of a single component in a complex mixture without concern for the other components presents a challenging problem that can be solved by rank annihilation factor analysis (RAFA).[16] The technique involves the use of sensors with linear response with respect to concentration of the components, such that the response of the mixture, m_{ik}, has the form

$$m_{ik} = \sum_{j=1}^{n} x_{ij}\alpha_j y_{jk} \tag{6.22}$$

Here x_{ij} and y_{jk} are variables of the response of component j, and α_j is the amount of j present in the mixture; α_j is the quantity that we are interested in determining.

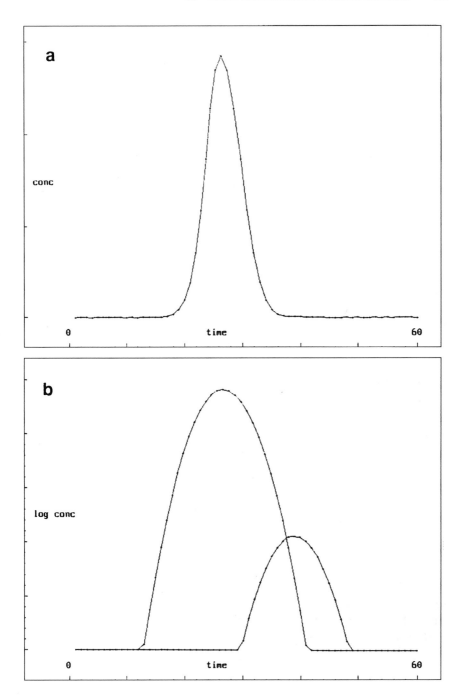

Figure 6.13   (*a*) Simulated chromatogram of a two-component mixture (0.2% impurity, R_s 1.0), and (*b*) log (concentrations) of the substances (0.2% impurity, R_s 1.0). [Reprinted with permission from H. R. Keller and D. L. Massart, *Anal. Chim. Acta*, **246**, 379 (1991). Copyright 1991 Elsevier Science.]

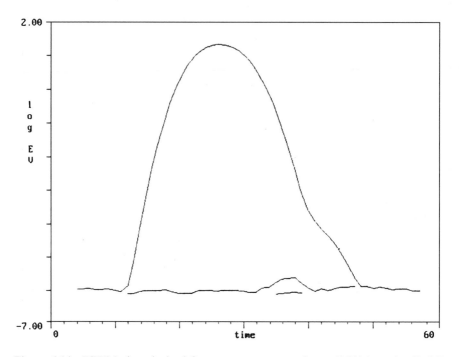

Figure 6.14   FSWM plot, obtained for a two-component mixture (0.2% impurity, R_s 1.0). [Reprinted with permission from H. R. Keller and D. L. Massart, *Anal. Chim. Acta*, **246**, 379 (1991). Copyright 1991 Elsevier Science.]

This expression applies to fluorescence spectra where m_{ik} is the fluorescence intensity, x_{ij} is related to the excitation wavelength, i, and y_{jk} is related to the emission wavelength, k. It also applies to chromatography with multiwavelength detection, where x_{ij} is related to the spectra of component j and y_{jk} is related to the chromatographic profile of the component. In matrix form (6.22) becomes

$$\mathbf{M} = \mathbf{X}\alpha\mathbf{Y}' \tag{6.23}$$

Consider a standard solution containing a known amount of the pure component of interest. If the standard solution is subjected to the same process as the mixture, rank annihilation will subtract the correct amount of a standard present in the mixture. More specifically, if $\mathbf{M}$ is the mixture matrix and $\mathbf{N}$ is the pure component matrix, then there exists a residual matrix, $\mathbf{L}$, with a rank one unit less than the rank of $\mathbf{M}$, defined by the following equation:

$$\mathbf{L} = \mathbf{M} - \tau\mathbf{N} \tag{6.24}$$

where

$$\tau = \alpha_j / \beta_j \tag{6.25}$$

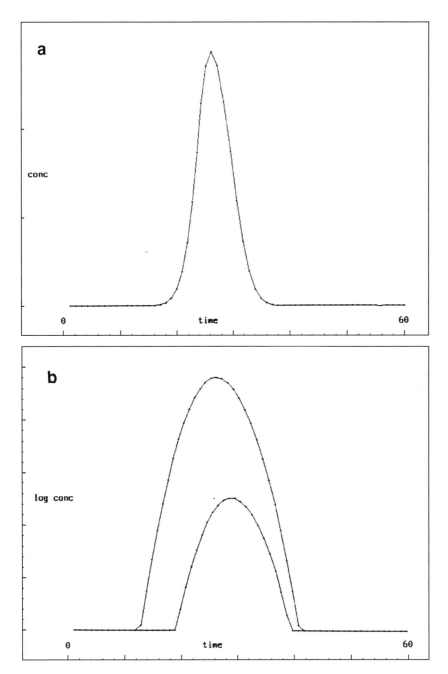

Figure 6.15 (*a*) Simulated chromatogram of a two-component mixture (0.5% impurity, R_s 0.5), and (*b*) (b) log (concentrations) of the substances (0.5% impurity, R_s 0.5). [Reprinted with permission from H. R. Keller and D. L. Massart, *Anal. Chim. Acta*, **246**, 379 (1991). Copyright 1991 Elsevier Science.]

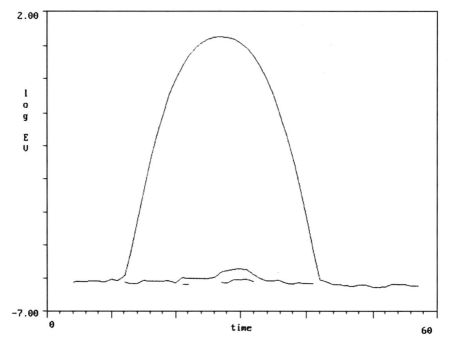

Figure 6.16 FSWM plot, obtained for a two-component mixture (0.5% impurity, R_s 0.5). [Reprinted with permission from H. R. Keller and D. L. Massart, *Anal. Chim. Acta*, **246**, 379 (1991). Copyright 1991 Elsevier Science.]

In (6.25) β_j is the concentration of j in the standard. It must be understood, of course, that **N** represents the responses of a single component, and that any solvent response has been appropriately corrected.

Ho et al.,[16] the originators of RAFA, determined the true value of τ by plotting the eigenvalues of the residual matrix as a function of τ. As shown in Figure 9.10 where $c_k/c_k^0 = \tau$, the nth eigenvalue reaches a minimum at the correct ratio. Unfortunately, this requires repeated eigenanalyses of the residual matrices for each increment of τ, a time-consuming process.

6.3.1 Direct Solution

Lorber[17] showed that τ can be calculated directly. The results of his derivation are summarized here. First, **N** is decomposed into an outer product of vectors, **x** and **y**:

$$\mathbf{N} = \mathbf{x}\mathbf{y}' \qquad (6.26)$$

This is possible if the pure component matrix is composed of only one factor, as can be expected for standards that are free from chemical anomalies, impurities, and background effects. Factor analysis of **N** should yield only one eigenvalue, one row factor, **x**, and one column factor, **y**.

Factor analysis of **M** yields the number of spectroscopically visible components, n, the abstract row-factor matrix $\bar{\mathbf{R}}$, and the abstract column-factor matrix $\bar{\mathbf{C}}$. The value of τ is simply the reciprocal of the inner product between vectors **a** and **b**,

$$\tau = 1/\mathbf{a}'\mathbf{b} \tag{6.27}$$

where

$$\mathbf{a} = [\bar{\mathbf{R}}\Lambda^{-1}]'\mathbf{x} \tag{6.28}$$

$$\mathbf{b} = \bar{\mathbf{C}}\mathbf{y} \tag{6.29}$$

Direct calculation of τ not only reduces the computational time enormously but also overcomes the difficulty in deciding which eigenvectors to scrutinize for a minimum in the eigenvalue-versus-τ plots. Another advantage of the direct calculation is the fact that the method is relatively insensitive to an overestimation of the factor space as shown in Table 6.2, concerning unresolved liquid chromatographic data.[18,19] Note that the calculated concentration ratio is fairly constant when the factor level is three or greater, in agreement with the fact that the mixture contains three components.

Malinowski[19] adapted the Clifford method[20] to estimate the standard deviation in τ, s_τ, by the expression

$$s_\tau = \sqrt{v_{ab}}\,\tau^2 \tag{6.30}$$

Here v_{ab} is the variation in the dot product of **a** and **b** and is calculated as follows:

$$v_{ab} = \ddot{\mathbf{a}}'\mathbf{V}\ddot{\mathbf{y}} \tag{6.31}$$

TABLE 6.2 Direct Solution RAFA of Unresolved LC Peaks of Ethylbenzene in a Single Mixture of Ethylbenzene, *o*-Xylene, and *p*-Xylene[a]

n	τ	s_τ
1	2.433	0.035
2	2.286	0.098
3	0.364	0.003
4	0.365	0.003
5	0.362	0.006
6	0.362	0.009

[a] Based on UV absorbances of LC fractions taken from the doctoral thesis of M. M. McCue (Stevens Institute of Technology, Hoboken, NJ, 1982). For a discussion of the experimental details see M. M. McCue and E. R. Malinowski, *J. Chromatogr. Sci.*, **21**, 229 (1983).

where $\ddot{\mathbf{a}}$ and $\ddot{\mathbf{y}}$ are vectors whose elements are the squares of the elements of $\mathbf{a}$ and $\mathbf{y}$, respectively, and $\mathbf{V}$ is the fractional variance–covariance matrix defined by Eq. (4.146). Typical results are shown in Table 6.2.[19] Note that the standard deviation in the concentration ratios tends to a minimum at $n = 3$. The highest degree of accuracy is attained when the correct number of factors are employed. Overspanning or underspanning the factor space tends to increase the uncertainty.

6.3.2   RAFA with Incomplete Information

The application of RAFA to situations where only one vector component, either $\mathbf{x}$ or $\mathbf{y}$, of the pure component matrix, $\mathbf{N}$, is known has been investigated.[21] For example, in chromatographic/spectral measurements the analyst may want to subtract the spectra, $\mathbf{s}$, of a component from the mixture matrix in order to remove its contributions. This requires a judicious choice of the unknown component chromatographic profile, $\mathbf{p}$, such that

$$\mathbf{L} = \mathbf{M} - \mathbf{s}\mathbf{p}' \tag{6.32}$$

where $\mathbf{L}$ has a rank that is one unit less than $\mathbf{M}$. Note that τ does not appear in (6.32) because this constant is contained within the profile vector.

A unique solution to this problem is not possible without the imposition of further constraints. One criterion for a solution is based on subtracting the maximum amount of component spectrum mathematically permissible. This can be achieved by satisfying the following equation, as closely as possible:

$$\mathbf{M} = \mathbf{s}\mathbf{p}' \tag{6.33}$$

The least-squares solution to (6.33) involves the pseudoinverse of $\mathbf{s}$:

$$\mathbf{p} = \mathbf{M}'\mathbf{s}(\mathbf{s}'\mathbf{s})^{-1} = \mathbf{M}'\mathbf{s}^{+} \tag{6.34}$$

The pseudoinverse estimate of $\mathbf{p}$ not only removes all the spectral character of the component but also removes those portions of the other components that are nonorthogonal to the component spectra. Thus the method overestimates the profile, yielding an upper bound estimate. The total amount of spectra removed depends on the spectral overlap between the other components and the component in question. If there were no overlap (6.34) would give the correct profile. Such situations are not likely to occur in practice.

Equation (6.34) could also lead to negative regions in the profile. One way to correct for such misbehavior is to invoke an error function that penalizes for negative terms. Simplex optimization can be used to adjust the elements of $\mathbf{p}$ to minimize the error function.

Another criterion is to require $\mathbf{p}$ to have a minimum overlap with the residual matrix $\mathbf{L}$, which contains the profiles of the remaining components. This will remove the least amount of component spectra from the mixture matrix, yielding a lower bound estimate of the profile. The following expression for the overlap, O, has been suggested:[21]

$$O = \sum_i \sum_k |l_{ik}|s_i p_k \Big/ \left(\sum_i \sum_k l_{ik}^2 \right) \sum_i \sum_k (s_i p_k)^2 \qquad (6.35)$$

where l_{ik}, s_i, and p_k are components of **L**, **s**, and **p**, respectively, the sums being taken over all rows and columns of the data matrix. Simplex optimization can be used to obtain the best estimate of the component profile that minimizes (6.35). Unfortunately, however, this requires a fairly good initial estimate of the profile, otherwise the simplex tends to find a local minimum rather than the global minimum.

Figure 6.17 illustrates the profiles obtained by the three criteria described above, for a three-component mixture subjected to high-performance thin-layer chromatography with a multichannel imaging spectrometer. The minimum overlap constraint was found to give the best results.

Rank Annihilation by Evolving Factor Analysis (RAEFA). A clever technique for quantifying an individual component in a mixture when only the spectrum of the component is available has been devised by Gampp and co-workers.[22] The method combines information gleaned from evolving factor analysis (see Section 6.2) with the general principle of rank annihilation.

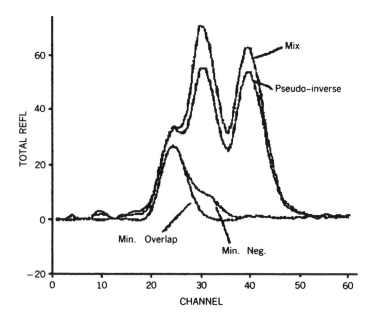

Figure 6.17 Reconstructed chromatogram by pseudoinverse, minimum negativity, and minimum overlap. [Reprinted with permission from D. H. Burns, J. B. Callis, and D. G. Christian, *Anal. Chem.*, **58**, 2805 (1986). Copyright 1986 American Chemical Society.]

Consider a matrix **D** generated by LC/UV measurements, in which the chromatograph is unresolved. Figure 6.18 represents a hypothetical chromatograph of four overlapping components, one of which we wish to quantify. Evolving factor analysis can be used to locate the "windows" of the chromatograph where the analyte is visible, although mixed with the other components. The AFA of the complete data matrix yields n visible components and the abstract matrices portrayed in Figure 6.19. These matrices can be divided into several regions, the analyte contributing only to the shaded portions. **D** can be separated into two matrices, $\mathbf{D}_k$, which concerns the shaded region, and $\mathbf{D}_u$, which concerns all other regions that exclude the analyte. RAEFA is based on the premise that all components except the analyte are represented in $\mathbf{D}_u$. If this is true, the rank of $\mathbf{D}_u$ is one unit less than the rank of **D**. The AFA of $\mathbf{D}_u$ will yield an abstract spectral matrix, $\mathbf{E}_u$, composed of linear combinations of the true spectra of all components except the analyte. A new spectral matrix **S** is formed by adding the known spectrum of the analyte, $\mathbf{A}_k$, to this matrix, increasing its rank by one unit. The following pseudoinverse calculation involving **S** and $\mathbf{D}_k$ produces a concentration matrix **C** that contains the chromatographic profile, $\mathbf{c}_k$, of the analyte in the last column:

$$\mathbf{C} = \mathbf{D}_k \mathbf{S}^+ \tag{6.36}$$

The need for methods such as RAEFA are required for studies involving chemical equilibria where it is impossible to generate a concentration profile of the species by independent experiments.

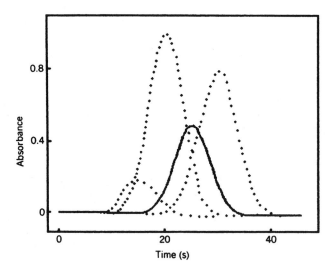

Figure 6.18 Quantification of a single species (—) from an unresolved four-component chromatogram using RAEFA. Profiles of the other species are shown as dotted (...) curves. [Reprinted with permission from H. Gampp, M. Maeder, C. J. Meyer, and A. D. Zuberbuhler, *Anal. Chim. Acta*, **193**, 287 (1987). Copyright 1987 Elsevier Science.]

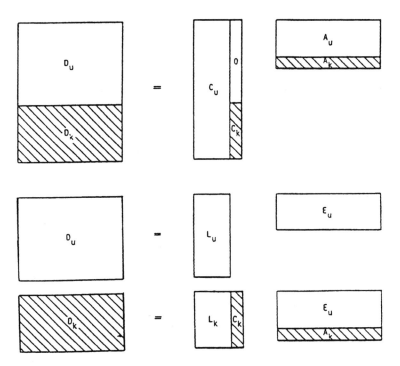

Figure 6.19 Symbolic diagram of the RAEFA methodology. [Reprinted with permission from H. Gampp, M. Maeder, C. J. Meyer, and A. D. Zuberbuhler, *Anal. Chim. Acta*, **193**, 287 (1987). Copyright 1987 Elsevier Science.]

Window Factor Analysis. Window factor analysis (WFA)[23,24] is mathematically equivalent to RAEFA but differs in theoretical development as well as the procedure used to determine the window of the component. Recall that the window is the region along the evolutionary axis that accounts for the total amount of the component of interest. All other components must have signals outside the window, although they may also have signals inside the window.

Consider **D** to be a matrix whose columns contain spectra recorded at different time intervals along the evolutionary axis. If there are n components, then

$$\mathbf{D} = \sum_{i=1}^{n} \mathbf{s}_i \mathbf{c}_i' \tag{6.37}$$

where $\mathbf{s}_i$ represents a vector containing the spectral absorptivities and $\mathbf{c}_i$ represents a vector containing the concentration profiles of component i. Submatrix $\mathbf{D}^0$ is created by deleting the window of component n; that is, by deleting all data columns containing signals of component n. If all components except n are contained in the submatrix, then abstract factor analysis of $\mathbf{D}^0$ will yield abstract matrices $\mathbf{S}^0$ and $\mathbf{C}^0$.

Matrix $\mathbf{S}^0$ contains $n - 1$ orthonormal, abstract spectral vectors $\mathbf{s}_j^0$. Matrix $\mathbf{C}^0$ contains $n - 1$ orthogonal, abstract concentration profiles $\mathbf{c}_j^0$ that are not normalized.

$$\mathbf{D}^0 = \sum_{j=1}^{n-1} \mathbf{s}_j^0 \mathbf{c}_j^0 = \mathbf{S}^0 \mathbf{C}^0 \tag{6.38}$$

The real spectral vectors, $\mathbf{s}_i$, of the $n - 1$ components are linear combinations of the abstract vectors because both sets of vectors occupy the common subspace, hence

$$\mathbf{s}_i = \sum_{j=1}^{n-1} \beta_{ij} \mathbf{s}_j^0 \tag{6.39}$$

where β_{ij} is a linear coefficient.

The real spectrum of component n lies inside the n-dimensional space, with projections inside the $(n - 1)$-dimensional hyperplane and perpendicular to the hyperplane, as expressed, respectively, by (6.40)

$$\mathbf{s}_n = \sum_{j=1}^{n-1} \beta_{nj} \mathbf{s}_j^0 + \beta_{nn} \mathbf{s}_n^0 \tag{6.40}$$

Inserting (6.39) and (6.40) into (6.37) and rearranging the result gives

$$\mathbf{D} = \sum_{j=1}^{n-1} \mathbf{s}_j^0 \left(\sum_{i=1}^{n} \beta_{ij} \mathbf{c}_i' \right) + \beta_{nn} \mathbf{s}_n^0 \mathbf{c}_n' \tag{6.41}$$

Multiplying both sides of (6.41) by $\mathbf{s}_j^{0\prime}$ and recalling that the abstract spectral vectors are orthonormal leads to

$$\mathbf{D} = \sum_{j=1}^{n-1} \mathbf{s}_j^0 \mathbf{s}_j^{0\prime} \mathbf{D} + \beta_{nn} \mathbf{s}_n^0 \mathbf{c}_n' \tag{6.42}$$

Equation (6.42) can be rearranged to

$$(\mathbf{I} - \mathbf{S}^0 \mathbf{S}^{0\prime}) \mathbf{D} = \beta_{nn} \mathbf{s}_n^0 \mathbf{c}_n' = \mathbf{X}_n \tag{6.43}$$

In this equation $\mathbf{I}$ is the identity matrix. Matrix $\mathbf{X}_n$ is calculated from $\mathbf{I}$, $\mathbf{S}^0$, and $\mathbf{D}$. Because $\mathbf{X}_n$ has unit rank, vector $\mathbf{s}_n^0$ and $\beta_{nn} \mathbf{c}_n'$ are determinable. Vector $\mathbf{s}_n^0$ is not the true spectral vector of the component. It represents the part of the vector that is orthogonal to the subspace of the other components. However, most importantly, $\beta_{nn} \mathbf{c}_n'$ is directly proportional to the concentration of the component.

Because the constant β_{nn} is unknown, the absolute concentration profile, $\mathbf{c}_n$, cannot be determined from (6.43) alone. Calibration can be accomplished by multiplying the true spectrum, $\mathbf{s}_n$, by the abstract spectrum, $\mathbf{s}_n^0$, because

$$\mathbf{s}_n' \mathbf{s}_n^0 = \beta_{nn} \tag{6.44}$$

Thus, by specifying the window, the concentration profile of a component can be deduced without recourse to any information concerning the other components. If the spectrum of the component is known, then quantification is possible.

Locating the window of totally unknown components can be done by considering every possible window, extracting the concentration profiles and examining them visually or computationally. Profiles that exhibit large negative values or double maximums are obviously inconsistent with the evolutionary constraints and are readily discarded. Examples of this procedure are illustrated in Figure 6.20, concerning the contour of $BiCl^{2+}$. Figure 6.20a is the "best" contour based on a window designated as 108 to 166 s. If the window is expanded to the right, Figure 6.20b results. Expansion to the left produces Figure 6.20c. Contraction on the left yields Figure 6.20d. Contraction on the right gives Figure 6.20e. Contraction on both sides gives Figure 6.20f. Notice that Figure 6.20a is the only profile that fits nicely inside the specified window. The other profiles are ruled out because they exhibit negative concentrations, multiple peaks, or concentrations outside the window.

WFA has more resolving power than EFA. When the flow injection analysis data in Figure 6.7, between 40 and 160 s, was processed by EFA, only six concentration profiles were obtained (see Figure 6.8). When the same data was subjected to WFA, the profiles of all seven bismuth species were extracted (see Figure 6.21).

For pedagogical reasons, a simple algorithm, written in MATLAB code, is provided below.

```
function [Xn,cn]=wfamodel(w1,w2)

                          %w1=left window edge of
                          component n
                          %w2=right window edge of
                          component n
conc(1,:)=[3 2 1 0 0];    %Create concentration profile of
                          component 1.
conc(2,:)=[0 1 2 1 0];    %Create concentration profile of
                          component 2.
conc(3,:)=[0 0 1 4 6];    %Create concentration profile of
                          component 3.
spex=rand(10,3);          %Create spectra of the three
                          components.
D=spex*conc;              %Create the data matrix.
D0=D;                     %Copy the data matrix.
```

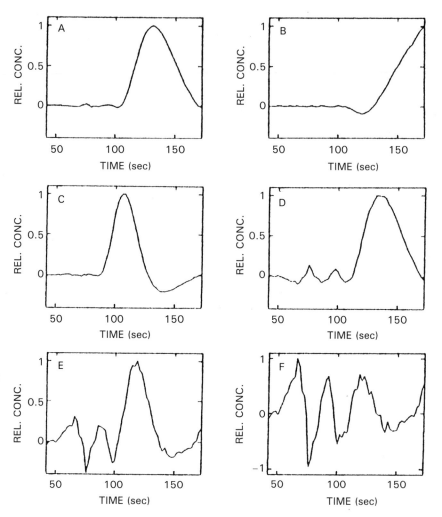

Figure 6.20 Concentration profiles of $BiCl^{2+}$ from WFA of Figure 6.7, using seven factors based on windows: A, 108–166; B, 108–172; C, 90–166; D, 116–166; E, 108–156; F, 120–140. [Reprinted with permission from E. R. Malinowski, *J Chemometrics*, **6**, 29 (1992). Copyright 1992 John Wiley & Sons, Ltd.]

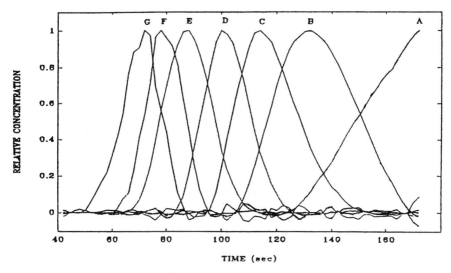

Figure 6.21   Concentration profiles of bismuth chloride complexes extracted by WFA of flow injection data based on seven factors: A, Bi^{3+} (128–172); B, $BiCl^{2+}$ (108–166); C, $BiCl_2^+$ (98–150); D, $BiCl_3$ (86–122); E, $BiCl_4^-$ (66–114); F, $BiCl_5^{2-}$ (62–92); G, $BiCl_6^{3-}$ (42–84) (values in parentheses designate the component window). [Reprinted with permission from E. R. Malinowski, *J. Chemometrics*, **6**, 29 (1992). Copyright 1992 John Wiley & Sons, Ltd.]

```
D0(:,w1:w2)=[ ];          %Remove the window of n from
                          the copy
[U0,S0,V0]=svd(D0,0);     %Subject D0 to singular value
                          decomposition.
Xn=D-U0*U0'*D;            %Apply Eq. 6.43.
cn=sum(Xn)                %Obtain uncalibrated
                          concentration profile of n.
```

The uncalibrated concentration profiles of the three modeled components can be extracted from D by running the program using three separate commands: wfamodel(1,3), wfamodel(2,4), and wfamodel(3,5). The numbers inside the parentheses designate the left and right window edges, w1 and w2, under consideration. Because Xn has rank one, the rows of Xn are mutually proportional to the concentration profile of n. Hence, the sum of the elements in each column is a valid representation of the concentration profile. In the algorithm the concentration profiles of the three species obey the necessary requirements for WFA. For example, by removing the first three data columns [wfamodel(1,3)] we have created a matrix, D0, that has *no* signal from the first component but has signals from *all* other components. Similar situations exist for wfamodel(2,4) and wfamodel(3,5). This is the fundamental principle upon which WFA is based.

A data-preprocessing algorithm for WFA has been suggested by Shao, Cai, and Pan.[25] They demonstrated how wavelet transform can be used to remove high-frequency noise and low-frequency baseline drift, leading to improved quantification.

It should be noted that the orthogonal projection resolution (OPR) method of Liang and Kvalheim[26] is mathematically equivalent to WFA even though the methods differ in derivation and algorithmic procedure.[27]

Automatic Window Factor Analysis. Determining the windows by EFA or by WFA requires a great deal of operator interaction, a task that is tedious and time consuming. In order to eliminate the need for such interaction, an algorithm called automatic window factor analysis (AUTOWFA) was designed by Malinowski.[28] AUTOWFA makes use of principal factor analysis (PFA), iterative key set factor analysis (IKSFA), the uniqueness test (UNIQ), and window factor analysis (WFA). A summary of the steps involved in the algorithm is given below.

1. Absorbance spectra from an evolutionary process are digitally recorded, corrected for baseline, and assembled into a matrix where each column represents a spectrum at a given time interval and each row represents a digitized wavelength.

2. The number of absorbing species, n, is determined by PFA (see Section 4.3).

3. Iterative key set factor analysis (IKSFA) (see Section 3.4.5), based on n factors, is used to locate n key spectra along the evolutionary axis where each species is the most pure.

4. Uniqueness vectors (see Section 3.4.3), each consisting of unity for the key column and zeros for all other columns, are designed and separately target tested. Each target test yields a predicted profile that is a better representation of the true concentration profile than the target vector.

5. All points inside a window must be positive. The first points that fall below zero on the left and right sides of the maximum of a predicted profile represent the first approximation of the window edges.

6. Each window in step 5 is individually subjected to WFA. The window is improved by locating the edges of each window as done in step 5. However, in this step, the edges are designated by the first points lying to the right and left of the profile maximum that reaches a specified baseline noise level. If the baseline noise is set too high, the window will be too small. If the baseline noise is set too low, the window will be too large.

7. For each window, step 6 is repeated until either the window no longer changes or the number of iterations exceeds some designated limit.

When the bismuth data (Figure 6.7) was subjected to AUTOWFA, the computational time was reduced to minutes instead of hours as required by conventional WFA. Although AUTOWFA is efficient and easy to use, the method has not been fully optimized. Improvements can often be made by specifying more factors, nf, in step 6 than n, the number of species. This procedure compensates for baseline bias and instrumental artifacts that produce "ghost" factors, factors that have no concentration profiles but fictitiously increase the overall factor space.

Subwindow Factor Analysis. The method of Manne and co-workers,[29] called subwindow factor analysis (SFA), extracts component spectra directly from overlapping chromatograms without the need to resolve the concentration profiles. SFA is particularly useful when complete resolution cannot be attained and component identification is the prime goal of the analysis.

The underlying theory of SFA is best understood by referring to Figure 6.22. Interfering substances on both sides flank the concentration profile of the component that lies in the region L + M + R. In the middle subwindow, M, all three components coexist. In the left subwindow, L, only the left interferent and the analyte coexist. In the right subwindow, R, the right interferent and the analyte coexist. Subwindow assignments must be made so that a specific interferent is not present in both the left and right subwindows. A variety of techniques such as EFA, WFA, FSWEFA, and so on can be employed for this purpose.

Principal factor analysis is used to determine the number of components, m and n, in the left and right subwindows as well as the orthonormal spectral bases sets, $\mathbf{E} = [\mathbf{e}_1, \mathbf{e}_2, \dots, \mathbf{e}_m]$ and $\mathbf{F} = [\mathbf{f}_1, \mathbf{f}_2, \dots, \mathbf{f}_n]$. Because the analyte spectrum, $\boldsymbol{v}$, is common to both subwindows, it can be expressed as linear combinations of the basis vectors with coefficients $\mathbf{a}$ and $\mathbf{b}$.[30]

$$\boldsymbol{v} = \mathbf{Ea} \tag{6.45}$$

$$\boldsymbol{v} = \mathbf{Fb} \tag{6.46}$$

Furthermore, in accord with (3.128), the following is true:

$$\mathbf{EE'}\boldsymbol{v} = \boldsymbol{v} \tag{6.47}$$

$$\mathbf{FF'}\boldsymbol{v} = \boldsymbol{v} \tag{6.48}$$

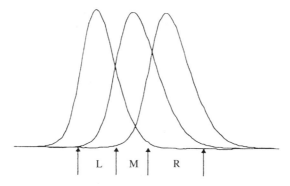

Figure 6.22 Illustration of the subwindows of the middle peak. [Reprinted with permission from R. Manne, H. Shen, and Y. Liang, *Chemom. Intell. Lab. Syst.*, **45**, 171 (1999). Copyright 1999 Elsevier Science.]

Inserting (6.45) into (6.48), and inserting (6.46) into (6.47) gives

$$\mathbf{E'FF'Ea = a} \tag{6.49}$$
$$\mathbf{F'EE'Fb = b} \tag{6.50}$$

According to (6.49) and (6.50), $\mathbf{a}$ and $\mathbf{b}$ are eigenvectors of their respective symmetric matrices with unit eigenvalues.

To demonstrate its utility, SFA was applied to a HPLC-DAD recording of an atmospheric sample at the Hong Kong Baptist University.[29] Although the sample contained 47 components, a time interval was selected where four unresolved components existed. The average elution profile (Figure 6.23) exhibits two large maximums and a weak central maximum. FSWEFA was used to determine the subwindows (see Table 6.3). Subwindow factor analysis produced the spectra portrayed in Figure 6.24. Because the spectra of all four components were found, the uncalibrated concentration profiles in Figure 6.24 were determined by a pseudoinverse calculation. To convert these profiles into true concentrations, it is necessary to identify the components and perform spectral calibrations based on pure materials.

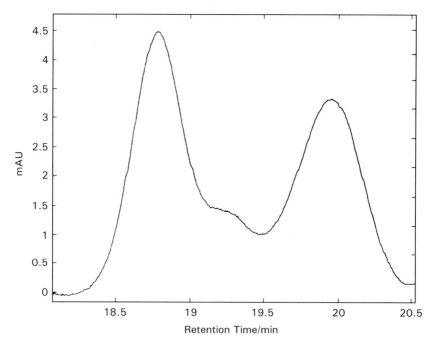

Figure 6.23 Average concentration profile of the analyzed data set. [Reprinted with permission from R. Manne, H. Shen, and Y. Liang, *Chemom. Intell. Lab. Syst.*, **45**, 171 (1999). Copyright 1999 Elsevier Science.]

TABLE 6.3   Subwindows Found by FSWEFA[a]

Component	Left Subwindow (min)	Right Subwindow (min)
1	18.166–18.897	18.166–19.359
2	18.904–19.319	19.490–19.801
3	19.325–19.622	19.808–20.270
4	19.491–20.484	20.277–20.848

[a] Reprinted with permission from R. Manne, H. Shen, and Y. Liang, *Chemom. Intell. Lab. Syst.*, **45**, 171 (1999).

6.3.3   Single Component with Multiple Rank

As a consequence of impurities, background, or chemical modifications caused by the process employed, the component matrix, **N** [required for application of (6.24)], may have a rank greater than one. Rank annihilation can be applied to this situation by expanding (6.26), (6.28), and (6.29) to full matrix form, applicable to multi-components, as follows:

$$\mathbf{N} = \mathbf{XY}' \tag{6.51}$$

$$\mathbf{A} = [\bar{\mathbf{R}}\Lambda^{-1}]'\mathbf{X} = \bar{\mathbf{U}}\bar{\mathbf{S}}^{-1}\mathbf{X} \tag{6.52}$$

$$\mathbf{B} = \bar{\mathbf{C}}\mathbf{Y} = \bar{\mathbf{V}}'\mathbf{Y} \tag{6.53}$$

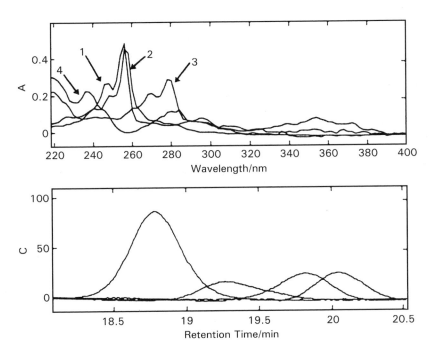

Figure 6.24   Resolved spectra and concentration profiles from SFA. [Reprinted with permission from R. Manne, H. Shen, and Y. Liang, *Chemom. Intell. Lab. Syst.*, **45**, 171 (1999). Copyright 1999 Elsevier Science.]

Lorber[31] showed that the direct solution to this problem yields

$$\tau = n/\text{trace}(\mathbf{A}'\mathbf{B}) \tag{6.54}$$

where n is the rank of $\mathbf{N}$ and trace( ) denotes the sum of the diagonal elements of the matrix inside the parentheses.

Combined with the standard addition method, (6.54) not only allows quantification of a target component without prior knowledge of the other species present but also eliminates the need for "zeroing" the instrument as required by conventional methods. Furthermore, there is no need to remove background contributions because the background is viewed as an additional component.

6.3.4   Generalized Rank Annihilation Factor Analysis

Sanchez and Kowalski[32−34] developed a generalized rank annihilation factor analysis (GRAFA) method, also known as GRAM, that permits simultaneous quantification of analytes in an unknown sample using only one bilinear calibration matrix gleaned from a mixture of standards. The full bilinear data of each individual analyte is not required. Instead, a single calibration sample containing all analytes of interest is prepared. In this case the calibration matrix is represented as

$$\mathbf{N} = \mathbf{X}\boldsymbol{\beta}\mathbf{Y}' \tag{6.55}$$

Here $\boldsymbol{\beta}$ is a diagonal matrix whose elements are the concentrations of the analytes in the standard solution. $\mathbf{X}$ and $\mathbf{Y}$ are the same matrices expressed in (6.23). It is important to recognize that $\boldsymbol{\beta}$ is a diagonal matrix whose elements are the concentrations of *all* components present in the sample. GRAFA requires the terms in (6.23) and (6.55) to have the same dimensions. Therefore the matrices on the right of both equations must be expanded to incorporate *all* components in the sample mixture as well as *all* components in the standard mixture. Components of the sample that have not been included in the calibration mixture will have zeros for their diagonal terms in $\boldsymbol{\alpha}$. Similarly, components of the calibration mixture that are not present in the sample will have zeros for their diagonal terms in $\boldsymbol{\beta}$.

GRAFA finds the "base" that satisfies (6.23) and (6.55) simultaneously. There is only one base that diagonalizes $\mathbf{M}$ and $\mathbf{N}$ simultaneously. This base contains elements in the diagonal only for those components common to both matrices.

To find the base, $\mathbf{N}$ is expressed as a function of $\mathbf{M}$:

$$\begin{aligned}
\mathbf{N} &= \mathbf{X}\boldsymbol{\beta}\mathbf{Y}' \\
&= \mathbf{X}\boldsymbol{\alpha}^{-1}\boldsymbol{\beta}\boldsymbol{\alpha}\mathbf{Y}' \\
&= \mathbf{X}(\boldsymbol{\alpha}^{-1}\boldsymbol{\beta})\mathbf{X}^{+}(\mathbf{X}\boldsymbol{\alpha}\mathbf{Y}') \\
&= \mathbf{X}(\boldsymbol{\alpha}^{-1}\boldsymbol{\beta})\mathbf{X}^{+}\mathbf{M}
\end{aligned} \tag{6.56}$$

Now define

$$\tau = \alpha^{-1}\beta \tag{6.57}$$

Postmultiplying both sides of (6.56) by M^+ yields

$$NM^+ = X\tau X^+$$
$$(NM^+)X = X\tau \tag{6.58}$$

This is the usual eigenvalue–eigenvector problem because the overall matrix between the parentheses is square; however, the eigenvectors, X, representing the spectra, are not orthogonal because the overall matrix is not symmetric. Eigenvalue matrix τ contains the ratios of the concentrations. Matrix Y can be determined because

$$Y' = \alpha^{-1}X^+M \tag{6.59}$$

Because (6.57) contains α_j in the denominator, these equations are valid only if the eigenvalues of the superset of M are all nonzero. This means that M must contain all the components that are present in the standard. A simple way to circumvent this problem is to create a new matrix, W, composed from the sum of the sample and the standard matrices:

$$W = M + N \tag{6.60}$$

By definition, this matrix includes all components represented by both matrices. Arguments similar to those described above lead to the following conclusions:

$$(NW^+)X = X\Lambda \tag{6.61}$$

where the eigenvalues are functions of the concentration ratios, namely,

$$\lambda_j = \beta_j/(\alpha_j + \beta_j) \tag{6.62}$$

Rearranging (6.62) gives

$$\alpha_j = \beta_j[(1 - \lambda_j)/\lambda_j] \tag{6.63}$$

Matrix Y, in this case, is determined by

$$Y' = \Lambda^{-1}X^+W \tag{6.64}$$

If component j has not been included in the calibration sample, β_j is zero, its eigenvalue is zero, and its concentration, α_j, in the sample is not determinable.

The algorithm can be improved by projecting the calibration and unknown sample matrices into the lower-dimensional, orthonormal, factor subspace.[35] The results obtained by this procedure are more stable.

An important advantage of GRAFA is the fact that preliminary testing with target factor analysis is not a necessary prerequisite. Furthermore, if several calibration matrices, consisting of various amounts of different compounds, have been prepared, an unknown sample can be analyzed by repeated eigenanalysis of the single mixture matrix combined with the calibration matrices. Alternatively, calibration standards can be added to the sample and **W** can be measured directly.

GRAFA offers a powerful method for analyzing complex mixtures. The analyst need not worry about interferences, contaminants, and background. GRAFA takes full advantage of all the data produced by bilinear instrumentation.

6.3.5   Direct Exponential Curve Resolution Algorithm

Windig and Antalek[36] developed a self-modeling method for resolving mixtures in which the component concentrations follow exponential decay pathways. The method, called direct exponential curve resolution algorithm (DECRA), is based on the generalized rank annihilation method (GRAM) (see Section 6.3.4.). However, unlike GRAM, which requires two matrices, DECRA requires only one matrix.

Two matrices of equal size are carved out of a single matrix by offsetting the time axis. An exponential decay curve thus split has the unique property that the two resulting curves are directly proportional to each other. Exponential curves belonging to different species have their own unique proportionality constant. The two resulting matrices, **M** and **N**, involve the same components, but in different relative proportions. These proportions are functions of the respective decay constants. The two matrices have the proper form for GRAM analysis. By inserting their transposed forms into (6.58) and invoking (6.59) the decay curves, **Y**, of the components are readily obtained.

DECRA has been used to resolve ultraviolet/visible spectra of reaction mixtures. An excellent example is the reaction of 3-chlorophenylhydrazonopropane dinitrile (A) with 2-mercaptoethanol (B) which forms an intermediate (C) which hydrolyzes to 3-chlorophenylhydrazonocyanoacetamid (D) and ethylenesulfide (E), a by-product.[37]

$$A + B \xrightarrow{k_1} C \xrightarrow{k_2} D + E \tag{6.65}$$

Because B is in large excess, the reactions are essentially first order as shown below:

$$C_{A,i} = C_{A,0} e^{-k_1 t_i} \tag{6.66}$$

$$C_{C,i} = \frac{k_1 C_{A,0}}{k_2 - k_1} (e^{-k_1 t_i} - e^{-k_2 t_i}) \tag{6.67}$$

$$C_{D,i} = C_{A,0} - C_{A,i} - C_{C,i} \tag{6.68}$$

In these expressions $C_{A,i}$, $C_{C,i}$, and $C_{D,i}$ are the concentrations of the species at time t_i, and $C_{A,0}$ is the initial concentration of A at the beginning of the reaction.

The reaction was followed by UV/VIS spectroscopy from 300 to 500 nm, regions where components B and E do not absorb any radiation. Figure 6.25 is a three-dimensional plot of the measured absorbances. Each spectrum in this figure represents the average of 10 replicate measurements. Because DECRA invokes the GRAM methodology, two matrices, **M** and **N**, are required, as dictated by (6.58). In accord with the DECRA prescription, matrix **M** was assembled from spectra 1 to 269, and matrix **N** was assembled from spectra from 2 to 270.

Equations (6.66), (6.67), and (6.68) can be expressed as a linear combination of exponential decay profiles:

$$C_{A,i} = f_1 e^{-k_1 t_i} \tag{6.69}$$

$$C_{C,i} = f_2 e^{-k_1 t_i} - f_2 e^{-k_2 t_i} \tag{6.70}$$

$$C_{D,i} = f_1 e^{-0 t_i} - f_1 e^{-k_1 t_i} - \left(f_2 e^{-k_1 t_i} - f_2 e^{-k_2 t_i} \right) \tag{6.71}$$

where $f_1 = C_{A,0}$ and $f_2 = k_1 C_{A,0}/(k_2 - k_1)$. These equations are appropriate for DECRA analysis because they express the concentrations as a linear sum of three different exponentials. It is important to notice that the initial concentration of A is expressed as an exponential with a decay constant equal to zero, $f_1 e^{-0 t_i}$.

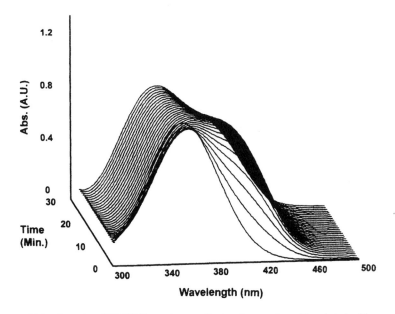

Figure 6.25   Sample of UV/VIS spectra monitoring the reaction. [Reprinted with permission from W. Windig, B. Antalek, L. J. Sorriero, S. Bijlsma, D. J. Louwerse, and A. K. Smilde, *J. Chemometrics*, **13**, 95 (1999). Copyright 1999 John Wiley & Sons, Ltd.]

The profiles generated by DECRA do not represent concentrations; they represent uncalibrated exponential decay functions. From these profiles the decay constants were found to be $k_1 = 0.314$, $k_2 = 0.027$, and 0.000, as expected. To obtain the concentration profiles, these constants must be inserted into (6.66), (6.67), and (6.68) and scaled by a method of least squares to reproduce the total signal (TSI) of the original spectra. The results of this procedure are displayed in Figure 6.26. The fact that the TSI is identical to the sum of the profiles lends credence to the exponential model.

By a pseudoinverse computation, the spectra of the three species were obtained. The resolved spectra of reactant A and product D were found to be identical to their reference spectra. There is no spectrum of the intermediate, C, available for comparison.

DECRA has also been used to resolve pulsed gradient spin echo nuclear magnetic resonance of polymer solutions (see Section 9.72) as well as magnetic resonance images of human brain (see Section 12.6).

6.4   TRANSMUTATION

Transmutation[38] is designed to convert broad overlapping chromatographic bands (see Figure 6.27) into narrow bands, sufficient to resolve and quantify a specific

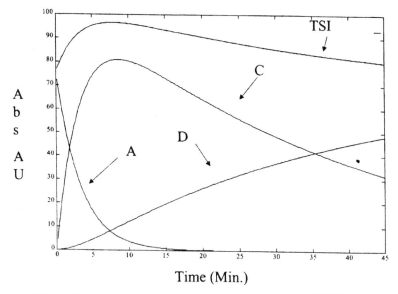

Time (Min.)

Figure 6.26   Extracted and scaled concentration profiles from DECRA. Reactant A, intermediate C, final product D, and total signal TSI. [Reprinted with permission from W. Windig, B. Antalek, L. J. Sorriero, S. Bijlsma, D. J. Louwerse, and A. K. Smilde, *J. Chemometrics*, **13**, 95 (1999). Copyright 1999 John Wiley & Sons, Ltd.]

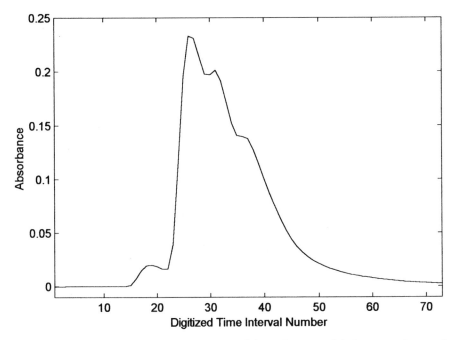

Figure 6.27 Chromatogram of a mixture containing toluene, naphthalene, *m*-xylene, and biphenyl, recorded at 295 nm. [Reprinted with permission from E. R. Malinowski, *Anal. Chem.*, **72(18)**, 4363 (2000). Copyright 2000 © American Chemical Society.]

component (see Figure 6.28). Unlike most chemometric methods, the transmutation method does *not* require two-dimensional arrays of data. It is particularly useful for a system equipped with a single detector, such as thermal conductivity, flame ionization, beta emission, gas density, and so forth. Mixtures of toluene, ethylbenzene, *m*-xylene, naphthalene, and biphenyl were analyzed by this technique.[39,40] As shown in Table 6.4, the determined concentrations have the same degree of accuracy as those obtained by window factor analysis, rank annihilation factor analysis, and matrix regression analysis, methods requiring two-dimensional data arrays.

A "transmutation function" is generated from the chromatogram of the component of interest, recorded with the same instrument and experimental conditions used to record the unresolved chromatogram of the sample mixture. Both chromatograms are digitized at equally spaced time intervals, and corrected for baseline drift. The time windows of the pure component and the mixture must be the same. Zero baseline regions must encompass the windows.

A chromatogram is a row vector composed of detector responses as a function of equally spaced digitized time intervals. The transmutation function, $\mathbf{T}$, is a matrix designed to convert a broad real chromatographic vector, $\mathbf{c}_{real}$, into a narrow vector, $\mathbf{c}_{ideal}$, called an ideal chromatographic vector.

$$\mathbf{c}_{real}\mathbf{T} = \mathbf{c}_{ideal} \tag{6.72}$$

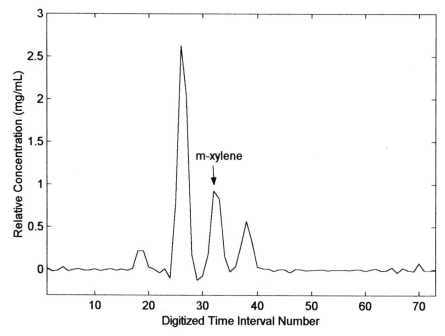

Figure 6.28 Concentration profiles obtained by transmutation of the chromatogram in Figure 6.27, based on the triangular function and *m*-xylene standard. [Reprinted with permission from E. R. Malinowski, *Anal. Chem.*, **72(18)**, 4363 (2000). Copyright 2000 © American Chemical Society.]

The most ideal chromatogram is a Dirac delta function, infinitely thin with a height directly proportional to the concentration of the component. Unfortunately, the delta function is not suitable for transmutation because it is sensitive to errors along the time axis. A triangular function serves as an excellent ideal chromatogram because it has a finite bandwidth and can tolerate a degree of error along the time axis.

The transmutation matrix, $\mathbf{T}$, is generated from two matrices, a standard matrix and an ideal matrix. The standard matrix, $\mathbf{C}_{std}$, is constructed from $\mathbf{c}_{std}$, the chromatogram of the pure component of interest, called the standard. $\mathbf{C}_{std}$ is obtained by expanding $\mathbf{c}_{std}$ threefold by adding time points with zero intensity to the front of the experimental time window as well as to the end of the time window. This is done to prevent truncation of the profiles at the edges of the experimental time window. The first row of $\mathbf{C}_{std}$ contains vector $\mathbf{c}_{std}$, shifted so its peak maximum coincides with the first time point of the expanded time scale. The second row of $\mathbf{C}_{std}$ contains $\mathbf{c}_{std}$, shifted so its peak maximum coincides with the second time point. The third, fourth, fifth, and so on rows are similarly constructed, shifting $\mathbf{c}_{std}$ so their maximums coincide with their appropriate time points. Responses that fall outside the expanded time window are deleted. As an example, if the experimental time window consisted of 73 time points, $\mathbf{C}_{std}$ would be a 219×219 matrix. An example

TABLE 6.4   Concentrations (mg/mL) of Components in Four Mixtures Determined by Triangular TRANSMUTATION, Window Factor Analysis (WFA), Rank Annihilation Factor Analysis (RAFA), and Matrix Regression Analysis (MATRA)[a]

Component	Expected[b]	TRANSMUTATION	WFA[b]	RAFA[b]	MATRA[b]
Mixture 1					
Toluene	0.976	0.954	1.073	1.025	0.917
Ethylbenzene	0.941	1.049	0.996	1.030	1.062
m-Xylene	1.002	1.091	0.912	1.078	1.042
Mixture 2					
Toluene	0.958	0.688	0.755	0.804	0.763
Naphthalene	0.093	0.072	0.088	0.086	0.007
m-Xylene	1.002	0.739	0.966	0.964	1.028
Mixture 3					
Toluene	0.958	0.955	0.931	1.056	0.948
Naphthalene	0.093	0.097[c,d]	0.070	0.081	0.098
Ethylbenzene	0.941	3.107[c,e]	0.640	1.617	1.003
m-Xylene	1.002	0.487[f]	1.066	1.393	1.116
Mixture 4					
Toluene	0.958	0.869	0.945	0.991	0.941
Naphthalene	0.093	0.091	0.073	0.108	0.094
m-Xylene	1.002	0.940	1.128	1.178	1.359
Biphenyl	0.038	0.034	0.042	0.043	0.039

[a] Reprinted with permission from E. R. Malinowski, *Anal. Chem.*, **72(18)**, 4363 (2000). Copyright 2000 © American Chemical Society.
[b] Taken from K. J. Schostack and E. R. Malinowski, *Chemometrics Intell. Lab. Syst.* **20**, 173–182 (1993).
[c] Insufficient chromatographic separation between naphthalene and ethylbenzene.
[d] Because naphthalene is a strong UV absorber, its signal dominates the signal of ethylbenzene.
[e] The weak signal of ethylbenzene is buried under the strong signal of naphthalene.
[f] The transmuted peak maximum does not coincide with the retention time of the standard.

of C_{std} is shown in Figure 6.29. This matrix was constructed from the chromatogram of m-xylene shown in Figure 6.30.

The ideal matrix, C_{ideal}, is similarly constructed, using an ideal chromatographic vector instead of c_{std}. For the delta function the ideal matrix is simply

$$C_{ideal} = Ic_0 \qquad (6.73)$$

Here c_0 is the concentration of the standard solution and I is the identity matrix. Multiplying the ideal shape function by the concentration of the standard converts the response units into concentration units. If c_0 is arbitrarily set equal to unity, then the transmutation will yield relative concentrations $c(\text{mixture})/c(\text{standard})$, where $c(\text{mixture})$ is the concentration of the component in the mixture and $c(\text{standard})$ is the concentration of the component in the pure standard.

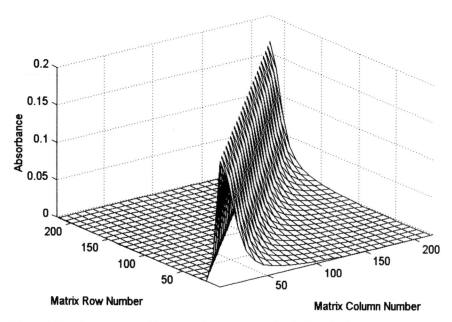

Figure 6.29 C_{std} constructed from the chromatogram of m-xylene displayed in Figure 6.30. [Reprinted with permission from E. R. Malinowski, *Anal. Chem.*, **72(18)**, 4363 (2000). Copyright 2000 © American Chemical Society.]

The transmutation matrix, **T**, is a matrix that transforms C_{std} into C_{ideal}, that is,

$$C_{std}T = C_{ideal} \tag{6.74}$$

Rearranging (6.74) gives

$$T = C_{std}^{-1}C_{ideal} \tag{6.75}$$

T is calculable because c_{std} and c_{ideal} are both known. The transmutation matrix not only converts C_{std} into C_{ideal} but also converts the detector response units into concentration units, *relative to the response of the standard solution of the component being quantified.*

Any arbitrary chromatographic vector, c_{arb}, can be transmuted into a linear sum of standard chromatographic vectors, c_{std}. This can be proven as follows. According to Eq. (6.72), c_{arb} can be transmuted into a delta function, c_{delta}, as shown in (6.76):

$$c_{arb}T = c_{delta} \tag{6.76}$$

Inserting (6.73) into (6.75) gives

$$T = C_{std}^{-1}C_{delta} = C_{std}^{-1}Ic_0 = C_{std}^{-1}c_0 \tag{6.77}$$

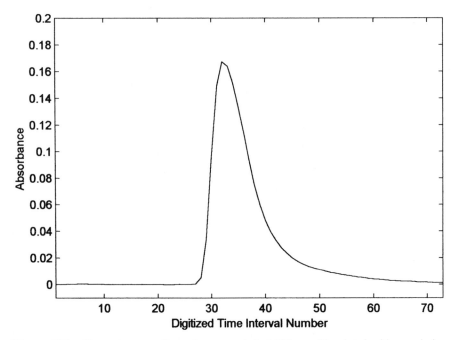

Figure 6.30   Chromatogram of *m*-xylene recorded at 295 nm. [Reprinted with permission from E. R. Malinowski, *Anal. Chem.*, **72(18)**, 4363 (2000). Copyright 2000 © American Chemical Society.]

Inserting (6.77) into (6.76) and postmultiplying both sides by $\mathbf{C}_{std}$ gives

$$\mathbf{c}_{arb}c_0 = \mathbf{c}_{delta}\mathbf{C}_{std} \tag{6.78}$$

According to (6.78)

$$\mathbf{c}_{arb} = \mathbf{d}\mathbf{C}_{std} \tag{6.79}$$

where

$$\mathbf{d} = \mathbf{c}_{delta}/c_0 \tag{6.80}$$

Equation (6.79) can be expressed as a linear sum of vectors.

$$\mathbf{c}_{arb} = \sum_j d_j\mathbf{c}_{std}(j) \tag{6.81}$$

The d_j values in (6.81), obtained from delta transmutation, represent a set of coefficients that express any arbitrary chromatogram as a linear sum of standard chromatograms, $\mathbf{c}_{std}(j)$. The profiles of every $\mathbf{c}_{std}(j)$ are exactly the same, but each vector is shifted so its peak maximum corresponds to time j. The sum is taken over all such vectors.

A mixed chromatographic vector exhibiting overlapping bands, $\mathbf{c}_{\text{real,mix}}$, can be expressed as the sum of component chromatograms and null chromatograms, $\mathbf{c}_{\text{real}}(j)$, each associated with a specific row labeled j. A null chromatogram is a vector whose elements are all statistically equal to zero. Accordingly, we may write

$$\mathbf{c}_{\text{real,mix}} = \sum_j m_j \mathbf{c}_{\text{real}}(j) \tag{6.82}$$

In (6.82) m_j is the amount of component with retention time j, in the mixture. For a null chromatographic vector, m_j equals zero. Postmultiplying both sides of (6.82) by $\mathbf{T}$ and invoking (6.81) leads to

$$\mathbf{c}_{\text{real,mix}} \mathbf{T} = \mathbf{c}_{\text{ideal,mix}} \tag{6.83}$$

where

$$\mathbf{c}_{\text{ideal,mix}} = \sum d_j \mathbf{c}_{\text{ideal}}(j) \tag{6.84}$$

Thus we see that a mixed chromatogram can be transmuted into a mixture of ideal chromatograms.

Transmutations based on the delta function ($\ldots$ 0 0 1 0 0 $\ldots$) are sensitive to errors along the time axis. Marked improvement can be achieved by using triangular transmutation function ($\ldots$ 0 0.5 1 0.5 0 $\ldots$). Triangular functions have some broadness that accommodates a degree of error along the time axis. For a triangular function, Eq. (6.73) should be replaced by (6.85).

$$\mathbf{C}_{\text{ideal}} = \mathbf{S} c_0 \tag{6.85}$$

The triangular shape matrix, $\mathbf{S}$, is obtained by adding 0.5 to each of the zero elements adjacent to the diagonal elements of $\mathbf{I}$, the identity matrix. From Eqs. (6.73), (6.75), (6.83), and (6.85) one can obtain the following relationship:

$$\mathbf{c}_{\text{ideal,mix}}^{\text{triangle}} = \mathbf{c}_{\text{ideal,mix}}^{\text{delta}} \mathbf{S} \tag{6.86}$$

where $\mathbf{c}_{\text{ideal,mix}}^{\text{triangle}}$ and $\mathbf{c}_{\text{ideal,mix}}^{\text{delta}}$ are the vector profiles after triangular and delta transmutation. The advantage of triangular transmutation is clearly seen by comparing Figure 6.28 to Figure 6.31. Figure 6.31 is the result of delta transmutation whereas Figure 6.28 is the result of triangular transmutation. Both figures are based on m-xylene standard solution.

The method has limitations. An excessive number of digitized time points produces an augmented transmutation matrix that is so large that present-day computer algorithms are unable to carry out the required inverse computation with sufficient accuracy. Furthermore, the delta coefficients resulting from large

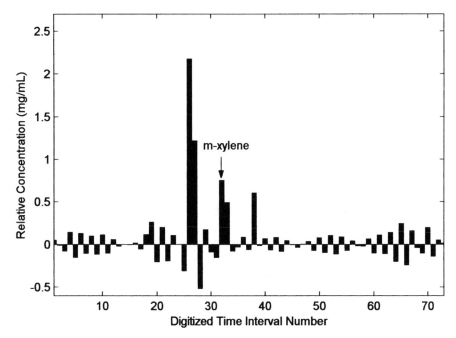

Figure 6.31   Result of delta transmutation of the mixture in Figure 6.27, using the chromatogram of *m*-xylene (Figure 6.30) as the standard.

chromatographic vectors will exhibit greater oscillations, that is, larger and more frequent positive and negative values, making triangular transmutation less effective. By reducing the number of digitized points in the chromatograms, successful transmutations can be achieved.[39]

REFERENCES

1. J. J. Kankare, *Anal. Chem.*, **42**, 1322 (1970).

2. R. I. Shrager and R. W. Hendler, *Anal. Chem.*, **54**, 1147 (1982).

3. R. I. Shrager, *SIAM J. Alg. Discuss. Meth.*, **5**, 351 (1984).

4. R. I. Shrager, *Chemometrics Intell. Lab. Syst.*, **1**, 59 (1986).

5. F. J. Knorr and J. M. Harris, *Anal. Chem.*, **53**, 272 (1981).

6. F. J. Knorr, H. R. Thorsheim, and J. M. Harris, *Anal. Chem.*, **53**, 821 (1981).

7. S. D. Frans, M. L. McConnell, and J. M. Harris, *Anal. Chem.*, **57**, 1552 (1985).

8. S. D. Frans and J. M. Harris, *Anal. Chem.*, **56**, 466 (1984).

9. S. D. Frans and J. M. Harris, *Anal. Chem.*, **57**, 1718 (1985).

10. P. J. Gemperline, *J. Chem. Inf. Comput. Sci.*, **24**, 206 (1984).

11. B. Vandenginste, W. Derks, and G. Kateman, *Anal. Chim. Acta*, **173**, 253 (1985).

12. H. Gampp, M. Maeder, C. J. Meyer, and A. D. Zuberbuhler, *Talanta*, **32**, 1133 (1985); *Chimia*, **39**, 315 (1985).

13. P. J. Gemperline and J. C. Hamilton, *J. Chemometrics*, **3**, 455 (1989).

14. M. Maeder, *Anal. Chem.*, **59**, 527 (1987).

15. H. R. Keller and D. L. Massart, *Anal. Chim. Acta*, **246**, 379 (1991).

16. C. N. Ho, G. D. Christian, and E. R. Davidson, *Anal. Chem.*, **52**, 1108 (1978).

17. A. Lorber, *Anal. Chim. Acta*, **164**, 293 (1984).

18. M. McCue and E. R. Malinowski, *J. Chromatogr. Sci.*, **21**, 229 (1983).

19. E. R. Malinowski, unpublished work.

20. A. A. Clifford, *Multivariate Error Analysis*, Applied Science Publishers, London. 1973.

21. D. H. Burns, J. B. Callis, and G. D. Christian, *Anal. Chem.*, **58**, 2805 (1986).

22. H. Gampp, M. Maeder, C. J. Meyer and A. D. Zuberbuhler, *Anal. Chim. Acta*, **193**, 287 (1987).

23. E. R. Malinowski, *J. Chemometrics*, **6**, 29 (1992).

24. W. Den and E. R. Malinowski, *J. Chemometrics*, **7**, 89 (1993).

25. X. Shao, W. Cai, and Z. Pan, *Chemometrics Intell. Lab. Syst.*, **45**, 249 (1999).

26. Y. Liang and O. M. Kvalheim, *Anal. Chim. Acta*, **66**, 43 (1994).

27. Q. Xu and Y. Liang, *Chemometrics Intell. Lab. Syst.*, **45**, 335 (1999).

28. E. R. Malinowski, *J. Chemometrics*, **10**, 273 (1996).

29. R. Manne, H. Shen, and Y. Liang, *Chemometrics Intell. Lab. Syst.*, **45**, 171 (1999).

30. H. Shen, R. Manne, Q. Xu, D. Chen, and Y. Liang, *Chemometrics Intell. Lab. Syst.*, **45**, 323 (1999).

31. A. Lorber, *Anal. Chem.*, **57**, 2395 (1985).

32. E. Sanchez and B. R. Kowalski, *Anal. Chem.*, **58**, 499 (1986).

33. E. Sanchez and B. R. Kowalski, *J. Chemometrics*, **2**, 247 (1988).

34. E. Sanchez and B. R. Kowalski, *J. Chemometrics*, **2**, 265 (1988).

35. B. E. Wilson, E. Sanchez, and B. R. Kowalski, *J. Chemometrics*, **2**, 493 (1989).

36. W. Windig and B. Antalek, *Chemometrics Intell. Lab. Syst.*, **37**, 241 (1997).

37. W. Windig, B. Antalek, L. J. Sorriero, S. Bijlsma, D. J. Louwerse, and A. K. Smilde, *J. Chemometrics*, **13**, 95 (1999).

38. E. R. Malinowski, *Anal. Chem.*, **72(18)**, 4363 (2000).

39. E. R. Malinowski, *J. Chemometrics*, accepted for publication.

40. K. J. Schostack and E. R. Malinowski, *Chemometrics Intell. Lab. Syst.*, **20**, 173 (1993).

7

Simple solutions require complicated explanations.

MULTIMODE FACTOR ANALYSIS

Up to this point we have been concerned with two-way arrays of data. However, many chemical problems involve multiway arrays (three way, four way, etc.). For example, high-performance liquid chromatography (HPLC) coupled with a spectro-photometric fluorescence detector yields a three-mode matrix composed of layers of two-way excitation–emission matrices that vary as a function of elution time. The fluorescent intensities depend upon the excitation wavelength, the emission wave-length, and elution time, variables representing three modes. Many chemical processes yield higher multimode data. For example, measurements of absorption spectra at different wavelengths during a kinetic experiment under various experi-mental conditions generate modes that can be represented by wavelength, time, temperature, initial concentration, and so on. These matrices contain a wealth of hidden information that can be unraveled by multimode factor analysis (MMFA).

Multiway analysis serves three primary functions: (1) curve resolution, (2) calibration, and (3) exploration. This chapter focuses on the most common techniques and their applications in these three areas. Although we will concentrate on three-way analysis, it is relatively easy to generalize the methodology to higher modes.

Multiway analysis originated from the psychometric studies of Tucker in the sixties.[1,2] Tucker expanded principal component analysis (PCA) and factor analysis

(FA) to incorporate three-way data arrays. In the early seventies, Carroll and Chang[3] developed a three-way model called canonical decomposition (CANDECOMP). At about the same time, Harshman[4] independently developed a method called parallel factor analysis (PARAFAC). The latter two methods are equivalent and are often labeled CP, giving credit to both research groups. In reality, the CP method is a special case of the Tucker model, yielding unique solutions under restricted conditions, removing the rotational ambiguity associated with two-way factor analysis and the generalized Tucker model. Because of the unique solutions, the CP method has become the most popular method used by chemometricians.

7.1   NOTATION

Multiway analysis is relatively new and is still undergoing evolutionary development from many different disciplines. As such, the emerging notation is quite inconsistent. Some practitioners favor array notation,[5] others advocate tensor algebra,[6,7] and some prefer modified matrix algebra.[8] Recently, Kiers[9] proposed a standard notation, terminology, and symbols based on earlier convention, with the intent to simplify the notation and reduce the confusion.

Rather than present a concise notation at the beginning of this chapter, it is more instructive to introduce the special notation of MMFA at appropriate stages in the exposition. Furthermore, we will make use of various notations (arrays, tensors, and algebra) best suited for clarity and understanding of the particular procedure being discussed. This should provide the reader with a broader background of the current language. Figure 7.1 illustrates how a three-way block of data with modes A, B, and C, can be cut into horizontal, lateral, and frontal slices (also called sheets or pages).

7.2   THREE-DIMENSIONAL RANK ANNIHILATION FACTOR ANALYSIS (3DRAFA)

Attempts have been made to apply rank annihilation to three-dimensional data arrays. A liquid chromatograph with a video fluorimeter detector (LC/VF) can produce such data. The data take the form m_{ikl}, which represent the fluorescence emission intensity at wavelength λ_i of the sample observed at elution time t_l when excited by light with wavelength λ_k. This response can be expressed in trilinear form:

$$m_{ikl} = \sum_{j=1}^{n} \alpha_j x_{ji} y_{jk} z_{jl} \qquad (7.1)$$

where z_{jl} is the chromatographic profile of component j, and the other terms are analogous to (6.22), the bilinear counterpart.

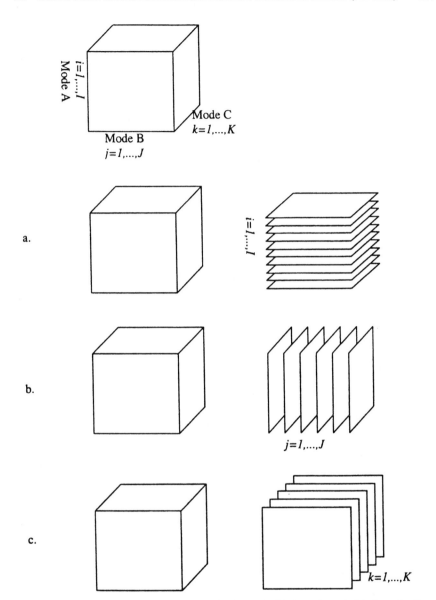

Figure 7.1 Three-way array, cut into (*a*) horizontal, (*b*) lateral, and (*c*) frontal slices. [Reprinted with permission from H. A. L. Kiers, *J. Chemometrics*, **14**, 105 (2000). Copyright 2000 © John Wiley & Sons, Ltd.]

Appelof and Davidson[10] carried out a simulation study of LC/VF, emulating a mixture of perylene, fluoranthene, teracene, and 9,10-dimethylanthracene. Three possible covariance matrices were constructed as follows:

$$\mathbf{Z}_x = \sum_k \sum_l \mathbf{m}_{ikl}\mathbf{m}_{i'kl}$$
$$\mathbf{Z}_y = \sum_i \sum_l \mathbf{m}_{ikl}\mathbf{m}_{ik'l} \qquad (7.2)$$
$$\mathbf{Z}_z = \sum_i \sum_k \mathbf{m}_{ikl}\mathbf{m}_{ikl'}$$

The number of components in the sample is estimated by the results of eigenanalysis of these matrices. The matrix with the smallest nonzero eigenvalue is less sensitive to errors in the data than the other two. This matrix is chosen as the basis for rank annihilation. Analogous to (6.24) and (6.25), an arbitrary amount, τ, of the three-dimensional array of the component standard is subtracted from the mixture array:

$$\mathbf{l}_{ikl} = \mathbf{m}_{ikl} - \tau\mathbf{n}_{ikl} \qquad (7.3)$$

A covariance matrix is formed from $\mathbf{l}_{ikl}$ in the same fashion used to construct the chosen basis matrix of the mixture. The value of τ that makes the smallest nonzero eigenvalue a minimum is determined by repeated eigenanalysis. This value is the best estimate of the concentration of the component in the sample relative to the standard solution.

Selecting the most appropriate covariance matrix is important. For example, if the spectrum of the analyte completely overlaps the spectra of the other components in one of the dimensions (e.g., the emission spectrum), changing τ will have little or no effect on the smallest nonzero eigenvalue of the corresponding covariance matrix. 3DRAFA permits selection of the most sensitive covariance matrix and could provide accurate results when two-dimensional analysis fails.

Sanchez and Kowalski[11] have considered extending the GRAFA technique to three-dimensional data. Currently, however, all trilinear decompositions are based on iterative minimizations of residuals and convergence is not always achieved. GRAFA attempts to extract the intrinsic vectors directly without iteration. Thus, at the present time, GRAFA can solve trilinear data by slicing the data into bilinear form.

7.3   SIMULTANEOUS ANALYSIS (ALS MCR)

Three-Way Analysis of Data with Common Spectra. Concentration profiles extracted by iterative target transformation factor analysis (ITTFA), evolving factor analysis (EFA), and window factor analysis (WFA) may contain ambiguities even when the constraints of nonnegativity, unimodality, and closure are applied. Such ambiguities, called "rotational" ambiguities often occur in titration experiments when one or more species exists at the beginning or at the end of the titration.

All of the above methods force the concentrations to zero at the edges of the titration windows.

Tauler and associates[12-14] have shown that rotational ambiguities can be removed by analyzing the spectra of several spectroscopic titrations, recorded at the same digitized wavelengths, using different initial concentrations of reagents. The first step in the procedure (labeled ALS MCR, alternating least-square multiple component regression) is to determine n, the number of species in each data set (see Section 4.3). The next step involves estimating the concentrations by subjecting each matrix, individually, to EFA. Although each matrix, D_1, D_2, ..., D_k, is factor analyzed separately, yielding different concentration profiles matrices, C_1, C_2, ..., C_k, the spectral absorptivity profile matrices, S_1, S_2, ..., S_k, should be the same because they represent the absorptivities of the chemical species. The spectral absorptivities are common among the data sets. In accord with Beers' law we expect

$$D_i = SC_i \qquad \text{where } i = 1, 2, \ldots, k \qquad (7.4)$$

Because each data matrix involves the same wavelengths, a single augmented matrix, D_{aug}, can be constructed by laying each matrix side by side as shown in (7.5):

$$D_{aug} = [\bar{D}_1 \ \bar{D}_2 \ \ldots \ \bar{D}_k] = S[C_1 \ C_2 \ \ldots \ C_k] = SC_{aug} \qquad (7.5)$$

In (7.5) $\bar{D}_1 \ \bar{D}_2 \ \ldots \bar{D}_k$ represent the reproduced data matrices based on n abstract factors. Since S is common to the augmented matrix, (7.5) represents the complete solution, where C_{aug} is a truer representation of the concentration profiles.

These profiles are assembled in accord with (7.5) and a new spectral matrix is determined by rearranging (7.5) to:

$$S = D_{aug}C_{aug}^+ \qquad (7.6)$$

A new estimate of C_{aug} is obtained by inserting the new S into (7.6):

$$C_{aug} = S^+D_{aug} \qquad (7.7)$$

New estimates for S and C_{aug} are obtained by repeated applications of (7.6) and (7.7) while subjecting the solutions to the following constraints:

Concentration Constraints

1. *Nonnegativity.* All concentrations must be equal to or greater than zero.
2. *Unimodality.* Each concentration profile must have a unimodal shape.
3. *Closure.* For, at least, one constituent, the sum of its concentrations must equal its total concentration at the titration point.

4. *Selectivity.* Species known to be absent in a specific titration experiment must have a zero concentration.

5. *Correspondence.* When the number of species differs between titrations, the algorithm must consider such differences.

Spectral Constraints

1. *Nonnegativity.* All absorbances must be equal to or greater than zero.

2. *Constant Spectra.* The species of a given component must be the same in all titration sets.

The results of a model study,[15] involving a three-component system, is presented in Figure 7.2. The solid lines represent the true concentration profiles of one of the three different data sets. The broken lines represent the profiles obtained from individual analyses of each data set. At high pH, the predicted concentrations depart greatly from the true values. The plus signs (+) represent the ALS MCR profiles obtained using augmented matrices after iterative application of (7.6) and (7.7). These profiles are in agreement with expectations.

This model-free method was used by Saurina et al.[16] to investigate the first-order decomposition of 1,2-naphthoquinone-4-sulphonate (NQS), a complicated double

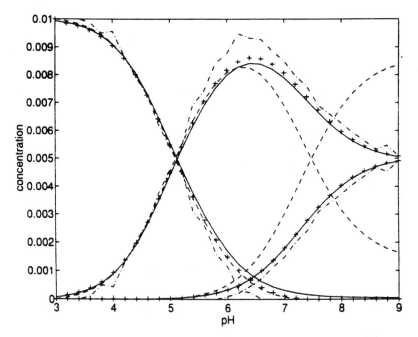

Figure 7.2   Comparison of recovered (- - individual analysis) (++ augmented data analysis) and true (—) concentration profiles pertaining to one of three different data matrices. [Reprinted with permission from R. Tauler, A. Smilde, and B. Kowalski, *J. Chemometrics*, **9**, 31 (1995). Copyright 1995 © John Wiley & Sons, Ltd.]

decomposition that involved four species in equilibria, as depicted in Figure 7.3. The kinetic process was carried out at three different pH (9.4, 10.5, and 13.3). Each run was conducted with a stopped flow system equipped with a diode array spectrophotometer that recorded spectra from 290 to 590 nm every 2 nm.

Individual analysis of each matrix (D_1, D_2, and D_3) yielded incorrect results due to rotational ambiguity. However, the three matrices share a common spectral matrix, **S**. When the data were analyzed using augmented matrices, the results shown in Figure 7.4 were obtained. From the three concentration matrices portrayed in Figure 7.4 (C_1, C_2, and C_3) the pK_a of NQS was determined to be 10.4, 10.7, and 10.4, respectively, values in excellent agreement with literature.

One of the most important features of this study is the fact that the data of matrix D_2, corresponding to pH 10.5, exhibits rank deficiency when factor analyzed separately. This is caused by closure. In this case, closure results from the double equilibria portrayed in Figure 7.3. These equilibria place a severe restriction on the concentrations of the four species. Namely,

$$K_{12} = \frac{c_2}{c_1} \quad \text{and} \quad K_{34} = \frac{c_4}{c_3} \tag{7.8}$$

Hence,

$$c_2 = K_{12}c_1 \quad \text{and} \quad c_4 = K_{34}c_3 \tag{7.9}$$

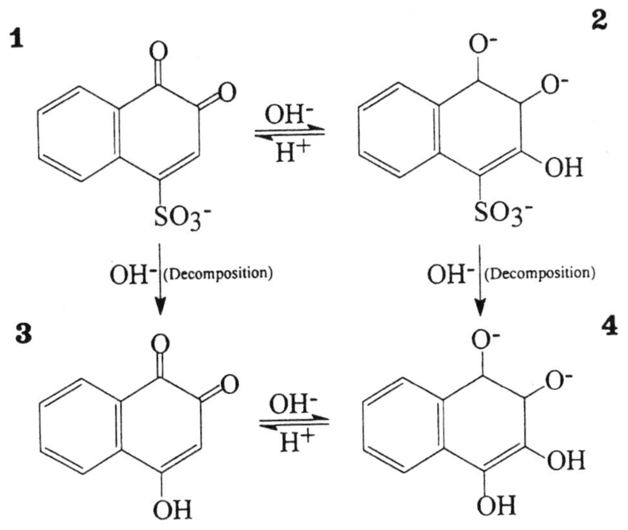

Figure 7.3   Kinetic scheme portraying the decomposition of NQS in basic media. [Reprinted with permission from J. Saurina, S. Hernández-Cassou, R. Tauler, and A. Izquierdo-Ridorsa, *J. Chemometrics*, **12**, 183 (1998). Copyright 1998 © John Wiley & Sons, Ltd.]

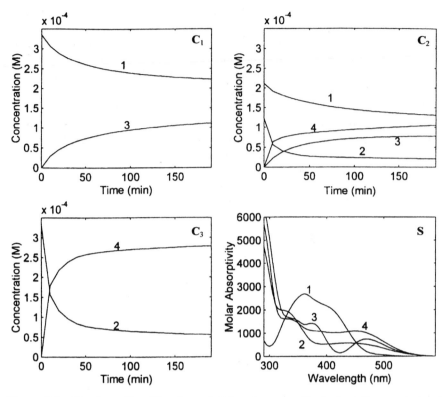

Figure 7.4 Kinetic profiles (C_1, C_2, and C_3) and spectra (S) obtained from analysis of augmented matrices. The species (1, 2, 3, and 4) are identified in Figure 7.3. [Reprinted with permission from J. Saurina, S. Hernández-Cassou, R. Tauler, and A. Izquierdo-Ridorsa, *J. Chemometrics*, **12**, 183 (1998). Copyright 1998 © John Wiley & Sons, Ltd.]

At pH 9.4 only species 1 and 3 coexist, hence D_1 has rank 2. At pH 13.3 only species 2 and 4 coexist, hence D_3 has rank 2. Because all four species are present at pH 10.5, one might expect the rank of D_2 to be 4, but the rank is 2 because the concentrations are not linearly independent, as shown in (7.9). In other words, although there are four species with four unique spectra, the data can be reduced to two independent variables as derived in (7.10):

$$d_2 = s_1 c_1 + s_2 c_2 + s_3 c_3 + s_4 c_4$$
$$= (s_1 + s_2 K_{12})c_1 + (s_3 + s_4 K_{34})c_3$$
$$= s_{12} c_1 + s_{34} c_3 \tag{7.10}$$

where s_{12} and s_{34} are hybrid spectra.

By combining the three data sets in the form of an augmented matrix the closure is broken, as clearly illustrated in this study.

Three-Way Analysis of Data with Common Concentrations. Tauler and co-workers[17] have developed and applied the ALS MCR technique to analyze three-way data that share common concentration profiles. The method was applied to a single acid–base titration of salicylic acid monitored by emission spectrofluorimetry at three different excitation wavelengths, thus generating three different data matrices with the same digitized row and column designations. Most importantly, these matrices have a unique trilinear structure. This unique situation permits the unambiguous extraction of the spectroscopic and concentration profiles of the underlying species as shown below.

To start, each individual matrix is subjected to rank analysis, evolving factor analysis and alternating least squares with natural constraints such as nonnegativity, unimodality, and closure. This provides initial estimates of the spectral profiles indigenous to the three matrices. These profiles, however, still suffer from rotational ambiguities.

The rotational ambiguities are removed by considering the trilinear nature of the data. The data matrices, as well as the spectral profiles, are assembled by stacking them one on top of the other to form augmented matrices, $\bar{\mathbf{D}}_{aug}$ and $\mathbf{S}_{aug}$.

$$\bar{\mathbf{D}}_{aug} = \begin{bmatrix} \bar{\mathbf{D}}_1 \\ \bar{\mathbf{D}}_2 \\ \dots \\ \bar{\mathbf{D}}_k \end{bmatrix} = \begin{bmatrix} \mathbf{S}_1 \\ \mathbf{S}_2 \\ \dots \\ \mathbf{S}_k \end{bmatrix} \mathbf{C} = \mathbf{S}_{aug}\mathbf{C} \tag{7.11}$$

This is possible because the column designations correspond to pH titration measurements while the row designations correspond to the emission intensities. In this case k equals three because there are three different excitation wavelengths (280, 297, and 314 nm). The three matrices share a common concentration matrix because they were generated by the same titration experiment.

Equation (7.11) is subjected to ALS optimization with constraints. These include nonnegativity, unimodality, selectivity, closure, and an additional constraint imposed by the trilinearity of the data. Figure 7.5 illustrates the mathematical details of the trilinearity constraint. Every column of $\mathbf{S}_{aug}(3NW \times NS)$ is folded into a new matrix, $\mathbf{SR}$, $(NW \times 3)$, where NW is the number of different emission wavelengths and NS is the number of titration points. Because the three columns that comprise $\mathbf{SR}$ share the same concentration distribution, this matrix should have rank one. For this reason only the score and loading of the first principal factor obtained by decomposition of $\mathbf{SR}$ have chemical significance. The outer product of this score and loading yields a better representation of $\mathbf{SR}$. The recalculated $\mathbf{SR}$ is then unfolded and inserted in $\mathbf{S}_{aug}$ in place of the original column.

Figure 7.6 shows the results obtained when this method was applied to acid–base titration of salicylic acid. Although three species, H_2sal, $Hsal^-$, and sal^{2-}, are

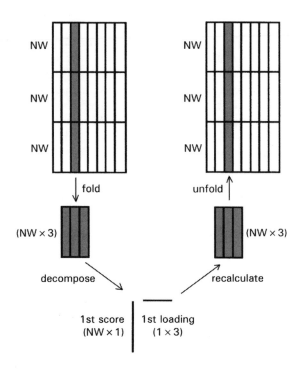

Figure 7.5 Trilinearity constraint. Three spectra in the third column of S_{aug} are folded to give **SR**, which is decomposed into scores and loadings. The first score and the first loading are used to recalculate **SR**, which is then unfolded and inserted into S_{aug} in place of the original column. [Reprinted with permission from R. Tauler, I. Marques, and E. Casassas, *J. Chemometrics*, **12**, 55 (1998). Copyright 1998 © John Wiley & Sons, Ltd.]

present, only two species are detected because the fully protonated species, H_2sal, is not fluorescent. The concentration profile of this species was obtained by mass balance and not mass action as required by traditional methods.

7.4   THREE-MODE FACTOR ANALYSIS

In 1964 Tucker[18] solved the problem of three-mode factor analysis (TMFA) by converting the data "box" into the conventional two-mode "plane." The terminology and notation of Tucker are particularly useful and are used throughout this exposition. Before discussing three-mode analysis, it is instructive to apply the Tucker notation to the two-mode case. Accordingly, the data matrix, labeled $_iX_j$, can be factored into a product of three matrices:

$$_iX_j \approx _i\tilde{X}_j = _iA_nG_nB_j \tag{7.12}$$

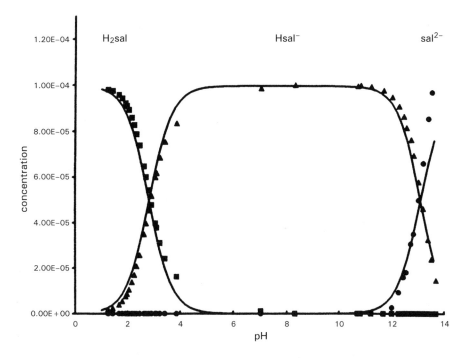

Figure 7.6   Concentration profiles of H_2sal, $Hsal^-$, and sal^{2-} (symbols) determined by ALS MCR compared to theoretical calculations (lines) based on $pK_{a1} = 2.82$ and $pK_{a2} = 13.09$. [Reprinted with permission from R. Tauler, I. Marques, and E. Casassas, *J. Chemometrics*, **12**, 55 (1998). Copyright 1998 © John Wiley & Sons, Ltd.]

This equation is equivalent to the singular value decomposition (SVD) expression (2.5),

$$\mathbf{D} \approx \bar{\mathbf{D}} = \bar{\mathbf{U}}\bar{\mathbf{S}}\bar{\mathbf{V}}' \qquad (2.5),\ (7.13)$$

where

$$\bar{\mathbf{D}} = {}_i\tilde{\mathbf{X}}_j \qquad \bar{\mathbf{U}} = {}_i\mathbf{A}_n \qquad \bar{\mathbf{S}} = {}_n\mathbf{G}_n \qquad \bar{\mathbf{V}}' = {}_n\mathbf{B}_j \qquad (7.14)$$

Here $\mathbf{A}$ is the $(i \times n)$ orthonormal eigenvector matrix of $\mathbf{XX}'$, $\mathbf{B}$ is the $(n \times j)$ orthonormal eigenvector matrix of $\mathbf{X}'\mathbf{X}$, and $\mathbf{G}$ is the $(n \times n)$ "core" matrix, a diagonal matrix composed of the singular values, that is, the square roots of the eigenvalues. The Tucker subscripts are designed to provide additional information. They are used to identify both the general mode and the mode to which an element belongs and as a variable for the elements of the mode. The presubscript designates the row mode. The postsubscript designates the column mode. Reversal of the subscripts indicates matrix transposition, for example, ${}_j\mathbf{X}_i$ is the transpose of ${}_i\mathbf{X}_j$.

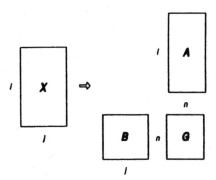

Figure 7.7   Illustration of matrix decomposition by principal component analysis of a two-mode matrix **X**.

The difference between two-mode and three-mode analysis is illustrated in Figures 7.7 and 7.8. In two-mode analysis the "plane" of data, **X**, is decomposed into two principal component matrices (**A** with dimensions $i \times n$, and **B** with dimensions $n \times j$) and a core matrix (**G** with dimensions $n \times n$). In three-mode analysis the "box" of data is decomposed into three two-way matrices, $_i\mathbf{A}_m$, $_j\mathbf{B}_p$, and $_k\mathbf{C}_q$, linked to a three-way core, $\mathbf{G}_{mpg}$.

An excellent exposition of TMFA has been presented by Zeng and Hopke,[19] which follows summarily. TMFA is based on expressing a data point, x_{ijk}, as a product function of factors associated with the three modes and a "core" value; namely,

$$x_{ijk} = \sum_m \sum_p \sum_q a_{im} b_{jp} c_{kq} g_{mpg} \tag{7.15}$$

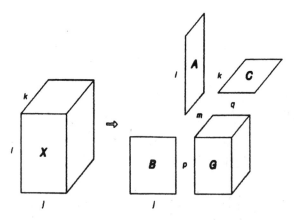

Figure 7.8   Illustration of matrix decomposition by three-mode principal component analysis of a three-mode matrix **X**.

In matrix notation (7.15) can be expressed as

$$_i\mathbf{X}_{(jk)} \approx {}_k\tilde{\mathbf{X}}_{(jk)} = {}_i\mathbf{A}_m\mathbf{G}_{(pq)}({}_p\mathbf{B}_j \otimes {}_q\mathbf{C}_k) \tag{7.16}$$

Matrix $_i\mathbf{X}_{(jk)}$ is the three-mode data table flattened into a two-way data table by slicing the data box into parallel sheets and unfolding the sheets as shown in Figure 7.9 and sequentially placing the k sheets, each with order i by j, side by side so the resulting array has order i by (jk).

The Kronecker product $_p\mathbf{B}_j \otimes {}_q\mathbf{C}_k$ is a supermatrix containing submatrices that are proportional to $_q\mathbf{C}_k$ and is defined as

$$_p\mathbf{B}_j \otimes {}_q\mathbf{C}_k \equiv \begin{bmatrix} (b_{11q}\mathbf{C}_k) & (b_{12q}\mathbf{C}_k) & \cdots \\ (b_{21q}\mathbf{C}_k) & (b_{22q}\mathbf{C}_k) & \cdots \\ \vdots & \vdots & \vdots \end{bmatrix} \tag{7.17}$$

The orthonormal eigenvector matrices $_i\mathbf{A}_m$, $_j\mathbf{B}_p$, and $_k\mathbf{C}_q$ are obtained by decomposing the respective covariance matrices $_i\mathbf{M}_i$, $_j\mathbf{P}_j$, and $_k\mathbf{Q}_k$ defined as

$$\begin{aligned} _i\mathbf{M}_i &\equiv {}_i\mathbf{X}_{(jk)}\mathbf{X}_i \\ _j\mathbf{P}_j &\equiv {}_j\mathbf{X}_{(ik)}\mathbf{X}_j \\ _k\mathbf{Q}_k &\equiv {}_k\mathbf{X}_{(ij)}\mathbf{X}_k \end{aligned} \tag{7.18}$$

Kroonenberg and DeLeeuw[20] developed the TUCKALS3 algorithm for decomposing these matrices by an alternating least-squares (ALS) approach that minimizes the mean-square loss function, f, defined as

$$f(\mathbf{A}, \mathbf{B}, \mathbf{C}, \mathbf{G}) \equiv \|\mathbf{x} - \tilde{\mathbf{x}}\|^2 \tag{7.19}$$

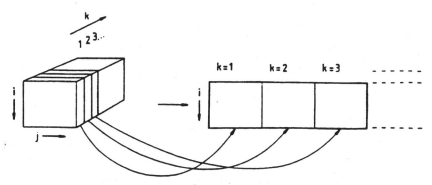

Figure 7.9 Unfolding a three-dimensional data matrix. [From S. Wold, P. Geladi, K. Esbensen, and J. Ohman, *J. Chemometrics*, **1**, 41 (1987). Copyright 1987 © John Wiley & Sons, Ltd.]

Because of the Kronecker product term, conventional methods cannot be used to minimize this function. The ALS method begins with arbitrary estimations for orthonormal matrices $\mathbf{A}$, $\mathbf{B}$, and $\mathbf{C}$, labeled $^0\mathbf{A}$, $^0\mathbf{B}$, and $^0\mathbf{C}$ (where the presuperscripts indicate the iteration number). The core matrix $\mathbf{G}$ is readily calculated by

$$\mathbf{G} = \mathbf{AX}(\mathbf{B} \otimes \mathbf{C}) \tag{7.20}$$

This equation results because the columns within matrices of $\mathbf{A}$, $\mathbf{B}$, and $\mathbf{C}$ are orthonormal. A new $\mathbf{A}$, labeled $^1\mathbf{A}$, is obtained by minimizing f while holding $^0\mathbf{A}$ and $^0\mathbf{B}$ constant (see A substep below). A new $\mathbf{B}$, labeled $^1\mathbf{B}$, is obtained by minimizing f while holding $^1\mathbf{A}$ and $^0\mathbf{C}$ constant (see B substep below). A new $\mathbf{C}$, labeled $^1\mathbf{C}$, is obtained by holding $^1\mathbf{A}$ and $^1\mathbf{B}$ constant (see C substep). This process is repeated again and again until convergence is achieved.

A substep:

$$^n\mathbf{M} = \mathbf{X}(^n\mathbf{B} \quad ^n\mathbf{B'} \otimes^n \mathbf{C} \quad ^n\mathbf{C})\mathbf{X'} \quad \text{with } \mathbf{X} = {}_i\mathbf{X}_{(jk)}$$
$$^{n+1}\mathbf{A} = {}^n\mathbf{M} \quad ^n\mathbf{A}(_n\mathbf{A'} \quad ^n\mathbf{M}^2 \quad ^n\mathbf{A})^{-1/2} \tag{7.21}$$

B substep:

$$^n\mathbf{P} = \mathbf{X}(^n\mathbf{C} \quad ^n\mathbf{C'} \otimes {}^{n+1}\mathbf{A} \quad ^{n+1}\mathbf{A})\mathbf{X'} \quad \text{with } \mathbf{X} = {}_j\mathbf{X}_{(ki)}$$
$$^{n+1}\mathbf{B} = {}^n\mathbf{P} \quad ^n\mathbf{B}(^n\mathbf{B'} \quad ^n\mathbf{P}^2 \quad ^n\mathbf{B})^{-1/2} \tag{7.22}$$

C substep:

$$^n\mathbf{Q} = \mathbf{X}(^{n+1}\mathbf{A} \quad ^{n+1}\mathbf{A'} \otimes {}^{n+1}\mathbf{B} \quad ^{n+1}\mathbf{B})\mathbf{X'} \quad \text{with } \mathbf{X} = {}_k\mathbf{X}_{(ij)}$$
$$^{n+1}\mathbf{C} = {}^n\mathbf{Q} \quad ^n\mathbf{C}(^n\mathbf{C'} \quad ^n\mathbf{Q}^2 \quad ^n\mathbf{C})^{-1/2} \tag{7.23}$$

Equation (7.19) can be expressed as

$$f = \sum_i \sum_j \sum_k (x_{ijk} - \tilde{x}_{ijk})^2 = \sum_i \sum_j \sum_k x_{ijk}^2 - \sum_i \sum_j \sum_k \tilde{x}_{ijk}^2 \tag{7.24}$$

These sums of squares (SS) can be represented in statistical notation as

$$\text{SS(residual)} = \text{SS(total)} - \text{SS(fit)} \tag{7.25}$$

where

$$\text{SS(fit)} = \sum_{m=1}^{s} \sum_{p=1}^{t} \sum_{q=1}^{u} g_{mpq}^2 \tag{7.26}$$

For each mode,

$$\text{SS(mode 1)} = \sum_{m=1}^{s} \alpha_m$$

$$\text{SS(mode 2)} = \sum_{p=1}^{t} \beta_p \qquad (7.27)$$

$$\text{SS(mode 3)} = \sum_{q=1}^{u} \gamma_q$$

where α_m, β_p, and γ_q are the eigenvalues of the respective modes, and the sums are taken over the number of factors (s, t, and u) retained in each respective mode. The sums in (7.27) are analogous to explained variance in conventional factor analysis, but the sum of the three equations in (7.27) does not equal the sum in (7.26).

The solutions of **A**, **B**, and **C** strongly depend on the number of factors assigned to the modes. If the number of factors assigned to one mode is increased by one unit, the solutions of **A**, **B**, and **C** may alter considerably. Care must be taken to use the correct number of factors for each mode, a task that is formidable compared to two-way factor analysis.

Interpreting the results requires rotation or transformation of the eigenvector axes. This is done by invoking square, nonsingular, transformation matrices $_m\mathbf{T}_m*$, $_p\mathbf{T}_p*$, and $_q\mathbf{T}_q*$ such that

$$_i\mathbf{A}_m\mathbf{T}_m* = {}_i\mathbf{A}_m*$$

$$_j\mathbf{B}_p\mathbf{T}* = {}_j\mathbf{B}_p* \qquad (7.28)$$

$$_k\mathbf{C}_q\mathbf{T}_q* = {}_k\mathbf{C}_q*$$

where $m*$, $p*$, and $q*$ signify the transformed modes. The core matrix becomes

$$_{m*}\mathbf{G}_{(p*q*)} = \left(_m\mathbf{T}_m*\right)_m^{-1}\mathbf{G}_{(pq)}\left[\left(_p*\mathbf{T}_p\right)^{-1} \otimes \left(_q*\mathbf{T}_q\right)^{-1}\right] \qquad (7.29)$$

After transformation, (7.16) becomes

$$_i\tilde{\mathbf{X}}_{(jk)} = {}_i\mathbf{A}_m*\mathbf{G}_{(p*q*)}\left(_p*\mathbf{B}_j \otimes {}_q*\mathbf{C}_k\right) \qquad (7.30)$$

Zeng and Hopke[19] applied TMFA to a simulated air pollution study based on Ontario, Canada. The three modes consisted of the following: mode 1 contained 20 chemical elements for each particulate sample; mode 2 listed 8 sampling sites; and mode 3 concerned 12 sampling time periods varying in wind directions and wind strengths. The data were generated from three particulate pollution sources, emulating the emissions from (1) local coal-fired power plants, (2) long-range power plants in the United States, and (3) a nickel smelter in Sudbury, Ontario. Thus the size of the data box was $20 \times 8 \times 12$.

The data were mean centered and scaled so the mean of each frontal plane equaled zero and the sum of squares per frontal plane equals unity. Each mode was examined individually after assigning a sufficiently large number of factors to the other modes in order to ensure that the other modes fit perfectly. Modes 1 and 3 required three factors while mode 2 required only two factors. The final SS(fit) accounted for 95% of the variance in the data.

To help interpret the factors, various abstract rotation schemes were tried. Varimax rotation without row normalization yielded the best agreement between the TMFA model and the known physical model.

The rotations portrayed in (7.28) require transformation matrices for the entire system, requiring simultaneous specifications for all sources of the factors. Unfortunately, at the present time, target testing for the existence of an individual factor has not been developed for TMFA.

Wold and co-workers[21] proposed that multimode matrices be decomposed by keeping one direction in the R-array distinct (called "objects") while treating the remaining $(R - 1)$ modes as direction variables. This can be accomplished by extracting vector components of the objects piecemeal, following the strategy of the nonlinear iterative partial least squares (NIPALS) algorithm (see Section 3.3.8). For an R-array $\mathbf{X}$:

$$\mathbf{X} = \mathbf{t}_1 \otimes \mathbf{P}_1 + \mathbf{t}_2 \otimes \mathbf{P}_2 + \mathbf{E} = \mathbf{T} \otimes \mathbf{P} + \mathbf{E} \qquad (7.31)$$

where the object vectors $\mathbf{t}_i$ are orthogonal to each other and the multimode matrices $\mathbf{P}_i$ are also mutually orthogonal. The object vectors are extracted until the norm of the residuals, $\|\mathbf{E}\|$, is sufficiently small as determined by cross validation or any other criteria.

$\mathbf{T}$ and $\mathbf{P}$ are R-mode analogs of the score and loading matrices. Note that if $R = 2$ then $\mathbf{X}$ is the usual two-mode matrix, $\mathbf{P}_i$ is a vector, and (7.31) reduces to the ordinary decomposition into a sum of outer products of two vectors.

Wold and co-workers[21] applied this technique to factor analyze six mixtures of anthracene and phenanthrene that were subjected to liquid chromatography with ultraviolet (UV) detection. Both of these components overlap in the two experimental domains. Ten wavelengths and 10 points from the time axis were selected, giving a $6 \times 10 \times 10$ three-way matrix. After variance scaling and subtracting the mean, two abstract factors described 93% of the variance whereas three factors described 99.6% of the variance.

Sanchez and Kowalski[22] and Burdick and co-workers[23] have independently shown that three-mode data arrays, which have the following trilinear form,

$$x_{ijk} = \sum_{m=1}^{n} a_{im} b_{jm} c_{km} \qquad (7.32)$$

have unique factor analytical decompositions that yield, directly, the true chemical factors responsible for the data. This is in marked contrast to two-mode factor analysis, which requires rotation of the abstract factors, a formidable task that is not always possible to achieve. The feasibility of this method was demonstrated with synthetic emission–excitation fluorescence spectra of three mixtures of anthracene, chrysene, and fluoranthene.

Burdick and co-workers[23] recorded and analyzed three-dimensional excitation–emission–frequency matrices generated from fluorescence spectra in which the measured emission intensity is a function of excitation wavelength, emission wavelength, and modulation frequency. Using data collected at three modulation frequencies (6, 18, and 30 MHz), they not only uniquely resolved the excitation–emission matrices of benzo[b]fluoranthene and benzo[k]fluoranthene from a single mixture containing only these two components, but they also obtained the fluorescence lifetimes of the two components.

7.5 PARAFAC

When a three-mode data array has the trilinear form depicted in Eq. (7.32), unique factors, representing true chemical factors, can be obtained directly by parallel factor analysis (PARAFAC)[3], also known as canonical decomposition (CANDECOMP).[4] This is in marked contrast to two-mode factor analysis and the Tucker models, which require rotation of the abstract factors, a formidable task that is not always possible to achieve.

A two-component three-way PARAFAC model is illustrated in Figure 7.10.[24] In this figure $\underline{\mathbf{X}}$ represents the data box; $\mathbf{a}_1$, $\mathbf{b}_1$, and $\mathbf{c}_1$ are the component vectors of the first factor; $\mathbf{a}_2$, $\mathbf{b}_2$, and $\mathbf{c}_2$ are the vectors associated with the second factor; and matrices $\mathbf{A}$, $\mathbf{B}$, and $\mathbf{C}$ are compilations of the respective vectors.

PARAFAC follows the *principle of parallel proportional profiles* proposed by Cattell in 1944.[25] According to this principle, a series of two-way data arrays that differ only in relative proportions will lead to a meaningful and unambiguous decomposition because the system has no rotational degrees of freedom. This constitutes the most important fundamental property of the PARAFAC algorithm.

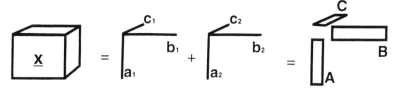

Figure 7.10   Two-component three-way PARAFAC model of a three-way array $\underline{\mathbf{X}}$. Vectors $\mathbf{a}_1$, $\mathbf{b}_1$, and $\mathbf{c}_1$ are component vectors of the first factor. Vectors $\mathbf{a}_2$, $\mathbf{b}_2$, and $\mathbf{c}_2$ are component vectors of the second factor. Matrices $\mathbf{A}$, $\mathbf{B}$ and $\mathbf{C}$ are compilations of the respective vectors. [Reprinted with permission from R. Bro, Multi-way Analysis in the Food Industry, Doctoral Thesis, University of Amsterdam (1998).]

For example, if matrix X_1 is composed of two factors, so that

$$X_1 = a_1 b_1' c_{11} + a_2 b_2' c_{12} \qquad (7.33)$$

and matrix X_2 is composed of the same two factors but proportionally different such that

$$X_2 = a_1 b_1' c_{21} + a_2 b_2' c_{22} \qquad (7.34)$$

where c_{11} is different from c_{21}, and c_{12} is different from c_{22}. The two matrices have the same (parallel) profiles but are proportionately different. These two matrices represent pages (sheets) of the box, which can be unfolded as shown in (7.35):

$$X = [X_1 \ X_2] = [a_1 \ a_2] \begin{bmatrix} b_1 c_{11} & b_1 c_{21} \\ b_2 c_{12} & b_2 c_{22} \end{bmatrix} \qquad (7.35)$$

Equation (7.36) is a general definition of the Khatri–Rao product.[26] It applies to two matrices that have the same number of columns, n.

$$C| \otimes |B = [c_1 \otimes b_1 \ c_2 \otimes b_2 \ \ldots \ c_n \otimes b_n] \qquad (7.36)$$

In our example, n equals two, $c_1 = [c_{11} \ c_{21}]'$ and $c_2 = [c_{12} \ c_{22}]'$. Using the Khatri–Rao notation, (7.35) can be written in a very compact form, namely,

$$X^{(I \times JK)} = A(C| \otimes |B)' \qquad (7.37)$$

Superscript $(I \times JK)$ specifies the direction of the unfolding as well as the size of the matrix. The data box can be unfolded in other directions, leading to

$$X^{(J \times IK)} = B(C| \otimes |A)' \qquad (7.38)$$
$$X^{(K \times IJ)} = C(B| \otimes |A)' \qquad (7.39)$$

Simultaneous solution of these three equations uniquely determines matrices A, B, and C. This can be accomplished by applying ALS to three unfolded matrices as follows:

Step 1: Determine n.

Step 2: Initialize B and C.

Step 3: Let $Z = (C| \otimes |B)'$. Find $A = X^{(I \times JK)} Z^+$.

Step 4: Let $Z = (C| \otimes |A)'$. Find $B = X^{(J \times IK)} Z^+$.

Step 5: Let $Z = (B| \otimes |A)'$. Find $C = X^{(K \times IJ)} Z^+$.

Step 6: Go to step 3 and repeat steps 3, 4, 5, and 6 until convergence.

The time required for convergence strongly depends upon the initial values of **B** and **C**. Poor estimates can also lead to a false solution because the ALS can converge to a local minimum instead of the global minimum. According to Harshman and Lundy,[27] the global solution can be found by repeating the algorithm using several different random initializations. If the same results are obtained, then the solution is most likely to be the global solution. Other starting points have been proposed.[28,29]

Kruskal[30] has shown that PARAFAC will yield unique solutions if

$$ n \leq \frac{n_A}{2} + \frac{n_B}{2} + \frac{n_C}{2} - 1 \qquad (7.40) $$

In (7.40) n is the number of PARAFAC components, and n_A, n_B, and n_C are the ranks of matrices **A**, **B**, and **C**. An excellent example of the Kruskal uniqueness rule is illustrated in Figure 7.11.[24] The emission–excitation fluorescence spectra of two samples containing different amounts of tryptophan, tyrosine, and phenylalanine were recorded and assembled into a three-way data box. When the data box was subjected to PARAFAC the emission spectra (bottom left) of the three components were obtained. These spectra coincide with the emission spectra of the pure

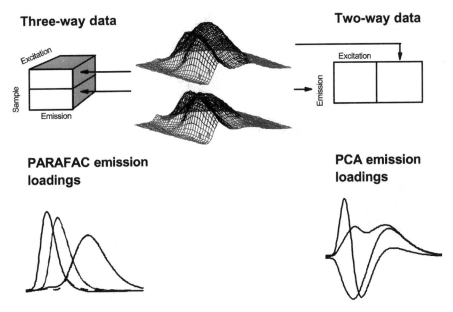

Three-way data

Excitation

Sample

Emission

Two-way data

Excitation

Emission

PARAFAC emission loadings

PCA emission loadings

Figure 7.11 Fluorescence spectra (excitation–emission) of two samples containing different amounts of tryptophan, tyrosin, and phenylalanine, assembled as a three-way data block and a two-way data plane. PARAFAC emission loadings from decomposition of the three-way data are shown in the bottom left corner. Orthogonal PCA emission loadings from decomposition of the two-way data are shown in the bottom right corner. [Reprinted with permission from R. Bro, Multi-way Analysis in the Food Industry, Doctoral Thesis, University of Amsterdam (1998).]

components (dashed lines). Two-way analysis of the same data produced the orthogonal set of emission spectra shown at the lower right. In this case, $n_A = 2$ (two samples), $n_B = 3$ (three emission components, and $n_C = 3$ (three excitation components). According to (7.40), PARAFAC will yield a unique solution when n is set equal to 3. Because n corresponds to the number of chemical components in the two samples, the results are reliable.

The uniqueness rule has been generalized by Sidiropoulos and Bro[31] to encompass any number of modes. For N-mode data, the rule is

$$n \leq \frac{1}{2}\left[1 - N + \sum_{i=1}^{N} n_i \right] \qquad (7.41)$$

For three-way data, $N = 3$, (7.41) reduces to (7.40).

The ranks of the modes can be deduced by the methods described in Section 4.3. However, because of the complexity of multimode data, these techniques often give ambiguous results. In order to determine the proper rank to be used in PARAFAC, the *spit-half analysis* of Harshman et al.[32] is very useful. If the data is split into two or more boxes, and each box is independently subjected to PARAFAC, the loadings should be the same because PARAFAC yields unique solutions. If too many or too few factors are employed, the loadings will be significantly different. To avoid unfortunate splittings, where some factor is missing in one group but present in another, the samples should be divided into two groups: I and II. Group III is prepared from the first half of I and II, and group IV is prepared from the last half of I and II. Because the four groups are pairwise independent, the correct solution can be established.

Problems. The PARAFAC algorithm has practical limitations. When two or more factors are equal or relatively equal, PARAFAC will lead to degenerate solutions that are unstable and unreliable. Degeneracy occurs when the factors are highly correlated, when too many factors are extracted, when the system is not truly trilinear, or when the data is improperly preprocessed. Degenerate solutions are characterized by two or more equally shaped loading vectors in all modes, with at least one negatively correlated.

PARAFAC can get caught in a *swamp*, a situation where an ALS algorithm moves extremely slowly.[33] Degeneracy often leads to a swamp. A swamp can occur when there is no degeneracy. For example, a poor starting approximation for matrices **B** and **C** (step 2) can lead to a parabolic localized least-squares minimum that oscillates rapidly between the true solution and the initial solution.

The techniques described in the following sections attempt to overcome these problems. Bro[34] has written an excellent tutorial that summarizes many of the aspects of PARAFAC not discussed in this monogram, such as preprocessing, restricted models, compressing the modes, initialization, stopping criterion, missing data points, leverage, residuals, iterative extrapolation, and so forth.

Constraining the Solution. The imposition of appropriate constraints on the PARAFAC algorithm has been used to confine the solution to a more realistic and easily interpretable output. Forcing the solution to nonnegativity effectively prevents negative correlations. Nonnegative solutions are based on theoretical and practical considerations. Certain types of chemical parameters are known to be nonnegative (e.g., pH, absorbance, emission, fluorescence, concentration, elution). Nonnegativity has been shown to remove degeneracy and to give unique solution models.[35] A fast nonnegativity-constrained least-squares algorithm has been introduced by Bro and De Jong.[36]

For chromatographic data, a unimodal constraint is applicable because each eluting component gives rise to a band with a single maximum. Furthermore, these bands are nonnegative, so a dual constraint is valid. A fast unimodal, nonnegative-constrained PARAFAC algorithm has been introduced by Bro and Sidiropoulos,[37] who demonstrated its utility in fluorescence spectroscopy and flow injection analysis.

Preprocessing. In comparison to two-way analysis, preprocessing three-way data is much more complicated.[38] *Single centering* is accomplished by subtracting the average value of all elements of the mode being centered. If more than one mode is to be centered, each mode is centered one at a time. Similarly, if scaling is desired, instead of scaling columnwise, the whole sheet must be scaled uniformly. If several modes are to be scaled, this must be done iteratively. However, the final results depend upon the order of the modes. Furthermore, convergence is not always achieved. Complications arising from centering and scaling can be avoided by using weighted regression.

7.5.1 Alternating Trilinear Decomposition

When two factors are similar in one mode but different in another mode, the PARAFAC algorithm may not converge to a chemically meaningful solution and the convergence may be extremely slow. Wu and co-workers[39] developed a method called alternating trilinear decomposition (ATLD) that overcomes these difficulties. Instead of unfolding $\underline{\mathbf{X}}$, they considered each page of each mode, developing the following set of equations appropriate for each mode:

$$\mathbf{X}_{i..} = \mathbf{B}\,\mathrm{diag}(\mathbf{a}_i^{\mathrm{T}})\mathbf{C}^{\mathrm{T}} \qquad i = 1,\,2,\,\ldots,\,I \qquad (7.42)$$

$$\mathbf{X}_{.j.} = \mathbf{C}\,\mathrm{diag}(\mathbf{b}_j^{\mathrm{T}})\mathbf{A}^{\mathrm{T}} \qquad j = 1,\,2,\,\ldots,\,J \qquad (7.43)$$

$$\mathbf{X}_{..k} = \mathbf{A}\,\mathrm{diag}(\mathbf{c}_k^{\mathrm{T}})\mathbf{B}^{\mathrm{T}} \qquad k = 1,\,2,\,\ldots,\,K \qquad (7.44)$$

The notation $\mathrm{diag}(\mathbf{a}_i^{\mathrm{T}})$ denotes a diagonal matrix of order $n \times n$ with diagonal elements $\mathbf{a}_i^{\mathrm{T}}$. In these expressions $\mathbf{X}_{i..}$ is the ith page, a $J \times K$ matrix. There are I pages in this mode. Similarly, $\mathbf{X}_{.j.}$ is the jth page, a $K \times I$ matrix. There are J pages in this mode. $\mathbf{a}_i^{\mathrm{T}}$, $\mathbf{b}_j^{\mathrm{T}}$, and $\mathbf{c}_k^{\mathrm{T}}$ represent, respectively, the ith, jth and kth rows of $\mathbf{A}$, $\mathbf{B}$, and **C**. Each of these row vectors contains n elements, where n is the maximum rank determined by (7.45).

$$n \geq \text{rank}(\underline{\mathbf{X}}) = \max\left\{\text{rank}(\mathbf{X}_P^I),\ \text{rank}(\mathbf{X}_P^J),\ \text{rank}(\mathbf{X}_P^K)\right\} \qquad (7.45)$$

In (7.45), $\mathbf{X}_P^I$, $\mathbf{X}_P^J$, and $\mathbf{X}_P^K$ represent two-way matrices obtained by unfolding $\underline{\mathbf{X}}$ as follows:

$$\mathbf{X}_P^I = [\mathbf{X}_{..1}\ \mathbf{X}_{..2}\ \ldots\ \mathbf{X}_{..K}] \qquad (7.46)$$

$$\mathbf{X}_P^J = [\mathbf{X}_{1..}\ \mathbf{X}_{2..}\ \ldots\ \mathbf{X}_{I..}] \qquad (7.47)$$

$$\mathbf{X}_P^K = [\mathbf{X}_{.1.}\ \mathbf{X}_{.2.}\ \ldots\ \mathbf{X}_{.J.}] \qquad (7.48)$$

The ATLD algorithm invokes the ALS principle and entails the following procedure based on (7.42), (7.43), and (7.44):

Step 1: Determine n.

Step 2: Initialize **B** and **C**.

Step 3: Find $\mathbf{a}_i^T = \text{diag}(\mathbf{B}^+\mathbf{X}_{i..}(\mathbf{C}')^+)$, for $i = 1,\ 2,\ldots,I$.

Step 4: Find $\mathbf{b}_j^T = \text{diag}(\mathbf{C}^+\mathbf{X}_{.j.}(\mathbf{A}')^+)$, for $j = 1,\ 2,\ldots,J$.

Step 5: Find $\mathbf{c}_k^T = \text{diag}(\mathbf{A}^+\mathbf{X}_{..k}(\mathbf{B}')^+)$, for $k = 1,\ 2,\ldots,K$.

Step 6: Go to step 3 and repeat steps 3, 4, 5, and 6 until convergence.

ATLD was found to converge 10 times faster than PARAFAC and to provide more accurate results.

7.5.2 Alternating Slicewise Diagonalization

The results gleaned from PARAFAC and ATLD are sensitive to the value of n used in the computations. When the model dimensionality, n, is incorrect, PARAFAC can get trapped in a computational swamp, giving meaningless results.[35] Jiang and co-workers[40] devised a technique that gives stable results provided that n used in the computations is equal to or greater than the actual number of components. The method, called alternating slicewise diagonalization (ASD), not only circumvents the two-factor degeneracy problem that arises in PARAFAC but also has a much faster convergence rate.

The underlying principle of ASD is "to find two matrices such that the matrix slices along some order, when multiplied by these two matrices at two sides, are all fitted to diagonal form in the least-squares sense." A slice, in this study, refers to a page with dimensions $I \times J$ of $\mathbf{X}_{..k}$. The method assumes that matrices **A** and **B** have full column rank and that the profiles of each component in **C** are linearly independent. This is stronger than the Kruskal condition displayed in (7.40), but it is the most common situation explored in chemistry.[41] Because the algorithm is quite intricate, those interested in learning the details should consult the original study.[39]

To demonstrate the advantages of ASD, Jiang and co-workers[40] studied a $31 \times 29 \times 6$ data array containing the excitation–emission fluorescence of six samples containing three components: (1) acridine red, (2) sodium fluorescein, and (3) rhodamine B. The resolved profiles of the excitation and emission processes are illustrated in Figure 7.12. Although there were only three components, the analysis was based on four components. The PARAFAC results are shown as solid lines in Figures 7.12*a* and 7.12*b*. The experimental profiles are shown as dotted lines. Clearly, an excessive number of factors produces unsatisfactory profiles. The ASD results are shown in Figures 7.12*c* and 7.12*d* as solid lines. The extracted profiles (solid lines) of the three chemicals correspond closely to the actual profiles (dotted lines). The fourth profile represents a ghost component that does not exist but is simply an aberration produced by overspecifying the factor space. The ghost is easily identified by the large negative values.

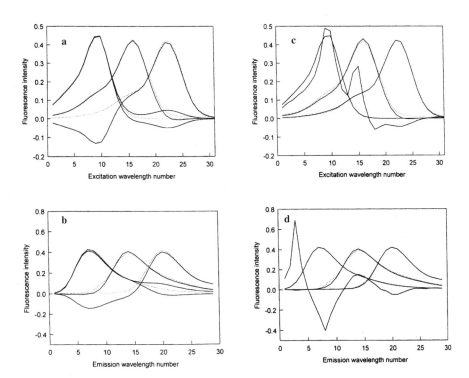

Figure 7.12   Spectral profiles of real fluorescence mixtures resolved by PARAFAC (*a* and *b*) and by ASD (*c* and *d*) when the component number was set to four. Dotted lines are the experimental measurements of the pure components. [Reprinted with permission from J. Jiang, H. Wu, Y. Li, and R. Yu, *J. Chemometrics*, **14**, 15 (2000). Copyright 2000 © John Wiley & Sons, Ltd.]

7.5.3   Pseudo–Alternating Least Squares

Because ATLD requires symmetry in the three modes, ATLD assigns the maximum rank to each mode [see (7.45)]. Furthermore, ATLD can lead to multiple solutions caused by convergence to local optimums. To overcome these deficiencies, Chen, Li, and Yu[42] developed an improved algorithm called pseudo-alternating least squares (PALS). By combining the PARAFAC loss function with objective functions based on (7.42), (7.43), and (7.44), the problem reduces to finding the minimum in functions such as:

$$\min \sum_{k=1}^{K} (\|\mathbf{X}_{..k} - \mathbf{A} \operatorname{diag}(\mathbf{c}_k^T \mathbf{B}^T\| + \lambda \|\mathbf{X}_{..k}(\mathbf{B}^T)^+ - \hat{\mathbf{A}} \operatorname{diag}(\mathbf{c}_k^T)\|) \tag{7.49}$$

where λ is an adjustable parameter. Such functions do not suffer from multiple optimums nor any symmetry constraint.

Because there is no straightforward optimization procedure, three different objective functions with interconnecting relationships were alternately minimized:

$$S(\mathbf{C}) = \sum_{k=1}^{K} (2\|\mathbf{X}_{..k} - \mathbf{A} \operatorname{diag}(\mathbf{c}_k^T)(\mathbf{B}^T)^+\| + \lambda(\|\mathbf{X}_{..k}(\mathbf{B}^T)^+ - \mathbf{A} \operatorname{diag}(\mathbf{c}_k^T)\|$$
$$+ \|\mathbf{A}^+\mathbf{X}_{..k} - \operatorname{diag}(\mathbf{c}_k^T)\mathbf{B}^T\|)) \tag{7.50}$$

Minimize $S(\mathbf{C})$, holding $\mathbf{A}$ and $\mathbf{B}$ constant, to find $\mathbf{C}$.

$$S(\mathbf{A}) = \sum_{k=1}^{K} (\|\mathbf{X}_{..k} - \mathbf{A} \operatorname{diag}(\mathbf{c}_k^T)(\mathbf{B}^T)^+\| + \lambda \|\mathbf{X}_{..k}(\mathbf{B}^T)^+ - \mathbf{A} \operatorname{diag}(c_k^T)\|) \tag{7.51}$$

Find $\mathbf{A}$ by minimizing $S(\mathbf{A})$ while holding $\mathbf{B}$ and $\mathbf{C}$ constant.

$$S(\mathbf{B}) = \sum_{k=1}^{K} (\|\mathbf{X}_{..k} - \mathbf{A} \operatorname{diag}(\mathbf{c}_k^T)(\mathbf{B}^T)^+\| + \lambda(\|\mathbf{A}^+\mathbf{X}_{..k} - \operatorname{diag}(\mathbf{c}_k^T)\mathbf{B}^T\|) \tag{7.52}$$

Find $\mathbf{B}$ by minimizing $S(\mathbf{B})$ while holding $\mathbf{A}$ and $\mathbf{C}$ constant.

In accord with the above objective functions, the pseudo-alternating least squares (PALS) algorithm is the following:

Step 1: Randomly generate loading matrices $\mathbf{A}(I \times N)$ and $\mathbf{B}(J \times N)$.

Step 2: Find $\mathbf{C}_1 = \sum_{i}^{I}(\mathbf{X}_{i..}^{T}(2\mathbf{B} + \lambda(\mathbf{B}^T)^+)\operatorname{diag}(\mathbf{a}_i^T)) + \lambda(\sum_{j}^{J}(\mathbf{X}_{j.}(\mathbf{A}^T)^+ \operatorname{diag}(\mathbf{b}_j^T)))$.

Step 3: Find $\mathbf{C}_2 = \sum_{i}^{I}(\operatorname{diag}(\mathbf{a}_i^T)(2\mathbf{B}^T\mathbf{B} + \lambda\mathbf{I})\operatorname{diag}(\mathbf{a}_i^T)) + \lambda(\sum_{j}^{J} \operatorname{diag}(\mathbf{b}_j^T)\operatorname{diag}(\mathbf{b}_j^T))$.

Step 4: Find $\mathbf{C} = \mathbf{C}_1 \mathbf{C}_2^+$.

Step 5: Find $\mathbf{A} = (\sum_{k}^{K}(\mathbf{X}_{..k}(\mathbf{B} + \lambda(\mathbf{B}^T)^+)\operatorname{diag})(\mathbf{c}_k^T))(\sum_{k}^{K}(\operatorname{diag}(\mathbf{c}_k^T)(\mathbf{B}^T\mathbf{B} + \lambda\mathbf{I})^+)$.

Step 6: Find $\mathbf{B} = (\sum_{k}^{K}(\mathbf{X}_{..k}(\mathbf{A} + \lambda(\mathbf{A}^T)^+)\operatorname{diag}(\mathbf{c}_k^T))(\sum_{k}^{K}(\operatorname{diag}(\mathbf{c}_k^T)(\mathbf{A}^T\mathbf{A} + \lambda\mathbf{I})^+)$.

Step 7: Go to step 2 until convergence or until the number of iterations reaches 3000.

In the examples studied by Chen and co-workers[42] the adjustable parameter λ was set equal to unity. The advantage of PALS, in comparison to other techniques, is the fact that PALS does not require an accurate estimation of the number of factors in the system, as long as the estimation is larger than the true number of factors. Furthermore, the method does not require symmetry in the three modes.

REFERENCES

1. L. R. Tucker in C. W. Harris (Ed.), *Problems in Measuring Change*, University of Wisconsin Press, Madison, 1963, pp. 122–137.

2. L. R. Tucker, *Psychometrika*, **31**, 279 (1966).

3. J. D. Carroll and J. J. Chang, *Psychometrika*, **35**, 283 (1970).

4. R. A. Harshman, *UCLA Working Papers in Phonics*, **16**, 1 (1970).

5. S. Leurgans and R. T. Ross, *Stat. Sci.*, **7**, 289 (1992).

6. D. S. Burdick, *Chemometrics Intell. Lab. Syst.*, **28**, 229 (1995).

7. E. Sanchez and B. R. Kowalski, *J. Chemometrics*, **2**, 247 (1988).

8. W. Wold, P. Geladi, K. H. Esbensen, and J. Ohman, *J. Chemometrics*, **1**, 41 (1987).

9. H. A. L. Kiers, *J. Chemometrics*, **14**, 105 (2000).

10. C. J. Appellof and E. R. Davidson, *Anal. Chim. Acta*, **146**, 9 (1983).

11. E. Sanchez and B. R. Kowalski, *J. Chemometrics*, **2**, 265 (1988).

12. R. Tauler, A. Izquierdo-Ridorsa, and E. Casassas, *Chemometrics Intell. Lab. Syst.*, **18**, 293 (1993).

13. R. Tauler and E. Casassas, *Anal. Chim. Acta*, **223**, 257 (1989).

14. R. Tauler, E. Casassas, and A. Izquierdo-Ridorsa, *Anal. Chim. Acta*, **248**, 447 (1991).

15. R. Tauler, A. Smilde, and B. Kowalski, *J. Chemometrics*, **9**, 31 (1995).

16. J. Saurina, S. Hernández-Cassou, R. Tauler, and A. Izquierdo-Ridorsa, *J. Chemometrics*, **12**, 183 (1998).

17. R. Tauler, I. Marques, and E. Casassas, *J. Chemometrics*, **12**, 55 (1998).

18. L. R. Tucker in N. Frederiksen and H. Gulliksen (Eds.), *Contributions to Mathematical Psychology*, Holt, Rinehart and Winston, New York, 1964, pp. 109–127.

19. Y. Zeng and P. K. Hopke, *Chemometrics Intell. Lab. Syst.*, **7**, 237 (1989).

20. P. M. Kroonenberg and J. DeLeeuw, *Psychometrika*, **45**(6), 9 (1980).

21. S. Wold, P. Geladi, K. Esbensen, and J. Ohman, *J. Chemometrics*, **1**, 41 (1987).

22. E. Sanchez and B. R. Kowalski, *J. Chemometrics*, **4**, 29 (1990).

23. D. S. Burdick, X. M. Tu, L. B. McGown, and D. W. Millican, *J. Chemometrics*, **4**, 15 (1990).

24. R. Bro, Multi-way Analysis in the Food Industry, Doctoral Thesis, University of Amsterdam (1998).

25. R. B. Cattell, *Psychometrika*, **9**, 267 (1944).

26. C. R. Rao and S. Mitra, *Generalized Inverse of Matrices and Its Applications*, Wiley, New York, 1971.

27. R. A. Harshman and M. E. Lundy, in H. G. Law, C. W. Snyder, J. A. Hattie, and R. P. McDonald (Eds.), *Research Methods for Multimode Data Analysis*, Praeger, New York, 1984 p. 120.

28. R. Sands and F. W. Young, *Psychometrika*, **45**, 39 (1980).

29. S. Li and P. J. Gemperline, *J. Chemometrics*, **7**, 77 (1993).

30. J. B. Kruskal, *Linear Algebra Appl.*, **18**, 95 (1977).

31. N. D. Sidiropoulos and R. Bro, *J. Chemometrics*, **14**, 229 (2000).

32. R. A. Harshman and W. S. de Sarbo in H. G. Law, C. W. Snyder, J. A. Hattie, and R. P. McDonald (Eds.), *Research Methods for Multimode Data Analysis*, Praeger, New York, 1984, p. 602.

33. B. C. Mitchell and D. S. Burdick, *J. Chemometrics*, **8**, 155 (1994).

34. R. Bro, *Chemometrics Intell. Lab. Syst.*, **38**, 149 (1997).

35. W. P. Krijnen and J. M. F. ten Berge, *Appl. Psych. Measurement*, **16**, 295 (1992).

36. R. Bro and S. De Jong, *J. Chemometrics*, **11**, 393 (1997).

37. R. Bro and N. D. Sidiropoulos, *J. Chemometrics*, **12**, 223 (1998).

38. R. Bro, J. Workman, Jr., P. Mobley, and B. R. Kowalski, *Appl. Spectrosc. Rev.*, **32**, 237 (1997).

39. H. Wu, M. Shibukawa, and K. Oguma, *J. Chemometrics*, **12**, 1 (1998).

40. J. Jiang, H. Wu, Y. Li, and R. Yu, *J. Chemometrics*, **14**, 15 (2000).

41. H. A. L. Kiers and A. K. Smilde, *J. Chemometrics*, **9**, 179 (1995).

42. Z. Chen, Y. Li, and R. Yu, *J. Chemometrics*, **15**, 149 (2001).

8

There are always more unknowns than knowns.

PARTIAL LEAST-SQUARES REGRESSION

How chemical and physical properties of compounds are interconnected is one of the prime goals of chemometrics. Such relationships are valuable for designing new materials, improving quality control, and predicting chemical behavior. Regression analysis is one of the most common ways for finding linear relationships between measurables and variables. This chapter describes several prominent factor analytical techniques that have been developed for the express purpose of solving regression problems.

223

8.1 MULTIPLE-REGRESSION ANALYSIS

Multiple-regression analysis (MRA) is used to express a matrix of *dependent* variables, **Y**, as a linear sum of *independent* variables, **X**,

$$\mathbf{Y} = \mathbf{XB} + \mathbf{E} \tag{8.1}$$

In (8.1) **B** is the set of linear coefficients that minimizes **E**, the error matrix. The sum of squares of the errors (**SSE**) in **E** is minimized by setting the derivative of **SSE** with respect to **B** equal to zero as shown in the derivation below:

$$\mathbf{SSE} = \mathbf{E'E}$$
$$= (\mathbf{Y} - \mathbf{XB})'(\mathbf{Y} - \mathbf{XB})$$
$$= \mathbf{Y'Y} - 2\mathbf{B'X'Y} + \mathbf{B'X'XB}$$
$$d\,\mathbf{SSE}/d\mathbf{B} = -2\mathbf{X'Y} + 2\mathbf{X'XB} = 0$$
$$\therefore \mathbf{B} = (\mathbf{X'X})^{-1}\mathbf{X'Y}$$
$$= \mathbf{X}^{+}\mathbf{Y} \tag{8.2}$$

Ordinary least squares (OLS) is a special case of MRA, where the dependent variable is a vector, **y**. In this case, the coefficient is a vector, **b**, and (8.1) takes the form:

$$\mathbf{y} = \mathbf{Xb} + \mathbf{e} \tag{8.3}$$

8.2 PRINCIPLE COMPONENT REGRESSION

A serious problem arises with MRA and OLS when the independent variables that comprise **X** are not independent but are colinear. In such cases the model parameters are more sensitive to noise, causing a loss of full rank. Principal component regression (PCR) and partial least squares (PLS) circumvent the colinearity problem because the eigenvectors (called latent variables) derived from the independent block are constrained to be orthogonal.

In PCR, **X** is replaced by $\mathbf{X}_{PCA}$, the abstract reproduced counterpart (**D** in Figure 3.1). $\mathbf{X}_{PCA}$ is obtained by deleting the error eigenvectors after subjecting **X** to abstract factor analysis (AFA). $\mathbf{B}_{PCR}$ is determined by substituting $\mathbf{X}_{PCA}$ into (8.2), giving

$$\hat{\mathbf{Y}} = \mathbf{X}_{PCA}\mathbf{B}_{PCR} \tag{8.4}$$

This procedure reduces the colinearity caused by experimental error.

A regression model that amalgamates OLS and PCR has been introduced by Iglarsh and Cheng.[1] They suggest using a weighted principal component (WPC) estimator, defined as:

$$\mathbf{b}_{WPC} \equiv (1 - \alpha)\mathbf{b}_{OLS} + \alpha\mathbf{b}_{PCR} \qquad (8.5)$$

Although WPC has been used in econometrics, as of this writing, it has not been used in chemometrics.

8.3 RELATIVE STANDARD DEVIATION PRINCIPAL COMPONENT REGRESSION

Principal component regression is usually carried out with eigenvectors ordered in accord with their eigenvalues, the largest eigenvalue considered to be most important and the smallest the least important. This top-down (TPCR) approach deletes all eigenvalues below a determined value, discarding weak signals of minor components. This procedure yields good predictions for the major components but poor predictions for the minor components.

This problem can be circumvented by reordering the eigenvectors. One widely used method, labeled CPCR, has been introduced by Xie and Kalivas.[2] The method is based on the correlation between the eigenvectors and vector $\mathbf{y}$,[2] placing more emphasis on the eigenvectors with greater correlations. The number of retained eigenvectors is based on a prediction error criterion. CPCR has several advantages over TPCR. It usually requires a smaller number of eigenvectors and has better predictability.

Unfortunately, CPCR will accept eigenvectors that have a high correlation with $\mathbf{y}$ by random chance. To overcome this, Fairchild and Kalivas[3] introduced a criterion, labelled (RSDPCR) based on the relative standard deviation of the absolute correlations obtained from a leave-one-out cross-validation process. If the correlation with the kth eigenvector is authentic, then each time a sample is deleted from $\mathbf{X}$ and $\mathbf{y}$ the correlation will change only slightly. However, if the correlation is purely coincidental, then the correlation will vary dramatically after each sample is deleted.

To illustrate the advantages of the RSDPCR criterion, they examined, among other things, the correlation between the octane numbers of 45 gasoline samples and their near-infrared (NIR) diffuse reflectance spectra. In order of importance, TPCR selected six eigenvectors: 1, 2, 3, 4, 5, and 6; CPCR selected the following six eigenvectors: 4, 1, 3, 7, 5, and 19; and RSDPCR selected five eigenvectors: 1, 4, 7, 5, and 10. These results were based on a cross-validation RSD cutoff of 8%. When the cutoff was decreased to 5%, only eigenvectors 1 and 4 were deemed important by CPCR and RSDPCR.

8.4 PARTIAL LEAST SQUARES

Partial least squares (PLS) is similar to PCR in its ability to extract orthogonal vectors, thus avoiding the problem of colinearity that plagues MRA. However, the principal components from PLS are computed to model X and to correlate Y, whereas the principal components from PCR are computed to model X only. For this reason, the PLS solution should have better predictive power when the calibration set (the training set) does not involve all of the constituents of the system.

PLS expresses *objects* (e.g., concentrations) as a linear function of *variables* (e.g., spectral absorbance). PLS is similar to PCR because both methods involve matrix decomposition. However, PCR extracts the eigenvectors solely from the matrix of variables, whereas PLS extracts the eigenvectors that are common to both the matrix of objects as well as the matrix of variables. The eigenvectors obtained from PLS are quite different from those of PCR. By path modeling with latent variables, PLS removes redundant information from the regression much in the same manner as PCR but only with respect to the objects of interest. The attractive feature of PLS is the fact that it yields viable solutions when the number of factors responsible for the object matrix is less than the number of factors responsible for the variable matrix.

Applications and techniques of PLS in chemistry have grown so vast since its inception that it is impossible to present this methodology in a single chapter. To provide an overview of the main features of PLS we will take a direct approach, using simplified notation. For more details, the serious reader should peruse the references provided at the end of the chapter.

Partial least squares is a factor analytical technique that is useful when the target matrix does not contain the full model representation, that is, when there are more factors in the data matrix than in the target matrix. If all factors were included in the target training set, there would be little or no difference between the results obtained with PCR or PLS.

The history and nature of PLS are presented nicely in a work by Geladi.[4] The method is the brainchild of Herman Wold[5] who began developing it in the early 1970s and completed it in 1977.[6] The first chemical application of PLS appeared in 1979 in a study by Gerlach, Kowalski and Herman Wold.[7] Svante Wold[8] (son of Herman Wold) and Naes and Martens[9] were also early propounders of the methodology in chemistry. An informative tutorial outlining the classical algorithm has been prepared by Geladi and Kowalski.[10]

In PLS the combination step is amalgamated with the decomposition step so that the eigenvectors of the data matrix are extracted in a sequence congruent with the eigenvectors of the target matrix. As an example of the need for PLS, consider the quantification of mixtures by infrared spectroscopy using a target matrix that only contains information about the constituents of interest. The major principal components of the mixture, as determined by PCA, may contain little information about these particular constituents. In fact, the largest principal component may be dominated by matrix effects or by other constituents, not of interest. Because PLS searches for the factor space most congruent to both matrices, its predictions are superior to PCR.

The PLS method involves regression between the scores of two matrices, **X** and **Y**. For spectroscopic analysis of mixtures, **X** and **Y** may represent, respectively, spectral and concentration matrices of the mixtures. PLS seeks a calibration model such that

$$\underset{k \times p}{\mathbf{Y}} = \underset{k \times m}{\mathbf{X}} \; \underset{m \times p}{\mathbf{B}_{\mathrm{PLS}}} \tag{8.6}$$

where $\mathbf{B}_{\mathrm{PLS}}$ is the best set of calibration constants for the system. For our example, the rows of **X** and **Y** correspond to the sample mixtures. The columns of **X** represent the spectral wavelengths. And the columns of **Y** are the concentrations of the constituents of interest, although the samples may contain many constituents. In general, the rows are called *samples* or *objects*. The columns of **X** are the *independent variables*; they represent the measurables. The columns of **Y** are the *dependent variables*; they represent the properties we want to predict. The properties may include esoteric phenomena such as octane numbers, molecular descriptors, quantitative structure activities, strengths of materials, sweetness, and so on, phenomena intrinsically imbedded in the measurement matrix, **X**. The rows and columns of $\mathbf{B}_{\mathrm{PLS}}$ are associated with the wavelengths and the concentrations, the independent and dependent variables. In (8.6) there are m measurables and p properties. Matrices **X** and **Y**, used to determine $\mathbf{B}_{\mathrm{PLS}}$, are collectively called the "training set."

The purpose in developing (8.6) is to predict the properties (e.g., concentrations) $\hat{\mathbf{Y}}_{\mathrm{new}}$, of new samples from a knowledge of their measurables (e.g., spectra) $\mathbf{X}_{\mathrm{new}}$. If there are q new samples, then, according to (8.6),

$$\underset{q \times p}{\hat{\mathbf{Y}}_{\mathrm{new}}} = \underset{q \times m}{\mathbf{X}_{\mathrm{new}}} \; \underset{m \times p}{\mathbf{B}_{\mathrm{PLS}}} \tag{8.7}$$

Partial least squares requires a *training set* consisting of matrices **X** and **Y**, which are used to determine $\mathbf{B}_{\mathrm{PLS}}$. It also requires an independent *test set* of measurables and properties that can be used to test the accuracy and validity of the developed PLS model.

There are several types of PLS regressions. The simplest type, called PLS1, applies to the situation where **Y** is a single, column vector, **y**. PLS2 applies to the situation where **Y** is a two-dimensional matrix. Multiblock PLS (MBPLS) makes use of several **X** blocks to obtain the regression coefficients. Multilinear PLS models, called N-PLS, applies to three-way, four-way, and so forth data blocks.

The classical notation developed in the early work is prevalent throughout the literature and represents the accepted standard. For the neophyte, however, this notation is difficult to absorb because it lacks mnemonics. Coupled with tricky concepts, this often leads to a sea of confusion. For pedagogical reasons we will use a simplified notation, expressing the factors as scores (**sx** and **sy**) and loadings (**lx** and **ly**) consistent with Section 1.2. At the end of the presentation a translation glossary will be provided.

Partial least squares is based on nonlinear iterative partial least squares (NIPALS) decomposition (Section 3.3.8), which, in turn, is based on (8.8):

$$\mathbf{X} = \mathbf{Sx}\,\mathbf{Lx}' + \mathbf{Ex} = \Sigma\mathbf{sx}\,\mathbf{lx}' + \mathbf{Ex} \tag{8.8}$$

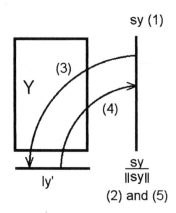

Figure 8.1   Diagram illustrating the iterative steps in the NIPALS algorithm.

In this expression **Ex** represents the residual error based on the number of factors used in the analysis. The underlying principal of NIPALS is illustrated in Figure 8.1. The NIPALS algorithm is given below. It is important to note that the NIPALS algorithm extracts orthogonal sets of eigenvectors. The score vectors are mutually orthogonal but not normalized. The loading vectors are mutually orthogonal and normalized. These orthogonality conditions are not preserved during PLS analysis.

NIPALS Algorithm

Principle: Substitute an approximation for **sy** in (A) and solve for **ly**. Then substitute **ly** into (A) and solve for **sy**. Repeat this procedure until convergence is achieved. Calculate the residual of **Y**. Extract the next set of eigenvectors substituting the residual for **Y**.

$$\mathbf{Y} = \mathbf{sy}\ \mathbf{ly}' \qquad (A)$$

Initialization:

    (1) **sy** =arbitrary   Create a column vector with n arbitrary numbers.

    (2) $\mathbf{sy} = \mathbf{sy}/\|\mathbf{sy}\|$ Normalize **sy**.

Y block:

    (3) $\mathbf{ly} = \mathbf{Y}'\mathbf{sy}$ Use (A) to find **ly**.

    (4) $\mathbf{sy} = \mathbf{Y}\ \mathbf{ly}$ Use (A) to find **sy**.

    (5) $\mathbf{sy} = \mathbf{sy}/\|\mathbf{sy}\|$ Normalize **sy**.

    (6) Repeat steps (3) to (5) until **sy** converges.

Calculate the residuals:

(7) $\mathbf{Ey} = \mathbf{Y} - \mathbf{sy\ ly'}$ Calculate the error matrix.

(8) $\mathbf{Y} = \mathbf{Ey}$ Let **Y** equal the residual error.

(9) SSE = trace($\mathbf{Ey'\ Ey}$)   Calculate SSE, the sum of squared errors.

(10) Go to step (1).   Extract the next set of eigenvectors if SSE is too large.

Partial least squares is based on the presumption that **X** and **Y** are linked together by a set of latent factors called scores, **s**. Hopefully, all of the **Y** scores are contained in **X**, although all of the **X** scores may not be contained in **Y**. In other words, **X** may have a larger factor space than **Y**. PLS involves finding individual, outer relations for **X** and **Y** and an inner relation linking both matrices. The outer relations can be expressed as

$$\underset{k\times m}{\mathbf{X}} = \underset{k\times n}{\mathbf{Sx}}\ \underset{n\times m}{\mathbf{Lx'}} + \underset{k\times m}{\mathbf{Ex}} = \Sigma\mathbf{sx\ lx'} + \mathbf{Ex} \tag{8.9}$$

$$\underset{k\times p}{\mathbf{Y}} = \underset{k\times n}{\mathbf{Sy}}\ \underset{n\times p}{\mathbf{Ly'}} + \underset{k\times p}{\mathbf{Ey}} = \Sigma\mathbf{sy\ ly'} + \mathbf{Ey} \tag{8.10}$$

The sums are taken over all n factors that are common to **X** and **Y**. The inner relationship involves converting **sx** into **sy** and **sy** into **sx** so that both score vectors are equal and common to **X** and **Y**. When the inner and outer relations are combined, (8.9) and (8.10) become

$$\underset{k\times m}{\mathbf{X}} = \underset{k\times n}{\mathbf{S}}\ \underset{n\times m}{\mathbf{Lx'}} + \underset{k\times m}{\mathbf{Ex}} = \Sigma\mathbf{s\ lx'} + \mathbf{Ex} \tag{8.11}$$

$$\underset{k\times p}{\mathbf{Y}} = \underset{k\times n}{\mathbf{S}}\ \underset{n\times p}{\mathbf{Ly'}} + \underset{k\times p}{\mathbf{Ey}} = \Sigma\mathbf{s\ ly'} + \mathbf{Ey} \tag{8.12}$$

where $\mathbf{s} = \mathbf{sx} = \mathbf{sy}$ and $\mathbf{S} = \mathbf{Sx} = \mathbf{Sy}$.

In PLS both $\|\mathbf{X} - \mathbf{S\ Lx'}\|$ and $\|\mathbf{Y} - \mathbf{S\ Ly'}\|$ are minimized. This involves finding eigenvectors **s**, **lx**, and **ly** for each factor level by the NIPALS routine. Figure 8.2 illustrates the sequence of steps involved. In this case, however, the iteration alternates between **X** and **Y**, searching for the solution that simultaneously minimizes $\|\mathbf{Ex}\|$ and $\|\mathbf{Ey}\|$ at each factor level.

In accord with the NIPALS routine, the outer products of the first set of eigenvectors are subtracted from **X** and **Y**, yielding residual matrices. To obtain the second principal components the iteration is repeated after replacing **X** and **Y** by their residuals. The third, fourth, and so forth eigenvectors are similarly extracted until the residual sum of squares reaches a satisfactory minimum. Cross validation can be used to determine the appropriate factor cutoff level.[6] In any case the cutoff should be equal to or less than the true factor space. If all of the factors are employed, the resulting coefficients will be identical to those obtained by simple regression analysis, defeating the purpose of PLS.

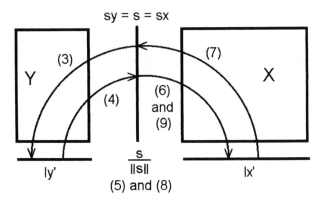

Figure 8.2   Diagram illustrating the iterative process in PLS.

Although **X, Y, Lx,** and **Ly** have different sizes, **Sx** and **Sy** are exactly the same size, $k \times n$. PLS searches for a mutual set of factor axes that are common to **Sx** and **Sy** so that **Sx** equals **Sy**. By setting **Sx = Sy = S** and deleting the uncommon residuals, **Ex** and **Ey**, (8.11) and (8.12) become

$$\mathbf{X} = \mathbf{S}\,\mathbf{Lx}' \tag{8.13}$$

$$\hat{\mathbf{Y}} = \mathbf{S}\,\mathbf{Ly}' \tag{8.14}$$

Inserting (8.13) and (8.14) into (8.6) and rearranging gives

$$\mathbf{B}_{\text{PLS}} = (\mathbf{S}\,\mathbf{Lx}')^{+}(\mathbf{S}\,\mathbf{Ly}') \tag{8.15}$$

Thus, by finding the common scores and their associated loadings, the calibration coefficients, expressed in (8.6) and (8.7), can be determined. If the scores are orthonormal, (8.15) reduces to

$$\mathbf{B}_{\text{PLS}} = (\mathbf{Lx}')^{+}\mathbf{Ly}' \tag{8.16}$$

It is important to note that, in PLS, the columns of **X** and **Y** are autoscaled to unit length; that is, they are mean centered and normalized. This is done to place equal statistical weight on each variable. This is the first preliminary step in all PLS analyses. Consequently, to predict the properties of the unknowns, the new measurement variables must be standardized using the mean values and normalization constants employed in autoscaling the training sets. At the end of the analysis the predictions are corrected for autoscaling.

8.4.1 PLS1

In PLS1 the property matrix, **Y**, is a single column vector, **y**, of size $k \times 1$. The NIPALS procedure expresses this vector as a $k \times n$ matrix, **S**, times an $n \times 1$ vector, **ly**. The classical PLS1 algorithm is given below.

Classical PLS1 Algorithm

Principle: To start, let **s** equal **y**, thus accounting for a maximum of **y** in **X**. Apply the NIPALS routine to obtain the factor components expressed by Eqs. (A) and (B). After each factor is extracted, calculate the residuals, label them **X** and **y**, and apply the NIPALS routine to extract the next factor components. When an appropriate number of factors have been obtained, compile the vectors into matrices and calculate $\mathbf{B}_{\text{PLS1}}$. (Note that $\mathbf{B}_{\text{PLS1}}$ is a column vector.)

$$\mathbf{X} = \mathbf{s} \, \mathbf{lx}' \tag{A}$$

$$\mathbf{y} = \mathbf{s} \, ly \tag{B}$$

Preliminary steps:

   (1) Autoscale **X** and **y**.

Initialization:

   (2) $\mathbf{s} = \mathbf{y}/\|\mathbf{y}\|$ To start, let **s** be **y** normalized.

X block:

   (3) $\mathbf{lx}' = \mathbf{s}'\mathbf{X}$ Use (A) to find $\mathbf{lx}'$.
   (4) $\mathbf{s} = \mathbf{X} \, \mathbf{lx}$ Use (A) to find **s**.
   (5) $\mathbf{s} = \mathbf{s}/\|\mathbf{s}\|$ Normalize **s**.

Y block:

   (6) $ly = \mathbf{s}' \, \mathbf{y}$ Use (B) to find scalar ly.

Complete X block:

   (7) $\mathbf{lx}' = \mathbf{s}' \, \mathbf{X}$ Use (A) to find **lx**.

Calculate the residuals:

   (8) $\mathbf{Ex} = \mathbf{X} - \mathbf{s} \, \mathbf{lx}'$ Calculate the residual in **X**.
   (9) $\mathbf{Ey} = \mathbf{y} - \mathbf{s} \, ly$ Calculate the residual in **Y**.
   (10) $\mathbf{X} = \mathbf{Ex}$ Label the residual **X**.

(11) $\mathbf{Y} = \mathbf{Ey}$ Label the residual $\mathbf{Y}$.

(12) $\mathrm{SSEy} = \mathbf{Ey}' \, \mathbf{Ey}$ Calculate the sum of squares of the residual error in $\mathbf{Y}$.

(13) Go to step (2).   Extract the next set of eigenvectors if SSEy is too large.

Calculate the coefficients:

(14) $\mathbf{B}_{\mathrm{PLS1}} = (\mathbf{S} \, \mathbf{Lx}')^{+}(\mathbf{S} \, \mathbf{ly}')$ Assemble the score and loading matrices and invoke (8.15).

It is interesting to note that the columns of $\mathbf{S}$ are not mutually orthogonal. Each extracted factor represents a "part of the least squares," hence the name "partial least squares." When all possible factors are included in step (14), the result is equivalent to classical least-squares regression. By stopping the analysis at the proper factor level, redundant information is removed from the procedure, leading to an improvement in predictability. In some situations the improvement may be substantial. In other cases the improvement may be negligible.

8.4.2 PLS2

Classical PLS2 follows the iteration procedure illustrated in Figure 8.2. Because the property matrix, $\mathbf{Y}$, consists of two or more columns, the iteration at each factor level is continued until the score vector $\mathbf{s}$ no longer changes. At this point the iteration is stopped, the loading vector $\mathbf{lx}$ is calculated, the residual matrices are calculated, and the next set of eigenvectors are extracted. Details of the computations are given in the algorithm below.

Classical PLS2 Algorithm
Principle: Substitute $\mathbf{s}$ in Eqs. (A) and (B), alternately, until converge is achieved.

$$\mathbf{X} = \mathbf{s} \, \mathbf{lx}' \qquad \text{(A)}$$
$$\mathbf{Y} = \mathbf{s} \, \mathbf{ly}' \qquad \text{(B)}$$

Preliminary steps:

(1) Autoscale $\mathbf{X}$ and $\mathbf{Y}$.

Initialization:

(2) $\mathbf{s} = \mathbf{y}/\|\mathbf{y}\|$ To start, let $\mathbf{s}$ be any normalized column of $\mathbf{Y}$.

Y block:

(3) $\mathbf{ly}' = \mathbf{s}'\mathbf{Y}/(\mathbf{s}'\ \mathbf{s})$ Use (B) to find $\mathbf{ly}'$.

(4) $\mathbf{s} = \mathbf{Y}\ \mathbf{ly}/(\mathbf{ly}'\ \mathbf{ly}$ Use (B) to find $\mathbf{s}$.

(5) $\mathbf{s} = \mathbf{s}/\|\mathbf{s}\|$ Normalize $\mathbf{s}$.

X block:

(6) $\mathbf{lx}' = \mathbf{s}'\ \mathbf{X}/(\mathbf{s}'\ \mathbf{s})$ Use (A) to find $\mathbf{lx}'$.

(7) $\mathbf{s} = \mathbf{X}\ \mathbf{lx}$ Use (A) to find $\mathbf{s}$.

(8) $\mathbf{s} = \mathbf{s}/\|\mathbf{s}\|$ Normalize $\mathbf{s}$.

Check convergence:

(9) Repeat steps (3) to (8) until $\mathbf{s}$ in step (5) converges.

Complete the X block:

(10) $\mathbf{lx}' = \mathbf{s}'\ \mathbf{X}$ Use (A) to find $\mathbf{lx}$.

Calculate the residuals:

(11) $\mathbf{Ex} = \mathbf{X} - \mathbf{s}\ \mathbf{lx}'$ Calculate the residual error in $\mathbf{X}$.

(12) $\mathbf{Ey} = \mathbf{Y} - \mathbf{s}\ \mathbf{ly}'$ Calculate the residual error in $\mathbf{Y}$.

(13) $\mathbf{X} = \mathbf{Ex}$ Label the residual $\mathbf{X}$.

(14) $\mathbf{Y} = \mathbf{Ey}$ Label the residual $\mathbf{Y}$.

(15) $\mathrm{SSEy} = \mathrm{trace}(\mathbf{Ey}'\ \mathbf{Ey})$ Calculate the sum of squares of the residual error in $\mathbf{Y}$.

(16) Go to step (2).   Extract the next set of eigenvectors if SSEy is too large.

Calculate the coefficients:

(17) $\mathbf{B}_{\mathrm{PLS}} = (\mathbf{Lx}')^{+}\mathbf{Ly}'$.   Assemble the score and loading matrices, and invoke (8.16).

It is interesting to note that the columns of $\mathbf{S}$ are orthonormal, and the columns of $\mathbf{Lx}$ and $\mathbf{Ly}$ are neither orthogonal nor normalized.

8.4.3 Kernel Methods

Because classical PLS involves iterative sequential computations, the classical algorithm is inefficient when the number of objects, k, or the number of variables,

m, in the training matrix, $\mathbf{X}(k \times m)$ is large. Lindgren and co-workers[11] developed an algorithm to handle many objects and few variables $(k \gg m)$. The method is based on the "kernel" matrix $\mathbf{X'YY'X}$. The kernel matrix is a square matrix of condensed size $(m \times m)$ that, according to Höskuldsson,[12] follows the eigenvalue–eigenvector equation:

$$\mathbf{lx} \, \lambda = (\mathbf{X'YY'X})\mathbf{lx} \tag{8.17}$$

where λ is the eigenvalue associated with $\mathbf{lx}$, the normalized eigenvector. A simplified kernel algorithm (KPLS) that illustrates the use of (8.17) is presented below.

Kernel Algorithm (Many Objects and Few Variables)

Principle: Use the kernel matrix to determine the loading vector of $\mathbf{X}$. Use this vector to determine the common score vector. Use the score vector to determine the loading vector of $\mathbf{Y}$.

$$\mathbf{lx} \, \lambda = (\mathbf{X'YY'X})\mathbf{lx} \tag{A}$$
$$\mathbf{X} = \mathbf{s} \, \mathbf{lx'} \tag{B}$$
$$\mathbf{Y} = \mathbf{s} \, \mathbf{ly'} \tag{C}$$

Preliminary steps:

(1) Autoscale $\mathbf{X}$ and $\mathbf{Y}$.

Use the kernel matrix to determine $\mathbf{lx}$:

(2) $\mathbf{lx} =$ any normalized row of $\mathbf{X}$.   Estimate $\mathbf{lx}$.
(3) $\mathbf{lx} \, \lambda = (\mathbf{X'YY'X})\mathbf{lx}$ Use (A) to find $\mathbf{lx} \, \lambda$.
(4) $\mathbf{lx} = \mathbf{lx} \, \lambda / \|\mathbf{lx} \, \lambda\|$ Find $\mathbf{lx}$ by normalizing $\mathbf{lx} \, \lambda$.
(5) Repeat steps (3) and (4) until $\mathbf{lx}$ converges.

Determine $\mathbf{s}$, normalized:

(6) $\mathbf{s} = \mathbf{X} \, \mathbf{lx}$ Use (B) to find $\mathbf{s}$.
(7) $\mathbf{s} = \mathbf{s}/\|\mathbf{s}\|$ Normalize $\mathbf{s}$.

Calculate $\mathbf{lx}$ and $\mathbf{ly}$:

(8) $\mathbf{lx} = \mathbf{X'} \mathbf{s}$ Use (B) to find $\mathbf{lx}$.
(9) $\mathbf{ly} = \mathbf{Y'} \mathbf{s}$ Use (C) to find $\mathbf{ly}$.

Calculate the residuals:

(10) $\mathbf{Ex} = \mathbf{X} - \mathbf{s}\,\mathbf{lx}'$ Calculate the residual error in $\mathbf{X}$.

(11) $\mathbf{Ey} = \mathbf{Y} - \mathbf{s}\,\mathbf{ly}'$ Calculate the residual error in $\mathbf{Y}$.

(12) $\mathbf{X} = \mathbf{Ex}$ Label the residual $\mathbf{X}$.

(13) $\mathbf{Y} = \mathbf{Ey}$ Label the residual $\mathbf{Y}$.

(14) $\mathrm{SSEy} = \mathrm{trace}(\mathbf{Ey}'\,\mathbf{Ey})$ Calculate the sum of squares of the residual error in $\mathbf{Y}$.

(15) Go to step (2).   Extract the next set of eigenvectors if SSEy is too large.

Calculate the coefficients:

(16) $\mathbf{B}_{\mathrm{PLS}} = (\mathbf{Lx}')^{+}\mathbf{Ly}'$ Assemble the score and loading matrices and invoke (8.16).

Rännar and co-workers[13] developed a kernel algorithm to handle many dependent and independent variables, and few objects ($p \gg k$) and ($m \gg k$). The method is based on the kernel matrix $\mathbf{XX}'\mathbf{YY}'$, which is a square, nonsymmetric matrix condensed to size ($k \times k$) and follows the eigenvalue–eigenvector equation:[12]

$$\mathbf{sx}\,\lambda = (\mathbf{XX}'\mathbf{YY}')\mathbf{sx} \tag{8.18}$$

An algorithm, applicable to $k \gg p$, can also be obtained by employing the kernel matrix $\mathbf{Y}'\mathbf{XX}'\mathbf{Y}$, which is a square, symmetric matrix of size ($p \times p$) that follows:[12]

$$\mathbf{ly}\,\lambda = (\mathbf{Y}'\mathbf{XX}'\mathbf{Y})\mathbf{ly} \tag{8.19}$$

The KPLS algorithm presented below makes use of (8.19).

Kernel Algorithm (Many Variables and Few Objects)
Principle: Use the kernel matrix to determine the loading vector of $\mathbf{Y}$. Use this vector to determine the common score vector. Use the score vector to determine the loading vector of $\mathbf{X}$.

$$\mathbf{ly}\,\lambda = (\mathbf{Y}'\mathbf{XX}'\mathbf{Y})\mathbf{ly} \tag{A}$$
$$\mathbf{X} = \mathbf{s}\,\mathbf{lx}' \tag{B}$$
$$\mathbf{Y} = \mathbf{s}\,\mathbf{ly}' \tag{C}$$

Preliminary steps:

(1) Autoscale $\mathbf{X}$ and $\mathbf{Y}$.

Use the kernel matrix to determine **ly**:

(2) **ly** = any normalized
row of **Y**.   Estimate **ly**.
(3) **ly** λ = (**Y'XX'Y**)**ly**   Use (A) to find **ly** λ.
(4) **ly** = **ly** λ/‖**ly** λ‖   Find **ly** by normalizing **ly** λ
(5) Repeat steps (2) and (3) until **ly** converges.

Determine **s** normalized:

(6) **s** = **Y ly**   Use (B) to find **s**.
(7) **s** = **s**/‖**s**‖   Normalize **s**.

Calculate **lx** and **ly**:

(8) **lx** = **X' s**   Use (B) to find **lx**.
(9) **ly** = **Y' s**   Use (C) to find **ly**.

Calculate the residuals:

(10) **Ex** = **X** − **s lx'**   Calculate the residual error in **X**.
(11) **Ey** = **Y** − **s ly'**   Calculate the residual error in **Y**.
(12) **X** = **Ex**   Label the residual **X**.
(13) **Y** = **Ey**   Label the residual **Y**.
(14) SSEy = trace(**Ey' Ey**)   Calculate the sum of squares of the residual error in **Y**.
(15) Go to step (2).   Extract the next set of eigenvectors if SSEy is too large.

Calculate the coefficients:

(16) $\mathbf{B}_{\text{PLS}} = (\mathbf{Lx'})^{+}\mathbf{Ly'}$ Assemble the score and loading matrices and invoke (8.16).

The advantage of the kernel algorithms, in comparison to the classical algorithm, is the speed and efficiency without loss in precision. Kernel algorithms that are even more efficient than those originally proposed[11,12] have been designed by Dayal and MacGregor.[14] Because the kernel algorithms extract the eigenvectors in a stepwise manner, the classical parameters can be determined at each step in the process. All of the classical parameters are intimately connected by their associated eigenvectors, **s**, **lx**, and **ly**. These relationships are described in Section 8.4.4.

8.4.4 Classical Notation

The symbols in the PLS1, PLS2, and kernel algorithms given above were employed for pedagogical reasons. By replacing these symbols with the classical analogs presented in the glossary of symbols, one can express these algorithms in classical form. The classical work defines other score and loading vectors, such as **t**, **p**, **w**, **u**, **q**, **c**, and b_{inner}. All of these quantities can be expressed as functions of **s**, **lx**, and **ly**, as shown in the glossary. They differ only by the nature of the desired normalization conditions. In classical notation Eqs. (8.9) and (8.10) are expressed as:

$$\mathbf{X} = \mathbf{TP}' + \mathbf{E} \tag{8.20}$$
$$\mathbf{Y} = \mathbf{UQ}' + \mathbf{F} \quad \text{or} \quad \mathbf{Y} = \mathbf{TC}' + \mathbf{F} \tag{8.21}$$

In the literature you are likely to find (8.15) expressed in its classical forms:

$$\mathbf{B}_{PLS} = \mathbf{P}(\mathbf{W}'\mathbf{P})^{-1}\mathbf{Q}' \quad \text{or} \quad \mathbf{B}_{PLS} = \mathbf{W}(\mathbf{P}'\mathbf{W})^{-1}\mathbf{C}' \tag{8.22}$$

GLOSSARY OF SYMBOLS

Pedagogical	Classical	Description
k	k	Number of samples in training set **X** and **Y**
m	m	Number of measurables in training set **Y**
p	p	Number of properties in training set **Y**
n	a	Number of factors employed
X	**X**	Matrix of measured features ($k \times m$)
Y	**Y**	Matrix of responses ($k \times p$)
$\mathbf{B}_{PLS}$	**B**	Matrix of PLS regression coefficients ($m \times p$)
Ex	**E**	Matrix of residuals for the **X** block ($k \times m$)
Ey	**F**	Matrix of residuals for the **Y** block ($k \times p$)
$\mathbf{s}\|\mathbf{lx}\|$	**t**	Scores for the **X** block, so that $\mathbf{t}\,\mathbf{p}' = \mathbf{s}\,\mathbf{lx}'$, ($k \times 1$)
$\mathbf{lx}/\|\mathbf{lx}\|$	**p**	Normalized loadings for the **X** block ($m \times 1$)
$\mathbf{lx}/\|\mathbf{ly}\|$	**w**	Weightings for the **X** block ($m \times 1$)
$\mathbf{s}\|\mathbf{ly}\|$	**u**	Scores for the **Y** block, so that $\mathbf{u}\,\mathbf{q}' = \mathbf{s}\,\mathbf{ly}'$, ($k \times 1$)
$\mathbf{ly}/\|\mathbf{ly}\|$	**q**	Normalized loadings for the **Y** block ($p \times 1$)
$\mathbf{ly}/\|\mathbf{lx}\|$	**c**	Weightings for the **Y** block ($p \times 1$)
$\|\mathbf{ly}\|/\|\mathbf{lx}\|$	b_{inner}	Inner relation defined by $\mathbf{u} = b_{inner}\,\mathbf{t}(1 \times 1)$
—	**T**	Matrix containing the **t** vectors ($k \times n$)
—	**P**	Matrix containing the **p** vectors ($p \times n$)
—	**W**	Matrix containing the **w** vectors ($p \times n$)
—	**U**	Matrix containing the **u** vectors ($k \times n$)
—	**Q**	Matrix containing the **q** vectors ($m \times n$)
—	**C**	Matrix containing the **c** vectors ($m \times n$)

Several interesting relationships can be gleaned from the glossary. For example:

$$s \; \mathbf{lx}' = \mathbf{tp}' = \mathbf{tw}'b_{\text{inner}} = \mathbf{uw}' \qquad (8.23)$$

$$s \; \mathbf{ly}' = \mathbf{tc}' = \mathbf{tq}'b_{\text{inner}} = \mathbf{uq}' \qquad (8.24)$$

Unfortunately, not all authors adhere to the classical notation, often interchanging $\mathbf{p}$, $\mathbf{c}$, and $\mathbf{q}$ or $\mathbf{t}$, $\mathbf{w}$, and $\mathbf{u}$. This can lead to confusion if one is not careful.

8.4.5 Number of Latent Factors

Ordinary multiple regression minimizes the sum of squares of the error in y, $\text{SSEy} = \Sigma(y - \hat{y})^2$. In this case, the mean-squared error of prediction (MSEP) is simply

$$\text{MSEP} = \text{SSEy}/\text{df} = \text{SSEy}/(p - k) \qquad (8.25)$$

In this case the number of degrees of freedom, df, is equal to the number of random (independent) variables, p, minus the number of parameters, k, employed in the model. Unfortunately, SSEy in PLS is *not* a true minimum but a compromised minimum, linking $\mathbf{X}$ and $\mathbf{Y}$, based on using nonlinear estimators for the b coefficients. For PLS there is no simple way to determine the degrees of freedom. Consequently, the literature is permeated with a wide variety of naïve approaches.

For nonlinear models additional factors are required to account for the nonlinearities. Determining the appropriate number of latent factors represents an important facet for successful PLS. Using too few factors leads to underfitting and poor predictability. Using too many factors adds unwanted noise to the predicted values, defeating the prime purpose of PLS. Finding the optimum number of latent factors is an essential underpinning in PLS methodology.

A variety of methods involve the determination of SSEy (or MSEP) at each factor level, extracting latent variables until SSEy (or MSEP) falls below a specified threshold. Other methods invoke cross validation, a method described in Section 4.3.3. Denham[15] studied 12 different methods for estimating MSEP in PLS and concluded that cross validation and boot strapping were the most reliable.

8.4.6 Improvements

A number of improvements have been suggested for PLS. For example, Forina and co-workers[16] introduced iterative predictor weightings (IPW) to eliminate useless predictors. Gil and Romero[17] used statistical procedures for covariance matrix robustification. Canonical PLS, developed by de Jong and co-workers,[18] makes use of the smallest rank of $\mathbf{X}$. Goutis[19] introduced a fast computational method for extracting the loadings in PLS. Denham[20] uses prediction intervals instead of point predictions. These represent only a small sample of the various improvements recently suggested.

8.5 CYCLIC SUBSPACE REGRESSION

Cyclic subspace regression (CSR), designed by Kalivas[21] in 1999, is a single method that incorporates all of the features of PCR, PLS1, least squares (LS), as well as ridge regression techniques. PCR and PLS1 have an advantage over LS because these methods reduce the dimensionality of the regression space to a small set of orthogonal factors, thus removing redundant information from the regression. PLS is considered to be better than PCR because it removes a greater portion of the redundant information. The similarities, as well as the differences, between PCR and PLS1 can be explored by examining the eigenvectors obtained from CSR.

The CSR algorithm provides an extra loop in the PLS1 algorithm, modifying $\mathbf{X}$ and $\mathbf{y}$, so the number of eigenvectors can be varied. It also involves an additional step for expressing $\mathbf{y}$ as a linear combination of eigenvectors.

Cyclic Subspace Regression (CSR) Algorithm
Principle: As a preliminary step, use singular value decomposition to find the eigenvectors of $\mathbf{X}$. Modify $\mathbf{y}$ and use the NIPALS decomposition to extract the scores that are common to $\mathbf{X}$ and $\mathbf{y}$.

$$\mathbf{X} = \mathbf{s} \, \mathbf{l}\mathbf{x}' \tag{A}$$

$$\mathbf{y} = \mathbf{s} \, ly \tag{B}$$

Preliminary steps:

(1) Autoscale $\mathbf{X}$ and $\mathbf{y}$.

(2) $\mathbf{USV'} = \mathbf{X}$ ⟶ Decompose $\mathbf{X}$ (see Section 3.37 or 3.38)

Initialization:

(4) For $j = 1, \ldots, k$ and $l = j, \ldots, k$:

(5) $\mathbf{P}_l = [\mathbf{u}_1 \, \mathbf{u}_2 \, \ldots \, \mathbf{u}_l][\mathbf{u}_1 \, \mathbf{u}_2 \, \ldots \, \mathbf{u}_l]'$ ⟶ Create the rotation matrix.

(6) $\mathbf{X}_l^1 = \mathbf{X}$ ⟶ Insert appropriate indices.

(7) $\mathbf{y}_l^1 = \mathbf{P}_l \mathbf{y}$ ⟶ Modify $\mathbf{y}$.

Extraction:

(8) For $i = 1, \ldots, j$ perform the following:

(9) $\mathbf{s}_l^i = \mathbf{y}_l^i / \|\mathbf{y}_l^i\|$ ⟶ To start, let $\mathbf{s}_l^i$ be $\mathbf{y}_l^i$ normalized.

(10) $\mathbf{s}_l^i = \mathbf{X}_l^i \mathbf{X}_l^{i'} \mathbf{s}_l^i$ ⟶ Find $\mathbf{s}_l^i$ [combine steps (3) and (4) in PLS1].

(12) $\mathbf{s}_l^i = \mathbf{s}_l^i / \|\mathbf{s}_l^i\|$ ⟶ Normalize $\mathbf{s}_l^i$.

(13) $\mathbf{X}_l^{i+1} = \mathbf{X}_l^i - \mathbf{s}_l^i \mathbf{s}_l^{i'} \mathbf{X}_l^i$ ⟶ Calculate the residual [combine (7), (8), and (10) in PLS1].

(14) $\mathbf{y}_l^{i+1} = \mathbf{y}_l^i - \mathbf{s}_l^i \mathbf{s}_l^{i\prime} \mathbf{y}_l^i$ Calculate the residual [combine (6), (9), and (11) in PLS1].

(15) Go to step (8).

(16) Go to step (4).   Extract the next set of eigenvectors.

According to Kalivas[21] the CSR regression vectors are linear combinations of the l eigenvectors and can be expressed as:

$$\mathbf{b} = \sum_{i=1}^{l} \left(\frac{\mathbf{u}_i'\mathbf{y}_l^i}{\sigma_i} \right) \mathbf{v}_i = \sum_{i=1}^{l} \phi_i \mathbf{v}_i \qquad (8.26)$$

In (8.26), $\mathbf{y}_l^i$ are the values calculated from step (14), based on l eigenvectors and j factors. The term in the parenthesis, labeled ϕ_i, represents the amount (the weight) of eigenvector $\mathbf{v}_i$ that contributes to $\mathbf{b}$. In fact the eigenvector weights, $\boldsymbol{\phi}$, can be obtained by rearranging (8.26) to

$$\boldsymbol{\phi} = \mathbf{V}'\mathbf{b} \qquad (8.27)$$

The relationship between PCR, PLS, LS, and CSR can be deduced from the CSR algorithm. PCR regression vectors result when $l = j$. PLS regression vectors are produced when all k eigenvectors are included in the j-factor model, that is, when $l = k$. The LS regression vector is obtained when $l = k = j$.

By applying CSR to several data sets, Kalivas[21] concluded that PCR is too restrictive (less variance but more bias) and PLS is too flexible (more variance but less bias). PCR tends to use too few eigenvectors, whereas PLS tends to incorporate too many eigenvectors. This study shows that it is not the *number* of factors but the *weightings* of the eigenvectors that is important. CSR can be used to develop a hybrid model that yields the best regression solution.

8.6 MULTIBLOCK PLS

Multistep processes are described best by multiblock partial least squares (MBPLS). In continuous processes samples are examined at points along the process line, providing valuable information about the end product. By combining this information, the analyst can obtain a better indication of the quality of the final product. Disturbances can be detected earlier and corrections can be appropriately applied. In fact, by combining the blocks of information one can look at the overall process and also zoom into a particular block to pinpoint local problems and make necessary adjustments.

Westerhuis and Coenegracht[22] applied MBPLS to a two-step pharmaceutical tablet manufacturing process depicted in Figure 8.3. The first step involved wet granulation of a powder mixture, blended with a binding agent, to produce a

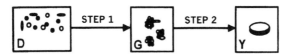

Figure 8.3 Two-step (granulation and compression) process of tablet manufacturing. [Reprinted with permission from J. H. Westerhuis and P. M. J. Coenegracht, *J. Chemometrics*, **11**, 379 (1997). Copyright 1997 © John Wiley & Sons, Ltd.]

homogeneous mixture. Block D was comprised from six descriptor variables: amount of water, amount of cellulose, amount of HPC (Aqualon) solution, granulation time, moisture of granules, and compression force. The second step involved compressing the powder into tablets, a process that involved many variables, such as drying time, temperature, pressure, sieving, size, mixing with other reagents, compression force, and so on. Block G was composed from some 14 variables. Block Y contained the crushing strength and disintegration time of 55 tablets, prepared in accord with a Box-Behnken design. The purpose of the analysis was to determine how the process variables affected the tablet properties.

A general MBPLS algorithm, incorporating any number of blocks, has been introduced by Wangen and Kowalski.[23] The two-block algorithm used by Westerhuis and Coenegracht[22] to model a two-step pharmaceutical process is presented below in classical notation and illustrated in Figure 8.4. Blocks D and G represent two matrices that share a common response block, **Y**. These three blocks contain the same number of rows (observations) but different columns (variables).

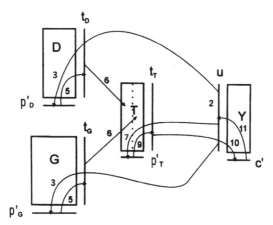

Figure 8.4 Diagram illustrating the iterative steps in the two-block PLS algorithm. [Reprinted with permission from J. H. Westerhuis and P. M. J. Coenegracht, *J. Chemometrics*, *11*, 379 (1997). Copyright 1997 © John Wiley & Sons, Ltd.]

Two-Block PLS Algorithm with Super Block Update

Principle: Use any estimate for **u** to find $\mathbf{t_D}$ and $\mathbf{t_G}$, independently. Combine $\mathbf{t_D}$ and $\mathbf{t_G}$, into a super block **T**. Use **u** to find $\mathbf{t_T}$, use $\mathbf{t_T}$ to find a better **u**. Repeat until converge is achieved. Calculate the residuals and then extract the next set of eigenvectors.

$$\mathbf{D} = \mathbf{t_D}\,\mathbf{p'_D} \qquad \text{(A)}$$
$$\mathbf{G} = \mathbf{t_G}\,\mathbf{p'_G} \qquad \text{(B)}$$
$$\mathbf{T} = \mathbf{t_T}\,\mathbf{p'_T} \qquad \text{(C)}$$
$$\mathbf{Y} = \mathbf{u}\,\mathbf{c'} \qquad \text{(D)}$$

Preliminary steps:

   (1) Autoscale **D**, **G**, and **Y**.

Initialization:

   (2) **u** = **y**   To start, let **u** be any column of **y**.

D and G blocks:

   (3) $\mathbf{p_D} = \mathbf{D'u}$ and   Let $\mathbf{t_D} = \mathbf{t_G} = \mathbf{u}$. Use (A) and (B) to find $\mathbf{p_D}$ and
       $\mathbf{p_G} = \mathbf{G'u}$ $\mathbf{p_G}$.

   (4) $\mathbf{p_D} = \mathbf{p_D}/\|\mathbf{p_D}\|$ and   Normalize $\mathbf{p_D}$ and $\mathbf{p_G}$.
       $\mathbf{p_G} = \mathbf{p_G}/\|\mathbf{p_G}\|$

   (5) $\mathbf{t_D} = \mathbf{D}\,\mathbf{p_D}$ and   Use (A) and (B) to find $\mathbf{t_D}$ and $\mathbf{t_G}$.
       $\mathbf{t_G} = \mathbf{G}\,\mathbf{p_G}$

T block:

   (6) $\mathbf{T} = [\mathbf{t_D}\ \mathbf{t_G}]$ Combine $\mathbf{t_D}$ and $\mathbf{t_G}$ into block T.
   (7) $\mathbf{p_T} = \mathbf{T'}\,\mathbf{u}$ Let $\mathbf{t_T} = \mathbf{u}$. Use (C) to find $\mathbf{p_T}$.
   (8) $\mathbf{p_T} = \mathbf{p_T}/\|\mathbf{p_T}\|$ Normalize $\mathbf{p_T}$.
   (9) $\mathbf{t_T} = \mathbf{T}\,\mathbf{p_T}$ Use (C) to find $\mathbf{t_T}$.

Y block:

   (10) $\mathbf{c} = \mathbf{Y'}\mathbf{t_T}/\mathbf{t'_T}\mathbf{t_T}$ Let $\mathbf{u} = \mathbf{t_T}$. Use (D) to find **c**.
   (11) $\mathbf{u} = \mathbf{Y}\,\mathbf{c}/\mathbf{c'}\mathbf{c}$ Use (D) to find **u**.

Check convergence:

(12) Repeat steps (2) to (11) until $\mathbf{u}$ in step (11) converges. Calculate the residuals:

(13) $b_{inner} = \mathbf{u}'\mathbf{t}_T/\mathbf{t}_T'\mathbf{t}_T$ Calculate the inner relation between $\mathbf{t}_T$ and $\mathbf{u}$.

(14) $\mathbf{Ey} = \mathbf{Y} - b_{inner}\mathbf{t}_T\mathbf{c}'$ Calculate the residual error in $\mathbf{Y}$.

(15) $\mathbf{Y} = \mathbf{Ey}$ Label the residual $\mathbf{Y}$.

(16) $\mathbf{p}_{DT} = \mathbf{D}_T'/\mathbf{t}_T'\mathbf{t}_T$ and   Use the super block to calculate $\mathbf{p}_{DT}$ and $\mathbf{p}_{GT}$.
$\mathbf{p}_{GT} = \mathbf{G}'\mathbf{t}_T/\mathbf{t}_T'\mathbf{t}_T$

(17) $\mathbf{D} = \mathbf{D} - \mathbf{t}_T\mathbf{p}_{DT}'$ and   Calculate and label the residuals $\mathbf{D}$ and $\mathbf{G}$.
$\mathbf{G} = \mathbf{G} - \mathbf{t}_T\mathbf{p}_{GT}'$

(18) $SSEy = trace(\mathbf{Ey}'\mathbf{Ey})$ Calculate the sum of squares of the residual error in $\mathbf{Y}$.

(19) Go to step (2).   Extract the next set of eigenvectors if SSEy is too large.

Predict new responses $\mathbf{y}_{new}$ using values based on new descriptor blocks $\mathbf{D}_{new}$ and $\mathbf{G}_{new}$:

(1') Autoscale $\mathbf{D}_{new}$ and $\mathbf{G}_{new}$ consistent with that used to calibration set $\mathbf{D}, \mathbf{G}, \mathbf{Y}$.

(2') $\mathbf{t}_{Dnew} = \mathbf{D}_{new}\mathbf{p}_D$ Use (A) and (B) to find $\mathbf{t}_{Dnew}$ and $\mathbf{t}_{Gnew}$.
$\mathbf{t}_{Gnew} = \mathbf{G}_{new}\mathbf{p}_G$

(3') $\mathbf{t}_{Tnew} = p_{T(D)}\mathbf{t}_{Dnew}$ Calculate $\mathbf{t}_{Tnew}$ associated with the super matrix.
$+ p_{T(g)}\mathbf{t}_{Gnew}$

(4') $\mathbf{u}_{new} = b_{inner}\mathbf{t}_{new}$ Calculate $\mathbf{u}_{new}$ using the inner relationship.

(5') $\mathbf{Y}_{new} = \mathbf{u}_{new}\mathbf{c}'$ Calculate the contribution to $\mathbf{Y}_{new}$.

(6') $\mathbf{D}_{new} = \mathbf{D}_{new} - \mathbf{t}_T\mathbf{p}_{DT}'$ Calculate the residuals.
$\mathbf{G}_{new} = \mathbf{G}_{new} - \mathbf{t}_T\mathbf{p}_{GT}'$

(7') Repeat steps (1') to (6') until all significant latent factors have been employed.

(8') $\mathbf{Y}_{new} = \Sigma\mathbf{Y}_{new}$ Add the contributions of each latent factor.

(9') Correct $\mathbf{Y}_{new}$ for autoscaling.

The super block update method is identical to ordinary PLS when the D and B blocks are amalgamated into a single X block. However, the separation of process blocks provides additional information about the nature of the separate blocks.

8.7   SERIAL PLS

Although serial partial least squares (S-PLS) is a multiblock model, the underlying philosophical principles are quite different from other multiblock methods. S-PLS treats the variable blocks in series rather than in parallel. The second block, X_2, is used to model the variation in Y that is not modeled by the first block, X_1. The underlying equations are

$$X_1 = T_1 P_1' + E_1$$
$$X_2 = T_2 P_2' + E_2 \qquad (8.28)$$
$$Y = T_1 C_1' + T_2 C_2' + F$$

Accordingly, the regression coefficients, B_1 and B_2, are intimately related to the response Y.

$$Y = X_1 B_1 + X_2 B_2 + F \qquad (8.29)$$

The algorithm developed for this model by Berglund and Wold[24] is given below.

S-PLS Algorithm
Principle: All intermediate steps employ classical PLS. Use the X_1 to model Y. Iterate until convergence is reached, using X_2 to model the residual of Y from the first block and X_1 to model the residual of Y from the second block and X_2.

Initialization:

(1) Autoscale X_1, X_2, and Y.
(2) Estimate the number of latent factors, n, by modeling Y with $X = [X_1 \ X_2]$, the combined blocks, using cross validation or any other technique.
(3) $F_2 = Y$ To start, let F_2 equal Y.

Use classical PLS to model F_2 with X_1 using n latent factors:

(4) $F_1 = Y - T_1 C_1'$ Calculate and label the residual F_1.

Use classical PLS to model F_1 with X_2 using n latent factors:

(5) $F_2 = Y - T_2 C_2'$ Calculate and label the residual F_2.

Repeat steps (3) to (5) until convergence is achieved.

Calculate the coefficients:

(6) $B_1 = W_1 (P_1' W_1)^{-1} C_1'$ and $B_2 = W_2 (P_2' W_2)^{-1} C_2'$

Because the blocks are treated separately by S-PLS, only variables having important information will contribute to the second block. Thus, it is possible to determine whether or not the second block has any significant modeling power. Such unique information cannot be gleaned by performing PLS on the combined blocks.

Two chemical examples, illustrating the value of S-PLS, have been nicely described by Berglund and Wold.[24] A multivariate characterization of "pulp" by NIR was successfully separated into the linear and nonlinear parts. The second study focused on the taste bitterness of 48 dipeptides. The first block contained five z scales to describe each amino acid.[25] The second block contained 22 parameters generated by the Molsurf program.[26] S-PLS demonstrated that the Molsurf parameters do not enhance the modeling power. This does not mean that the Molsurf parameters should not be used to model bitterness. It means that the five z scales sufficiently model the system.

8.8 MULTILINEAR PLS

In 1989, Ståhle[27] was the first to apply the principles of two-way PLS to three-way data. In 1996, Bro[28] developed a general multiway PLS regression model (N-PLS) that was optimally correct. N-PLS has several advantages over bilinear PLS. It is more stable and less perturbed by noise. It has greater predictive power and easier to interpret. In bilinear PLS the independent block, $\mathbf{X}$, contains one row for each object, each row containing the measured variables. In multilinear PLS the independent block, $\underline{\mathbf{X}}$, contains several rows for each object. For example, if $\underline{\mathbf{X}}$ is a three-way matrix ($I \times J \times K$) with elements x_{ijk} and $\mathbf{y}$ is a vector ($I \times 1$), with elements y_i, then N-PLS will decompose $\underline{\mathbf{X}}$ into a sum of single score vectors $\mathbf{t}(I \times 1)$ and two weight vectors $\mathbf{w}^J$ and $\mathbf{w}^K$. N-PLS is based on the contention that the measurable can be modeled, in a partial least-squares sense, by the following expression:

$$x_{ijk} = t_i w_j^J w_k^K + e_{ijk} \tag{8.30}$$

so that

$$t_i = \sum_{j=1}^{J} \sum_{k=1}^{K} x_{ijk} w_j^J w_k^K \tag{8.31}$$

Partial least squares is based on maximizing the covariance between $\mathbf{t}$ and $\mathbf{y}$ by varying w_j^J and w_k^K as expressed by

$$\max\left[\sum_{i=1}^{I} t_i y_i\right] = \max\left[\sum_{i=1}^{I} \sum_{j=1}^{J} \sum_{k=1}^{K} y_i x_{ijk} w_j^J w_k^K\right] \tag{8.32}$$

Because $\underline{\mathbf{X}}$ and $\mathbf{y}$ are known, it is advantageous to define $\mathbf{Z}(J \times K)$ with elements $z_{jk} = \Sigma y_i x_{ijk}$, so that (8.32) reduces to

$$\max\left[\sum_{i=1}^{I} t_i y_i\right] = \max\left[\sum_{j=1}^{J} \sum_{k=1}^{K} z_{jk} w_j^J w_k^K\right] \tag{8.33}$$

which, in matrix form, becomes

$$\max\left[\sum_{i=1}^{I} t_i v_i\right] = \max[(\mathbf{w}^J)'\mathbf{Z}\mathbf{w}^K] \tag{8.34}$$

Vectors $\mathbf{w}^J$ and $\mathbf{w}^K$ can be determined by principal factor analysis (PFA) or by singular value decomposition (SVD), as described in Sections 3.3.2 and 3.3.7. In SVD notation, this solution is expressed as

$$(\mathbf{w}^J, \; s, \; \mathbf{w}^K) = \text{SVD}(\mathbf{Z}, \; 1) \tag{8.35}$$

In (8.35) s is the singular value (the square root of the eigenvalue). SVD($\mathbf{Z}$, 1) symbolizes the first component of the singular decomposition of $\mathbf{Z}$. It is convenient to constrain the solution to unit weight vectors:

$$\|\mathbf{W}^J\| = \|\mathbf{w}^K\| = 1 \tag{8.36}$$

The tri-PLS1 algorithm developed by Bro[28,29] and extended by Smilde[30] is presented below. In N-PLS the recommended preprocessing practice involves mean centering each column of data by subtracting the mean of the respective column from every member of the column.

Tri-PLS1 Algorithm
Principle: Calculate matrix $\mathbf{Z}$ using $\underline{\mathbf{X}}$ and $\mathbf{y}$. Obtain weight vectors, $\mathbf{w}^J$ and $\mathbf{w}^K$, by SVD of $\mathbf{Z}$. Calculate residuals. Extract the next set of weight vectors. Iterate until convergence.

(1) Center $\underline{\mathbf{X}}$ and $\mathbf{y}$.	Subtract the mean of each respective column.
(2) $\mathbf{y}_0 = \mathbf{y}$	Store the original vector.
(3) $\mathbf{Z} = \mathbf{y}'\underline{\mathbf{X}}$	Calculate $\mathbf{Z}$ with elements $z_{jk} = \Sigma y_i x_{ijk}$.
(4) $(\mathbf{w}^J, \; s, \; \mathbf{w}^K) =$ SVD($\mathbf{Z}$, 1)	Determine the weights using SVD.
(5) $t_1 = \sum_{j=1}^{J}\sum_{k=1}^{K} x_{ijk} w_j^J w_k^K$	Calculate the elements of vector $\mathbf{t}_a$ ($i = 1$ to I).
(6) $e_{ijk} = x_{ijk} - t_i w_j^J w_k^K$	Calculate the residual errors.
(7) $x_{ijk} = e_{ijk}$	Label the residuals x_{ijk}.
(8) $\mathbf{w}_a = \mathbf{w}_k^K \otimes \mathbf{w}_j^J$	Unfold the blocks.
(9) $\mathbf{W} = [\mathbf{w}_1 \mathbf{w}_2 \; \dots \; \mathbf{w}_n]$	Insert vector $\mathbf{w}_a$ into matrix $\mathbf{W}$.
(10) $\mathbf{T} = [\mathbf{t}_1 \mathbf{t}_2 \; \dots \; \mathbf{t}_n]$	Insert vector $\mathbf{t}_a$ into matrix $\mathbf{T}$.
(11) $\mathbf{b} = (\mathbf{T}'\mathbf{T})^{-1}\mathbf{T}'\mathbf{y}_0$	Based on $\mathbf{y}_0 = \mathbf{T}\mathbf{b}$.
(12) $\mathbf{E}y = \mathbf{y}_0 - \mathbf{T}\mathbf{b}$	Calculate the error in $\mathbf{y}$.
(13) $\mathbf{y} = \mathbf{E}y$	Label the residual $\mathbf{y}$.

(14) $SSEy = trace(\mathbf{Ey'\ Ey})$ Calculate the sum of squares of the residual error.

(15) Go to step (3) if the SSEy is too large.

(16) $\mathbf{b}_{PLS} = \mathbf{Wb}$

Steps (8), (9), (10), (11), and (16) are based on expressing $\mathbf{y}$ as a function of the score matrix $\mathbf{T}$ and $\mathbf{X}^{(0)}$. $\mathbf{X}^{(0)}$ is the $I \times JK$ matrix obtained by unfolding $\underline{\mathbf{X}}$. Unfolding is necessary in order to permit application for predictive purposes. The interconnecting relationship is

$$\hat{\mathbf{y}} = \mathbf{Tb} = \mathbf{X}^{(0)}\mathbf{b}_{PLS} \qquad (8.37)$$

Smilde[30] has shown that

$$\mathbf{b}_{PLS} = [\mathbf{w}_1\mathbf{w}_2 \ \dots \ \mathbf{w}_n]\mathbf{b} = \mathbf{Wb} \qquad (8.38)$$

where $\mathbf{w}_a = \mathbf{w}_k^K \otimes \mathbf{w}_j^J$, the Kronecker product defined by Eq. (7.17).

Bro[29] has also developed a tri-PLS2 algorithm that is applicable when $\mathbf{Y}$ is a matrix containing two or more columns. N-way blocks greater than three cannot be decomposed by SVD as shown above. However, their solutions can be obtained by a one-component PARAFAC decomposition.[28,29] N-PLS, involving as many as five modes, has been used to investigate three-dimensional quantitative structural activity relationships (QSAR).[31]

8.9 SOME RECENT APPLICATIONS

Since its introduction in chemistry in 1979, partial least squares has been applied to almost every branch of chemistry. Thousands of publications attest to its importance. A few of the more recent applications presented at chemometric conferences are summarized in this section.

At the *First International Conference on Chemometrics in China* held in Zhangjiajie on October 17–22, 1997, a significant number of papers were devoted to PLS. These papers were compiled into a special issue of *Chemometrics and Intelligent Laboratory Systems*, volume 45. Ni and Jin[32] used PLS, as well as PCR and ITTFA, to analyze mixtures of Pb, Cu, Va, Cd, and Ni by differential pulse polarography. To analyze mixtures of Mn, Zn, Co, and Cd from UV measurements, Gao and Ren[33] employed kernel partial least squares (KPLS). Maeda and associates[34] investigated hydrogen bonding in alcohols via PLS of NIR spectra. Eleven regression methods, including several different PLS techniques, were applied to UV spectra by Liu and Wang[35] for the purpose of analyzing solutions of pharmaceuticals containing paracetamol, chloropromazine, promethazine, caffeine, phenacetin, phenobarbital, aminophenazone, and chlorophenamine. Lipsticks and mascaras contain different waxlike materials. Li[36] quantified the different waxes in commercial products by applying PLS to data gleaned from supercritical fluid extraction/supercritical fluid chromatography. Zhou and co-workers[37] used PLS to relate antiallergic activities of substituted benzamides with 3D-QSAR factors obtained from comparative molecular field analysis (CoMFA). Li and co-workers[38] combined nonlinear PLS with a numeric genetic algorithm (NPLSNGA) to investigate the relationship between QSAR and the fungal activity of certain drugs.

A special issue of the *Journal of Chemometrics* reported a selection of papers presented at the second *International Chemometrics Research Meeting* held in Veldhoven, The Netherlands, on May 24–28, 1998. These papers applied PLS to a variety of problems in chemistry. Based on Raman spectra, a universal calibration model for off-line and on-line predictions of poly(ethylene terephthalate) yarn shrinkage was reported by Sierenga and co-workers.[39] PLS regression was used by Masserschmidt and co-workers[40] to develop a relationship between organic matter content in soils and diffuse reflectance spectra in the mid-infrared region. Teppola and Minkkinen[41] applied PLS and fuzzy C-means clustering to model wastewater treatment. Multiblock serial PLS (S-PLS) was developed by Bergland and Wold[24] and applied to characterize wood pulp and to characterize bitter tasting dipeptides.

A large number of papers, devoted to PLS at the *Sixth Scandinavian Symposium on Chemometrics* held in Porsgrunn, Norway in August 1999, were published in a special issue of the *Journal of Chemometrics*. Seasonal fluctuations and low-frequency fluctuations in paper plant processes were investigated using PLS postprocessed and preprocessed by multiple resolution analysis (PLS-MRA and MRA-PLS), methods developed by Teppola and Minkkinen.[42] Three-way and two-way PLS techniques were used by Halstensen and Ebensen[43] and Ergon and Halstensen[44] to convert acoustical signals of powders during in-line batch process into particle size distribution measurements. The coordinates of the **X** data block represented sample number, time, and acoustical frequency. Peussa and co-workers[45] used PLS, preceded by orthogonal signal correction (OSC), to convert diffuse reflectance infrared Fourier transform spectra (DRIFTS) into hydroxyl group content in calcined silica. Multivariate image analysis (MIA) depend upon a number of variables such as channel, color (wavelength), and polarizing angle. MIA data were unfolded and subjected to kernel PLS by Lied and co-workers.[46] QSAR models were investigated by Eriksson and associates[47] and by Anderson and associates.[48] These studies, respectively, concerned the classification of 351 soil sorption coefficients and 205 heterogeneous primary amines. Leardi[49] demonstrated how a genetic algorithm (GA) can select spectral features to best characterize the system for PLS analysis. The method was applied to several sets of data including gasoline, soy oil, resorcinol, foodstuff, and wheat. PLS was used to improve the manufacturing process of a cellulose derivative, ethyl-hydroxy-ethyl-cellulose by Lindgre.[50] The bleaching of Norway spruce was studied by Malkavaara and co-workers.[51] Drift in gas sensors was studied by Artursson et al.[52] Lemberge and associates[53] applied PLS to convert electron probe X-ray microanalysis (EPXMA) and micro-X-ray fluorescence (μ-XRF) analysis of 16th- and 17th-century archeological glass samples into quantifications of Na_2O, SiO_2, K_2O, CaO, MnO, and Fe_2O_3. Among other techniques, PLS was employed by Wittrup[54] to classify 137 extracts of *Penicillium* and seven extracts of pure growth medium based on their fluorescence spectra.

The studies listed in this section attest to the wide variety of chemical problems that have been studied by partial least-squares techniques. Other examples are cited in Chapter 12. Altogether they represent only a tiny sample of the thousands of applications currently reported.

8.10 AUGMENTED CLASSICAL LEAST SQUARES

Quantitative multivariate spectral analysis by classical least squares (CLS) requires that the concentrations of all spectrally active constituents be known and included in the calibration step. Because of this severe limitation PCR and PLS methods have become popular during the past 20 years since their inception into chemistry. PCR and PLS yield significantly better predictions than CLS when all of the constituents are not included in the calibration. In 1985 Marten and Naes[55,56] developed a CLS model that involved augmenting the calibration step with information extracted from the residual errors. This method, however, received little attention until the year 2000 when Haaland and Melgaard[57−59] developed a family of augmented CLS techniques that have many advantages, yielding predictions with equal or better accuracy than PCR and PLS.

The newly developed augmented classical least-squares (ACLS) methods[57−60] can be used to quantify a single component, as well as multiple components, in situations where the identities and concentrations of the other components are unknown. ACLS has been applied to situations where effects of nonlinearities or nonchemical sources appear in the spectral calibration data. By combining various ACLS methods, successful quantifications have been achieved in the presence of unmodeled spectral components, spectrometer drift, and changes in spectrometers.

The original underlying principles of ACLS, as developed by Martens and Naes[55,56] and expounded by Haaland and Melgaard,[60] is outlined below. Consider the standard CLS model based on Beer's law:

$$\mathbf{A} = \mathbf{CK} + \mathbf{E_A} \tag{8.39}$$

where $\mathbf{A}$ is the spectral matrix of the multicomponent mixture, $\mathbf{C}$ is the concentration of the components, $\mathbf{K}$ is the response matrix, and $\mathbf{E_A}$ is the error matrix. Each row of $\mathbf{A}$ contains the spectra of a sample. Each column of $\mathbf{C}$ contains the concentration of a specific analyte. $\mathbf{E_A}$ is determined using $\hat{\mathbf{K}}$, the CLS estimate of $\mathbf{K}$,

$$\mathbf{E_A} = \mathbf{A} - \mathbf{C\hat{K}} \tag{8.40}$$

In (8.40) $\hat{\mathbf{K}}$ contains the CLS estimated responses (the spectra) of the known pure components, obtained by the pseudoinverse computation:

$$\hat{\mathbf{K}} = \mathbf{C^{+}A} \tag{8.41}$$

If all of the components are not included in $\mathbf{C}$, (8.41) will represent an approximation of $\mathbf{K}$.

If $\mathbf{C}$ is incomplete, or the model is nonlinear, the error matrix will include correlated errors. These errors contain information about the unspecified components as well as the component(s) of interest. Factor analysis can be used to further separate these errors into spectral components, correlated system errors and uncorrelated errors as expressed in (8.42):

$$\mathbf{E}_\mathbf{A} = \mathbf{SL} + \mathbf{E} \tag{8.42}$$

Here $\mathbf{S}$ and $\mathbf{L}$ represent, respectively, the score and loading matrices obtained from the factor analysis of $\mathbf{E}_\mathbf{A}$. $\mathbf{E}$ represents the random errors that compose the true null space, devoid of any useful information. For optimal prediction the true rank of $\mathbf{E}_\mathbf{A}$ must be accurately determined.

An augmented matrix $\tilde{\mathbf{K}}$ is created by inserting the loadings from (8.42) into $\hat{\mathbf{K}}$ so that

$$\tilde{\mathbf{K}} = \begin{bmatrix} \hat{\mathbf{K}} \\ \mathbf{L} \end{bmatrix} \tag{8.43}$$

The augmented CLS prediction estimate, $\tilde{\mathbf{C}}$, is then determined by (8.44):

$$\tilde{\mathbf{C}} = \mathbf{A}\tilde{\mathbf{K}}^+ \tag{8.44}$$

This augmentation improves the predictive ability of CLS because it takes into consideration the latent contributions of the unknown constituents. This procedure is referred to as spectral residual augmented classical least squares (SRACLS).

Alternatively, the score matrix, $\mathbf{S}$, obtained from (8.42) can be used to correct the CLS model. This method is referred to as the score-augmented classical least squares (SACLS).

The prediction augmented CLS (PACLS) method[57−60] is applicable when sources of spectral variation, neglected in the CLS model, can be identified or measured empirically. PACLS improves the predictive ability by accounting for new sources of variations in the spectral measurements not originally present in the calibration spectra. PACLS is particularly useful for updating spectral calibration models when changes, such as instrument drift, occur.

Concentration residual augmented CLS (CRACLS)[59] accounts for unmodeled sources of spectral variation without invoking PCR or PLS. CRACLS subjects concentration residuals to an iteration procedure involving a series of CLS analyses. The concentration residual matrix, $\mathbf{E}_\mathbf{C}$, is defined as:

$$\mathbf{E}_\mathbf{C} = \hat{\mathbf{C}} - \mathbf{C} \tag{8.45}$$

To accommodate unmodeled components, Beer's law takes the following form:

$$\mathbf{A} = \mathbf{CK} + \mathbf{C}_\mathbf{u}\mathbf{K}_\mathbf{u} + \mathbf{E} \tag{8.46}$$

where C_u and K_u represent the concentrations and pure spectra of the unmodeled components, and E represents the remaining error after removing unmodeled spectra. K_u can be decomposed into a sum of two terms:

$$K_u = DK + G \tag{8.47}$$

where DK represents the part of K_u that projects into the factor space spanned by K, and G is the part of K_u that is orthogonal to K. Inserting (8.47) into (8.46) gives

$$A = (C + C_u D)K + C_u G + E \tag{8.48}$$

Hence, the predicted concentration obtained by applying (8.44) is

$$\tilde{C} = C + C_u D \tag{8.49}$$

and the residual concentration error, defined in (8.45), is

$$E_C = C_u D \tag{8.50}$$

Although D is unknown, each column, e_c, of E_C approximates a linear combination of unknown concentrations. The original calibration matrix, C, can be augmented by inserting any one of the e_c vectors as an additional column in C. Only one such vector is required since all the other error vectors contain redundant information. An augmented pure-component spectral matrix, $\tilde{K}$, is obtained by applying an expression analogous to (8.41). The new $\tilde{K}$ is then used to predict $\tilde{C}$. A new E_C is computed, and a column vector of the new E_C is added to the previously augmented $\tilde{C}$. This process is repeated until no further improvement in the predictions occurs. A CRACLS algorithm, illustrating the steps in the iteration procedure, is provided below.

Concentration Residual Augmented Classical Least-Squares Algorithm
Principle: To start, use classical least squares to find an approximate spectral matrix, Kx, in accord with Eq. (A). Use this matrix to predict the concentrations, Cx, of the analyte(s), in accord with Eq. (A). Calculate the residual concentration errors, E, associated with the analyte(s) using Eq. (B). Augment the calibration concentration matrix by adding a single error vector, a column of the residual concentration error matrix, in accord with Eq. (C). Repeat this process until no further improvement is observed in the concentration reproductions.

$$A = Cx\ Kx \tag{A}$$

$$E = Cx - C \tag{B}$$

$$Cx = [Cx\ E(:, 1)] \tag{C}$$

(1) $Cx = C$ Store the analyte calibration data in C.
(2) $Kx = Cx^+ A$ Use (A) to find Kx.
(3) $Cxx = A\ Kx^+$ Use (A) to find Cxx.
(4) $E = Cxx(:, 1:nc) - C$ Use (B) to calculate the residual concentration errors.

(5) $Cx = [Cx\ E(:, 1)]$ Use (C) to augment the concentration matrix.
(6) Go to step (2).   Repeat until no further improvement occurs.
(7) $Kx = Cx^+A$ Use (A) to find Kx.

$Cxx(:, 1:nc)$ represents columns 1 through nc, where nc is the number of analytes involved in the calibration. $E(:, 1)$ represents the first column of matrix E, although any other column may be used.

To accommodate spectral variations that result from instrument drift, sample insertion and other instrumental artifacts, PACLS is used in combination with CRACLS or SRACLS. The usual outlier detection methods involving F ratios[61] and Mahalonobis distances[62] have been developed for ACLS.

A comparison of the results obtained by CLS, PLS, and CRACLS, based on mean-centered spectra, is shown in Table 8.1. This study involved the near-infrared measurements (from 7,500 to 11,000 cm^{-1}) of dilute aqueous solutions containing glucose, ethanol, and urea, four chemical constituents that absorb infrared radiation. The calibration models were based on 27 samples. The validation set also involved 27 samples based on the same sample design as the calibration set. The values in the table represent the standard errors of prediction (SEP) and the cross-validated standard errors of prediction (CVSEP). The PLS and CRACLS calibration models were developed using only one of the three analytes at a time. The CLS model required simultaneous fitting of all three analytes.

Although the spectral features of the water dominate the spectra, instrumental drift and temperature variations still remain after mean-centering the spectra. Consequently, as seen in Table 8.1, the fully specified CLS model is inadequate for quantification. However, note that both PLS and CRACLS, based on single-component calibration, inherently corrects for the unspecified constituents as well as some of the experimental variables. Unfortunately, PLS is sensitive to the number of latent factors employed in the computations, overfitting degrades the predictivity. Preliminary studies indicate that CRACLS does not suffer from overfitting. Most important, the CRACLS/PACLS combinations is a time and cost-efficient way to

TABLE 8.1   SEP for CLS and CVSEP for PLS and CRACLS for Prediction Set[a]

Component	CLS	PLS[b]	CRACLS[c]
Glucose (mg/dL)	287	21 (12)	20 (20)
Ethanol (mg/dL)	27	4 (13)	4 (20)
Urea (mg/dL)	42	4 (12)	4 (20)

[a] Based on Ref. 59. Reprinted with permission from D. K. Melgaard, D. M. Haaland, and C. M. Wehlburg, private communication.
[b] Values in parentheses represent the optimal number of PLS factors required.
[c] Values in parentheses represent the number of augmentation used in the calibration.

update previous calibration models. Recent studies of Wehlburg, Haaland, and Melgaard[63] have shown that CRACLS, in comparison to PLS, gives better qualitative information about the analytes, providing better estimates of the pure component spectra.

REFERENCES

1. H. J. Iglarsh and D. C. Cheng, *J. Statist. Comput. Simul.*, **10**, 103 (1980).
2. Y. L. Xie and J. H. Kalivas, *Anal. Chim. Acta*, **348**, 19 (1997).
3. S. Z. Fairchild and J. H. Kalivas, *J. Chemometrics*, **15**, 1 (2001).
4. P. Geladi, *J. Chemometrics*, **2**, 231 (1988).
5. H. Wold, *Eur. Econ. Revs.*, **5**, 67 (1974).
6. H. Wold, in C. Fornell (Ed.), *A Second Generation of Multivariate Analysis*, Preager, New York, 1982, p. 325.
7. R. Gerlach, B. R. Kowalski, and H. Wold, *Anal. Chim. Acta*, **112**, 417 (1979).
8. S. Wold, *Technometrics*, **20**, 397 (1978).
9. T. Naes and H. Martens, *Commun. Statist. Comput.*, **14**, 545 (1985).
10. P. Geladi and B. R. Kowalski, *Anal. Chim. Acta*, **185**, 1 (1986); **185**, 18 (1986).
11. F. Lindgren, P. Geladi, and S. Wold, *J. Chemometrics*, **7**, 45 (1993).
12. A. Höskuldsson, *J. Chemometrics*, **2**, 211 (1988).
13. S. Rännar, F. Lindgren, P. Geladi, and S. Wold, *J. Chemometrics*, **8**, 111 (1994).
14. B. S. Dayal and J. F. MacGregor, *J. Chemometrics*, **11**, 73 (1997).
15. M. C. Denham, *J. Chemometrics*, **14**, 351 (2000).
16. M. Forina, C. Casolino, and C. Pizarro Millan, *J. Chemometrics*, **13**, 165 (1999).
17. J. A. Gil and R. Romera, *J. Chemometrics*, **12**, 365 (1998).
18. S. de Jong, B. M. Wise, and N. L. Richer, *J. Chemometrics*, **15**, 85 (2001).
19. C. Goutis, *J. Chemometrics*, **11**, 33 (1997).
20. M. Denham, *J. Chemometrics*, **11**, 39 (1997).
21. J. H. Kalivas, *J. Chemometrics*, **13**, 111 (1999).
22. J. H. Westerhuis and P. M. J. Coenegracht, *J. Chemometrics*, **11**, 379 (1997).
23. L. E. Wangen and B. R. Kowalski, *J. Chemometrics*, **3**, 3 (1988).
24. A. Berglund and S. Wold, *J. Chemometrics*, **13**, 461 (1999).
25. J. Jonsson, L. Eriksson, S. Hellberg, M. Sjöström, and S. Wold, *Quant. Struct.-Act. Relat.*, **8**, 204 (1989).
26. P. Sjöberg, *Molsurf Version 2.1*, Qemist AB, Karlskoga (1997).
27. L. Ståhle, *Chemometrics Intell. Lab. Syst.*, **7**, 95 (1989).
28. R. Bro, *J. Chemometrics*, **10**, 47 (1996).
29. R. Bro, Multiway Analysis in the Food Industry, Doctoral Thesis, University of Amsterdam (1998).
30. A. K. Smilde, *J. Chemometrics*, **11**, 367 (1997).
31. J. Nilsson, S. D. Jong, and A. K. Smilde, *J. Chemometrics*, **11**, 511 (1997).

32. Y. Ni and L. Jin, *Chemometrics Intell. Lab. Syst.*, **45**, 105 (1999).
33. L. Gao and S. Ren, *Chemometrics Intell. Lab. Syst.*, **45**, 87 (1999).
34. H. Maeda, Y. Wang, Y. Ozaki, M. Suzuki, M. A. Czanecki, and M. Iwahashi, *Chemometrics Intell. Lab. Syst.*, **45**, 121 (1999).
35. S. Liu and W. Wang, *Chemometrics Intell. Lab. Syst.*, **45**, 131 (1999).
36. J. Li, *Chemometrics Intell. Lab. Syst.*, **45**, 385 (1999).
37. Y. Zhou, L. Xu, Y. Wu, and B. Liu, *Chemometrics Intell. Lab. Syst.*, **45**, 95 (1999).
38. T. Li, H. Mei, and P. Cong, *Chemometrics Intell. Lab. Syst.*, **45**, 177 (1999).
39. H. Swierenga, A. P. de Weijer, and L. Buydens, *J. Chemometrics*, **13**, 237 (1999).
40. I. Masserschmidt, C. J. Cuelbas, R. J. Poppi, J. C. de Andrade, C. A. de Abreu, and C. Davanzo, *J. Chemometrics*, **13**, 265 (1999).
41. P. Teppola and P. Minkkinen, *J. Chemometrics*, **13**, 445 (1999).
42. P. Teppola and P. Minkkinen, *J. Chemometrics*, **14**, 383 (2000).
43. M. Halstensen and K. Esbensen, *J. Chemometrics*, **14**, 463 (2000).
44. R. Ergon and M. Halstensen, *J. Chemometrics*, **14**, 617 (2000).
45. M. Peusso, S. Härkönen, J. Puputti, and L. Niinistö, *J. Chemometrics*, **14**, 501 (2000).
46. T. T. Lied, P. Geladi, and K. H. Esbensen, *J. Chemometrics*, **14**, 585 (2000).
47. L. Eriksson, E. Johansson, M. Müller, and S. Wold, *J. Chemometrics*, **14**, 599 (2000).
48. P. M. Anderson, M. Sjöström, S. Wold, and T. Lundstedt, *J. Chemometrics*, **14**, 629 (2000).
49. R. Leardi, *J. Chemometrics*, **14**, 643 (2000).
50. Å. Lindgren, *J. Chemometrics*, **14**, 657 (2000).
51. P. Malkavaara, J. P. Isoaho, R. Alén, and J. Soininen, *J. Chemometrics*, **14**, 693 (2000).
52. T. Artursson, T. Eklöv, I. Lunström, P. Mårtensson, M. Sjöström, and M. Holmberg, *J. Chemometrics*, **14**, 711 (2000).
53. P. Lemberge, I. de Raedt, K. H. Janssens, F. Wei, and P. J. van Espen, *J. Chemometrics*, **14**, 751 (2000).
54. C. Wittrup, *J. Chemometrics*, **14**, 765 (2000).
55. H. Martens and T. Naes, *Multivariate Calibration*, Wiley, Chichester, 1989.
56. H. Martens and T. Naes, in P. C. Williams and K. Norris (Eds.), *Near-Infrared Technology in Agricultural and Food Industries*, American Association of Cereal Chemists, St. Paul, MN, 1987, pp. 57–87.
57. D. M. Haaland and D. K. Melgaard, *Appl. Spectrosc.*, **54**, 1303 (2000).
58. D. M. Haaland and D. K. Melgaard, *Appl. Spectrosc.*, **55**, 1 (2001).
59. D. K. Melgaard, D. M. Haaland, and C. M. Wehlburg, submitted to *Appl. Spectrosc.*
60. D. M. Haaland and D. K. Melgaard, submitted to *Vibrational Spectroscopy*.
61. D. M. Haaland and E. V. Thomas, *Anal. Chem.*, **60**, 1193 (1988).
62. H. Mark, *Anal. Chem.*, **59**, 790 (1987).
63. C. M. Wehlburg, D. M. Haaland, and D. K. Melgaard, unpublished results.

9

There is no such thing as a simple solution.

COMPONENT ANALYSIS

In this chapter we describe how investigators have made use of the factor analytical approach for both qualitative and quantitative analysis. The discussion involves a variety of analytical methods such as absorption and emission spectroscopy, optical rotation, gas chromatography, mass spectrotometry, and nuclear magnetic resonance. Our intent is to illustrate the various factor analysis (FA) approaches that can be employed for component analysis.

9.1 ABSORPTION SPECTRA

9.1.1 Rationale

Factor analysis has been used in a powerful fashion for determining the number of components that contribute to the absorption spectra of multicomponent systems. This is to be expected because Beer's law, for a multicomponent system, involves a linear sum of product functions:

$$A = \sum_{j=1}^{n_c} \epsilon_j c_j \qquad (9.1)$$

Here A is the absorbance at a given wavelength measured in a cell of unit path length, ϵ_j the extinction coefficient of the jth component, and c_j the concentration of component j, the sum being taken over all n_c components in the mixture. Factor analysis is ideally suited to systems where the absorbance spectra of each individual component differ significantly.

For a series of solutions having the same species but different concentrations, the absorbances are best expressed as

$$A_{\lambda k} = \sum_{j=1}^{n_c} \epsilon_{\lambda j} c_{jk} \qquad (9.2)$$

Here λ refers to the wavelength, k refers to the particular solution, and j refers to the component. These equations can be written in a more compact form using matrix notation:

$$\mathbf{A} = \mathbf{EC} \qquad (9.3)$$

If there are n_c absorbing components in n_s solutions and the measurements are made at n_w wavelengths, then $\mathbf{A}$ is an $n_w \times n_s$ absorbance matrix, $\mathbf{E}$ is an $n_w \times n_c$ extinction coefficient matrix, and $\mathbf{C}$ is an $n_c \times n_s$ concentration matrix.

The absorbance data matrix is obtained by digitizing the spectra at intervals over a wide wavelength range. In this way, the entire band contour is analyzed rather than only the major spectral features. The use of a large number of digitized wavelengths is desirable since this will yield statistically more accurate results. The number of solutions of different composition to be used in the factor analysis study usually poses a more time-consuming problem. However, the larger the number of solutions involved, the more reliable the results. To ensure that the rank of the data matrix, as determined by factor analysis, will equal the number of components, both the number of digitized wavelengths and the number of solutions must not be less than the number of components. It is relatively easy to use several hundred digitized wavelengths, whereas 10 solutions are usually considered sufficient when there are about 6 or fewer components.

Because of experimental error, factor analysis will always yield a set of eigenvectors whose number will be equal to the number of columns or rows in the data matrix, whichever is smaller. However, eigenvectors associated with small eigenvalues simply reproduce experimental error and should be disregarded. A variety of tests, developed specifically to estimate the rank of a data matrix, are discussed in Chapter 4. It is good practice to use a combination of these methods to deduce the dimensionality of the factor space, which, in these problems, is equivalent to the number of components in the mixtures.

9.1.2 Pioneering Efforts

Although the early investigators did not use factor analysis to determine the number of components, their pioneering efforts laid the foundation for its ultimate use. For this reason we briefly review the early methodology in chronological order.

In 1960, Wallace[1] was the first to recognize that rank analysis of absorption spectra could be used to determine the number of components in a mixture (the first step of factor analysis). He recorded, at 25-nm intervals, the visible spectra of methyl orange and methyl red indicators, as well as mixtures of the indicators, in buffered solutions of known pH. To determine the rank of the **A** matrix, Wallace did not employ the sophisticated factor analysis procedure described in Chapter 3. Instead, he used a simple statistical criterion, which states that the determinant of a singular matrix is equal to its standard deviation. A square singular matrix is one in which the number of rows, or columns. equals the rank plus one. The rank of the data matrix equals the order of the largest nonsingular submatrix. The problem then is to find this submatrix.

The rank of each **A** matrix was deduced in the following way. First every possible 2×2 submatrix that could be formed from the **A** matrix was constructed. The value of the determinant, d, for each 2×2 submatrix was then compared to the standard deviation, σ. If the determinant was close to the standard deviation, the submatrix was deemed singular. If one of the 2×2 submatrices was nonsingular, as proved by $d > \sigma$, every possible 3×3 submatrix was constructed by adding a row and a column to this submatrix. This procedure of adding rows and columns was repeated until the largest singular matrix was found. The rank was taken to be the number of rows (or columns) in this matrix. Using this technique, Wallace concluded that solutions of supposedly pure methyl red and pure methyl orange each contain two absorbing components. This technique was also applied to data concerning mixtures of the two dyes. Rank analysis showed that there were four components present in the mixtures.

Although this method is readily applicable to small data matrices, it is unwieldy when large matrices are involved. Furthermore, the method leads to a dead end in the overall FA scheme because it can yield only one piece of information, the rank of the data matrix, and nothing more. However, this pioneering investigation brought to the attention of the chemist the fact that matrix rank analysis is a viable method for determining the number of absorbing components in a series of related mixtures.

Because this method is excessively tedious to carry out when large data matrices are involved, Wallace and Katz[2] developed the following alternative method. In addition to the absorbance matrix, **A**, an error matrix, **S**, was constructed. Each element of the error matrix is simply the estimated error associated with the corresponding element of the absorbance matrix. By using standard mathematical procedures,[3] the **A** matrix is reduced to a matrix whose elements are all zero below the principal diagonal. At each step in the reduction of **A**, appropriate operations, based on statistical theory, are performed on the error matrix, yielding a reduced **S** matrix. The rank of the **A** matrix is equivalent to the number of diagonal elements in the reduced **A** matrix, with absolute values greater than three times the corresponding elements in the reduced **S** matrix.

The original **A** matrix consisted of the absorbances of a methyl red solution as a function of wavelength and pH. Because the concentration of an absorbing ligand is a function of pH, varying the pH caused a change in absorbance. The original **S** matrix consisted of 64 identical elements, all equal to the experimental error, 0.003. Comparing the reduced **A** matrix to the reduced **S** matrix, Wallace and Katz[2] concluded that there were at least three absorbing components present, and possibly four, since one element in the reduced **A** matrix was approximately three times the corresponding element in the reduced **S** matrix.

Varga and Veatch,[4] relying on the computer method of Wallace and Katz,[2] investigated the nature and stabilities of hafnium chloranilic acid metallochrome. Two series of solutions were prepared, one containing chloranilic acid at various concentrations and the other containing both hafnium (IV) and chloranilic acid. The concentrations of chloranilic acid in the first set of 12 solutions were identical to the second set, which contained both hafnium and chloranilic acid. In the second set of 12 solutions, the total molar concentration of hafnium plus chloranilic acid was held constant. Absorbance measurements were made at 5-nm intervals from 260 to 360 nm over the region of maximum absorption. Thus two 21×21 individual data matrices were investigated. Their ranks were determined by comparing the reduced absorbance matrices with their reduced error matrices. They concluded that three absorbing species were present in the hafnium–chloranilic acid solutions.

In deducing the rank, a knowledge of the size of the experimental error is extremely important. Varga and Veatch[4] recognized this fact and made a systematic study of the effect of error on both the chloranilic acid and hafnium–chloranilic acid systems. When the overall absorbance error estimate was varied from 0.003 to 0.050, for chloranilic acid, the rank changed from 5 to 1. Since only one species, undissociated chloranilic acid, is expected to exist in $3\,M$ perchloric acid solution, the rank should be 1. This is consistent with the results if the error is estimated to be about 0.025 absorbance unit. The rank increases dramatically when the error is assumed to be less than 0.010 absorbance unit, reflecting the sensitivity to random fluctuations in absorbance measurements rather than to the number of absorbing species present.

A similar situation exists for the hafnium–chloranilic acid mixtures. The accepted rank is three if the error is estimated to be 0.025 unit. If the error is assumed to be less than 0.016 unit, the rank fallaciously increases. If it is assumed to be greater than 0.040, the rank fallaciously decreases.

Because of this sensitivity, Varga and Veatch[4] recognized that the assumption of a constant absorbance error could lead to incorrect conclusions. They calculated the absorbance error, $S_{\lambda k}$, from the photometric error in the transmittance, ΔT, and the measured absorbance, $A_{\lambda k}$:

$$S_{\lambda k} = 0.43429 \cdot \Delta T \cdot \text{antilog } A_{\lambda k} \qquad (9.4)$$

Using this equation together with the experimental absorbances for the hafnium–chloranilic acid solutions, they generated a series of error matrices based on different estimates of ΔT ranging from 0.001 to 0.050. Again the conclusions concerning the ranks of the matrices were found to depend on an accurate estimation of the photometric error. For the Beckman DU spectrophototometer employed, the photometric error was in the range between 0.3 and 0.5% ($\Delta T = 0.003$ to 0.005). This led to the conclusion that there were three species present.

Katakis[5] developed a computer method for determining the rank of an absorbance matrix, based on the Gauss process of elimination.[6] Here the **A** matrix is reduced by subtracting an appropriate matrix constructed from the elements of the row and column associated with the largest element in the matrix. This subtraction is repeated until all the elements of the residual matrix are less, in absolute value, than the corresponding elements of the error matrix. The rank of **A** equals n, the number of matrix subtractions required. Katakis applied this method to study the absorbance of Cr^{2+} solutions.

A graphical method for determining the rank of the absorbance matrix was developed by Coleman et al.[7] This method relies on the same numerical relationships used previously but does not require sophisticated computer analysis. The necessary computations can be done rapidly with a desk calculator. When applied to the chloranilic acid and methyl red spectra previously discussed, the same conclusions were reached.

All the methods just described represent variations in different rank analysis techniques. These studies are included here to stress the historical sequence and the importance of rank analysis in spectrophotometric studies of multicomponent systems. Rank analysis is the first step in factor analysis.

9.1.3 Factor Analyses

Kankare[8] recognized that factor analysis was an ideal mathematical tool for determining the number of components in a solution from its absorption spectrum. He was inspired by Simmonds,[9] who applied factor analysis to optical response in photography; by Reeves,[10] who used factor analysis to separate medium effects from concentration effects in dye solutions; and by Wernimont,[11] who used factor analysis to evaluate the performance of different spectrophotometers (see Section 12.4).

He recorded the absorption spectra of solutions containing 8×10^{-5} M bismuth ion, 1 M perchloric acid, and varying amounts of sodium chloride and sodium perchlorate to maintain a constant ionic strength. The spectra were recorded against blanks having the same composition but no bismuth. Absorbance measurements were made on 17 solutions between 230 and 360 nm at 5-nm intervals.

The rank of the absorbance matrix was determined by comparing the residual standard deviation with the estimated deviation. This method is described in Chapter 4. Since the residual standard deviation calculated from factor analysis must be less than the estimated deviation, Kankare concluded that seven absorbing species were present. These species were suspected to be Bi^{3+}, $BiCl^{2+}$, $BiCl_2^+$, $BiCl_3$, $BiCl_4^-$, $BiCl_5^{2-}$, and $BiCl_6^{3-}$.

To improve the data Kankare substituted factor analysis-regenerated data points for all points with excessive errors. A data point was considered to have an excessive error if the absolute value of the difference between the measured absorbance and the AFA-regenerated absorbance was greater than three times the standard deviation. Such errors were considered to be accidentally excessive. Thus data points with large errors were easily spotted and removed. The smoothed data matrix was then factor analyzed, yielding, hopefully, more reliable results.

Factor analysis of the absorbance matrix does not yield directly an extinction coefficient matrix E and a concentration matrix C in a true chemical sense, as portrayed by (9.3). Instead, it yields mathematical solutions that have basis axes (eigenvectors) that lie in the same chemical space but do not necessarily coincide with the chemical axes. These mathematical axes must be transformed into chemical axes. Additional information is required to perform such transformations.

Kankare[8] was able to obtain the concentration matrix by first speculating what the seven absorbing bismuth species were and then using, as a first approximation, the formation constants for these species as determined by other methods. He then developed a least-squares method for the purpose of adjusting these constants to give the best set compatible with the abstract matrices of factor analysis. The formation constants from the factor analysis study were considered to be more accurate and reliable.

Knowing A and having obtained C, Kankare calculated E. The elements of each column of E trace out the spectrum of one of the absorbing bismuth species, even though it is impossible to obtain the spectra of these species by direct spectrophotometric measurement. With the aid of factor analysis, the spectrum of each absorbing species in the complicated mixture was obtained.

Many years later the bismuth–chloride equilibrium system was reinvestigated by Gemperline and Hamilton[12] who employed an ion chromatograph coupled to an ultraviolet–visible (UV–VIS) diode array detector. Their data, illustrated in Figure 6.7, were subjected to evolving factor analysis (EFA), which revealed the concentration profiles as well as the spectra of the ionic species (see Figures 6.8 and 6.9), in agreement with Kankare's results. This is most impressive because the EFA method is self-modeling, requiring no assumptions, information, or modeling parameters as required in the earlier work.

Factor analysis of absorption spectra was used by Hugus and El-Awady[13] in their investigation of the hydrolytic depolymerization of certain binuclear cobalt (III) complexes. To test their detailed kinetic model, they needed to know how many absorbing species were present. Their data matrix consisted of the absorbances of 38 solutions measured at 9 wavelengths. Four criteria for deducing the number of species were employed: (1) trends in the eigenvalue, (2) the standard error in the eigenvalue, (3) the number of residuals greater than three times the estimated standard deviation, and (4) the chi-squared test. These criteria are fully discussed in Section 4.3. The results shown in Table 9.1 indicate that 3 species are present. As shown in the table, there is a severe drop in the eigenvalue when n is changed from 3 to 4, indicating three factors. The first 3 eigenvalues are greater than their respective standard errors, whereas the remaining 6 eigenvalues are much smaller than their standard errors. Of the 342 data points, 155 reproduced data points have an error greater than 3 times the estimated error when 2 factors are employed. When 3 factors are employed, all reproduced data points have errors less than 3 times the estimated error. The expectation value $[\chi_n^2(\text{expected}) = (r - n)(c - n)]$ using 2 factors is much smaller than the calculated χ_n^2, whereas the expectation value using 3 factors is closer to χ_n^2. All these criteria give evidence that 3 species are present.

By studying the imbedded error (IE) function, the factor indicator function (IND), and the significance level (%SL) (see Section 4.3), Malinowski[14-16] substantiated the conclusion that three components are present. As shown in Table 9.1, there is no further reduction in IE upon using more than three eigenvectors. Furthermore, the IND function reaches a minimum at $n = 3$, again indicating three factors. This conclusion is confirmed statistically by examining the significance levels obtained from the F test of the reduced eigenvalues. Reading from the bottom of the table upward, we see that the %SL values are highly significant (indicative of error eigenvalues) until we reach $n = 3$. For three factors, the real error (RE) is 0.00104 absorbance unit. Hence without recourse to any knowledge of the experimental error, as required by the previous criteria, Malinowski deduced not only the number of components but also the experimental error.

The data of Wallace and Katz[2] concerning methyl red solutions as a function of pH, discussed in Section 9.1.2, were factor analyzed by Hugus and El-Awady.[13] All four error criteria gave evidence that there were only three components. These calculations were based on the same reasonable error estimation as given by Wallace and Katz. The questionable fourth component mentioned by Wallace and Katz was clearly ruled out by this study. Malinowski,[14] however, presented evidence that the error estimation used to reach this conclusion was too large. He argued that there were four factors involved because the IND function exhibited a minimum at $n = 4$, yielding a real error of 0.00154 absorbance unit, considerably less than 0.003 used by the previous investigators.

Principal factor analysis of the Fourier transform infrared spectra, recorded and digitized from 500 to 3500 cm^{-1}, was used by Rasmussen and co-workers[17] to correctly identify the number of components in artificial mixtures of xylenes. However, the method was unsuccessful for determining the number of components in mixtures of alkanes because the infrared spectra of the individual components were not distinguishable. McCue and Malinowski[18] used target factor analysis of ultraviolet spectra in the region from 260 to 280 nm not only to chemically identify

TABLE 9.1 Factor Analysis Study of the Absorbances of a Co(III) Complex and Its Hydrolysis Products[a]

n	λ_n (Eigenvalue)	Standard Error in λ_n	3σ Misfit	χ^2_n Calculated	χ^2_n Expected	Imbedded Error, IE	Indicator Function, IND $\times 10^5$	Significance Level, %SL	Real Error, RE
1	1,627.301311	0.231	311	222,227	296	0.03219	150.90	0.0	0.09658
2	2.642417	0.092	155	8742	252	0.01083	46.87	0.0	0.02297
3	0.140080	0.111	0	35	210	0.00060	2.89	0.0	0.00104
4	0.000091	0.057	0	21	170	0.00060	3.62	29.5	0.00091
5	0.000057	0.063	0	14	132	0.00060	5.04	35.7	0.00081
6	0.000035	0.071	0	10	96	0.00061	8.32	49.3	0.00075
7	0.000027	0.056	0	5	62	0.00062	17.45	49.1	0.00070
8	0.000022	0.073	0	3	30	0.00063	62.80	55.5	0.00063
9	0.000015	0.052	0	0	0	—	—	—	—

[a]Reprinted with permission from Z. Z. Hugus, Jr., and A. A. El-Awady, *J. Phys. Chem.*, **75**, 2954 (1971), E. R. Malinowski, *Anal. Chem.*, **49**, 612 (1977), and E. R. Malinowski, *J. Chemometrics*, **3**, 49 (1988).

the components in mixtures of xylenes but also to determine the concentrations of the components.

Antoon et al.[19] used factor analysis to investigate the Fourier transform form-infrared (FT–IR) spectra of polymeric films. A factor analytical study of the fingerprint region (1100 to 1800 cm^{-1}) of seven films consisting of various proportions of atactic polystyrene and poly-2,6-dimethyl-1,4-phenyl oxide showed that three species were present, giving evidence that one of the constituent polymers undergoes a conformational change that is a function of the compositional blend. An FA–FT–IR study[19] of semicrystalline poly(ethylene terephthalate) yielded two components, corresponding to crystalline and amorphous phases, the trans and gauche conformers being indistinguishable. An FA–FT–IR study[19] of poly(vinyl chloride) films that have been subjected to various annealing treatments gave evidence for the existence of as many as eight components, due to a combination of configurational and conformational disorders in the chains.

Bulmer and Shurvell[20] employed factor analysis as a complement to band resolution studies of infrared spectra. Unfortunately, band resolution techniques require an a priori assumption concerning the general contour of the band. The shape is usually considered to be Lorentzian. A major advantage of factor analysis is that no such assumption is required.

They recorded and factor analyzed the infrared spectra of the carbonyl region of acetic acid–CCl_4 solutions. Their purpose was to study the monomer–cyclic dimer equilibrium. Their data matrix was constructed as follows. The spectra were digitized manually every 0.5 cm^{-1} from 1690.5 to 1790.0 cm^{-1}. The absorbance matrix consisted of 200 data points from each of nine solutions, ranging from 1.72×10^{-4} to 4.31×10^{-2} M acetic acid in CCl_4 solvent. The standard error in the eigenvalue, the 3σ misfit, and the chi-squared criteria clearly indicated that not two but four absorbing components were present. This conclusion was substantiated by Malinowski,[14,15] who studied the IE and IND functions. The F statistic,[16] however, shows $n = 2$ at the 5% significance level, and $n = 3$ at the 10% level. Considering the nature of the data, the 10% cutoff is favored. The results of these studies are discussed in Section 4.3 (sec Table 4.4). The previous approaches assumed only two species, a monomer and cyclic dimer of acetic acid. The factor analytical studies provide evidence for the existence of other hydrogen-bonded species.

Because of the success of the factor analysis technique in establishing the number of components in acetic acid solution, Bulmer and Shurvell[21] investigated trichloroacetic acid. They recorded the infrared spectra of nine solutions of trichloroacetic acid in CCl_4 solvent ranging from 0.61×10^{-3} to 0.16 M. At all concentrations, only two bands visually appeared in the spectrum of the carbonyl region. However, factor analysis gave evidence that, as in the case of acetic acid, four components were actually present. The detection of species other than monomer or dimer in this system cannot be gotten from simple band contour analysis nor from monomer–dimer equilibrium constants calculated from concentration studies.

The factor analysis technique was applied by Bulmer and Shurvell[22] to investigate the infrared spectra of solutions of $CDCl_3$ and di-n-butyl ether in CCl_4. Infrared spectra were recorded from 2290 to 2310 cm^{-1}, a region characteristic of the C—D stretching of $CDCl_3$. Nine solutions were prepared, in which the concentration of $CDCl_3$ was held constant throughout at approximately 2.2 M, and the concentration of di-n-butyl ether was varied from about 0.15 to 1.00 M. The spectra were digitized into 241 wavenumbers. Thus a 241×9 data matrix, containing 2169 points, was obtained.

The results of the eigenvalue studies[16,22] clearly showed that only two components were present. If this were so, then, in accord with the traditional approach, an isosbestic point should have occurred when the spectra were normalized to unit concentration and unit path length. No such intersection was obtained. Because factor analysis indicated only two components, Bulmer and Shurvell[22] searched for an explanation for the dilemma. Further investigation revealed that an isosbestic point occurred when the spectra were normalized to unit area as well as unit concentration and unit path length.

To obtain the extinction coefficient and concentration matrices, **E** and **C**, Bulmer and Shurvell[20] applied the method developed by Kankare[8] described earlier. First, a rough estimate of the equilibrium constant for the formation of the 1 : 1 complex between chloroform and di-n-butyl ether was made. The equilibrium constant was then varied until the abstract concentration matrix from FA was compatible with the corresponding matrix obtained using the equilibrium constant. The constant $0.0431\ M^{-1}$ obtained in this way agreed very well with the value $0.0456\ M^{-1}$ obtained from band resolution studies.

Korppi-Tommola and Shurvell[23] studied the complex formation between pentachlorophenol (PCP) and acetone in CCl_4 solution by factor analyzing separately the carbonyl and the hydroxyl stretching regions in the infrared. The stoichiometric concentration of PCP was fixed at 0.05 M for one study and at 0.10 M for another study, while the acetone concentration was varied from 0.005 to 1.0 M. The absorbance matrix for the carbonyl region consisted of 100 digitized wavelength intervals for each of 8 different solutions. For the hydroxyl region 401 digitized wavelength intervals per spectrum were involved.

Based on various criteria, such as chi-squared and residual standard deviation in the eigenvalues, factor analysis indicated three components for both types of data matrices. However, each region was interpreted quite differently. The hydroxyl region was believed to be the result of one monomer and two complexes, whereas the carbonyl region was believed to be due to one monomer, one combination band, and only one complex. Such an anomaly could occur if either one of the complexes did not significantly affect the carbonyl vibration or if both complexes contributed identically to the carbonyl vibration.

Although band resolution studies, based on Cauchy–Gauss product functions, for the hydroxyl region yielded a better spectral fit when four components were used, factor analysis gave clear evidence for only three absorbing components. Thus factor analysis served as an excellent complement to band resolution studies because it reduced the temptation to add an excessive number of bands in order to improve the fit.

Self-Modeling Spectral Isolation. Band resolution studies require some assumption of the band shape, such as Gaussian or Lorentzian. In actuality, the assumed shape may be incorrect. Lawton and Sylvestre[24] proposed a method, based on factor analysis, that requires no such assumption and that permits overlapped bands to be separated into their true shapes. This *self-modeling method* requires regions of the spectra where each component individually absorbs radiation.

Their study focused attention on a set of mixtures of standard dyes resulting from a production process. The absorption spectra of five mixtures, observed in the visible region, are shown in Figure 9.1. Principal factor analysis was performed on an absorbance matrix constructed by digitizing the wavelength scale every 10 nm from 410 to 700 nm. Since the first two principal eigenvectors accounted for the data within experimental error, it was concluded that the mixtures were composed of only two dyes.

A mathematical method for isolating the spectra of each dye was devised by Lawton and Sylvestre.[24] The rationale of the method can easily be visualized by examining Figure 9.2, which shows the results of the factor analysis. This diagram, analogous to Figure 5.2, shows the normalized principal axes, c_1 and c_2, and the five normalized data axes, labeled 1, 2, 3, 4 and 5, where each data axis represents one of the five mixtures. Each point on the diagram represents a specific wavelength. The chronological, oblique projection of these points onto any data axis virtually traces out the visible spectra of that mixture. According to Beer's law, the absorbance of a

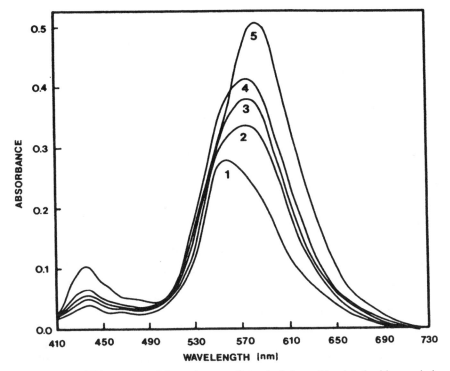

Figure 9.1 Visible spectra of five mixtures of standard dyes. [Reprinted with permission from W. H. Lawton and E. A. Sylvestre, *Technometrics*, **13**, 617 (1971).]

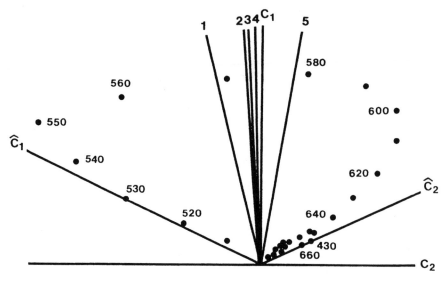

Figure 9.2 Results of principal factor analysis of spectra shown in Figure 9.1, after digitization. [Based on work of W. H. Lawton and E. A. Sylvestre, *Technometrics*, **13**, 617 (1971).]

mixture is a linear combination of the absorbance of its pure components. Because the absorbance of a pure component is a product of its absorptivity and concentration, both quantities being positive, all data points must lie in a region between a set of real axes representing the pure components. If the individual components absorb radiation of some wavelength uniquely, the pure component axes can be found because they will pass directly through the unique wavelength points. As seen in Figure 9.2, one such axis, $\hat{c}_1$, can be drawn through the 530-nm wavelength point, while the second, $\hat{c}_2$, can be drawn through the 430- and 660-nm points jointly. The spectrum of each pure component can now be generated by projecting, obliquely, the wavelength points onto the appropriate component axes. When this was done, the spectra shown in Figure 9.3 were resolved. Unfortunately, from mixed spectra alone it is impossible to determine the relative composition of the components.

Although the self-modeling method was originally restricted to two-component systems, Ohta[25] expanded the technique to a three-component system involving cyan, magenta, and yellow dyes, which are used in subtractive color photography. Although the spectral density distribution curves of the three dyes could not be determined uniquely, their ranges could be restricted to certain limits because the spectral densities and the spectral buildup of the mixtures is always nonnegative. Sylvestre et al.[26] showed that, under special circumstances involving chemical equilibria and reaction kinetics, the self-modeling technique could be applied to multicomponent systems beyond two or three components, yielding resolved spectra for each of the components.

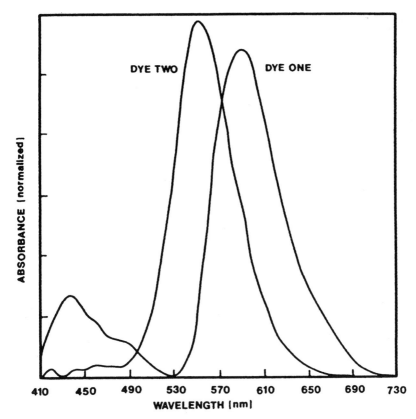

Figure 9.3 Spectra of pure components generated by projecting the wavelength points shown in Figure 9.2 onto the respective pure component axes. [Reprinted with permission from W. H. Lawton and E. A. Sylvestre. *Technometrics*, **13**, 617 (1971).]

When a polyelectrolyte is added to an aqueous solution of a staining dye, a new absorption band often appears at a lower wavelength than that of the free dye. A factor analytical study of this phenomenon, called *metachromasy*, was conducted by Yamaoka and Takatsuki.[27] The visible spectra of two metachromatic dyes, crystal violet and trypaflavine, in the presence of seven different polyelectrolytes, were measured at various concentration ratios of dye and polyanion. Factor analysis of the 14 data matrices revealed only 2 absorbing components in each case. The 2 components were interpreted to be the free dye and the dye molecule bound to the polyanion. The spectra of the bound dyes were generated by the spectral isolation method of Lawton and Sylvestre.[24]

Lin and co-workers[28,29] developed a factor analytical method, called *automated spectral isolation* (ASI), which they successfully used to isolate component spectra from the spectra of mixtures. The unique feature of ASI is that it does not require regions of spectral purity. The technique consists of the following. First each digitized spectrum is normalizd so that the absorbance of the maximum is 1.000.

Principal factor analysis (PFA) is then used to define the primary eigenvectors and, hence, the number of components. To find the spectral axes of the pure components, "prototype spectra" are target tested and the predicted vectors are judged by means of a *risk function*. The smaller the risk function, the closer the regenerated prototype spectrum is expected to resemble a pure component spectrum.

Prototype spectra for infrared are generated by a single-needle search that is similar to the uniqueness test but makes use of finite absorbance values of 0.08 instead of zeros. In other words, the test vectors consist of (1.000, 0.08, 0.08, ..., 0.08), (0.08, 1.000, 0.08, ..., 0.08), ..., (0.08, 0.08, 0.08, ..., 1.000). The purpose of using a base of 0.08 instead of 0.0 is to give better assurance that the absorbance values of the predicted target spectra will be positive rather than negative, which is physically unrealizable. The predicted target spectra that result from each of these prototype test vectors are considered as possible candidates for the pure component spectra. The risk value of each prototype spectrum is computed by the following defining equation

$$(\text{RISK})_j = \frac{1}{r}\sum_{i=1}^{r}(s_{ij} - a_{ij})^2 \sum_{i=1}^{r} s_{ij}^2 \tag{9.5}$$

Here a_{ij} is the ith absorbance value in the jth test vector and s_{ij} is the corrected ith absorbance $\hat{a}_{ij}$ in the predicted spectrum,

$$s_{ij} = \begin{cases} \hat{a}_{ij} & \text{if } \hat{a}_{ij} \geq 0 \\ 10\hat{a}_{ij} & \text{if } \hat{a}_{ij} < 0 \end{cases} \tag{9.6}$$

The sum is taken over all r components of the test vector. The reason for using s_{ij} instead of $\hat{a}_{ij}$ is to intuitively increase the risk value when negative absorbances appear in the target-reproduced prototype spectrum. Although there are r prototype spectra, only n of them, having the smallest risk values, will yield predicted spectra corresponding to the pure components.

Lin and Liu[28] recorded the infrared spectra of the ethylacetoacetate, lauryl alcohol, ethyl benzoate, and di-n-butyl ether. Using Beer's law, they generated artificial spectra of eight mixtures containing different amounts of these components. The eight-mixture data matrix was subjected to the ASI technique. Figure 9.4 shows the spectrum of each pure component and the isolated spectra obtained from the set of nonredundant prototype spectra having the smallest risk values.

Window Factor Analysis (WFA). Zhao and Malinowski[30] investigated the visible spectra of methylene blue in water over a very wide range of concentration, from 2.000×10^{-6} to 1.600×10^{-2} M (see Figure 9.5). Evidence for the existence of three species was found by applying abstract factor analysis based on *multiple sources of error* (see Section 4.3.3). The concentration profiles of the species,

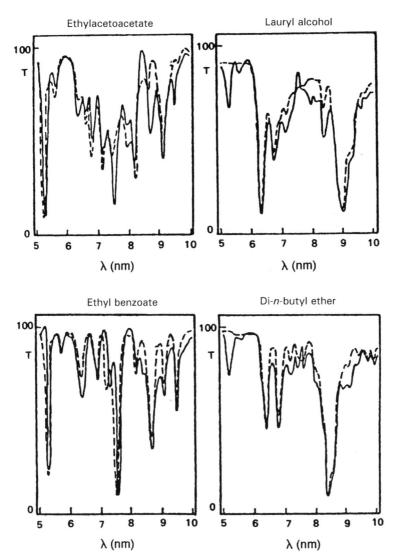

Figure 9.4 Isolated spectra (solid lines) and true component spectra (dashed lines). [Reprinted with permission from C. H. Lin and S. C. Liu, *J. Chin. Chem. Soc.*, **25**, 167 (1978).]

identified as monomer, dimer, and trimer, were obtained by WFA (see Section 6.2) and are portrayed as dotted lines in Figure 9.6. The solid lines represent predicted profiles based on the following equilibria:

$$(MB)_2^{2+} \rightleftharpoons 2MB^+ \quad \text{and} \quad (MB)_3^{3+} \rightleftharpoons 3MB^+ \tag{9.7}$$

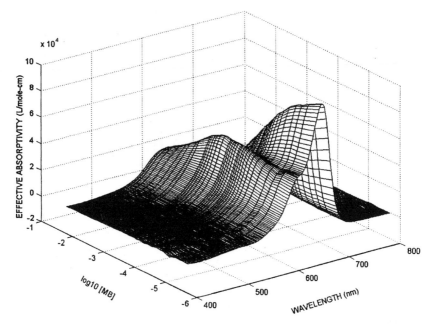

Figure 9.5 Three-dimensional plot of the absorptivity spectra of 33 aqueous solutions of methylene blue, ranging from 2.000×10^{-6} to 1.600×10^{-2} M. [Reprinted with permission from Z. Zhao and E. R. Malinowski, *J. Chemometrics*, **13**, 83 (1999). Copyright 1999 © John Wiley & Sons, Ltd.]

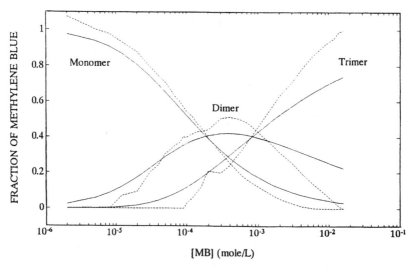

Figure 9.6 Concentration profiles of three methylene blue species extracted by WFA (broken lines) and predicted profiles (solid lines) (nonchlorinated trimer). [Reprinted with permission from Z. Zhao and E. R. Malinowski, *J. Chemometrics*, **13**, 83 (1999). Copyright 1999 © John Wiley & Sons, Ltd.]

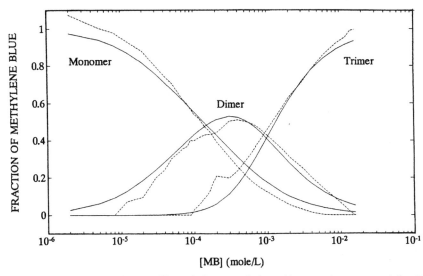

Figure 9.7   Concentration profiles of three methylene blue species extracted by WFA (broken lines) and predicted profiles (solid lines) (chlorinated trimer). [Reprinted with permission from Z. Zhao and E. R. Malinowski, *J. Chemometrics*, **13**, 83 (1999). Copyright 1999 © John Wiley & Sons, Ltd.]

As seen in Figure 9.6, the theoretical profile of the trimer does not satisfactorily correspond to the profile extracted by WFA. However, as shown in Figure 9.7, profiles in agreement with WFA were obtained by considering a trimer bonded to a chlorine atom that undergoes the following equilibrium:

$$(MB)_3Cl^{2+} \rightleftharpoons 3MB^+ + Cl^- \tag{9.8}$$

This example illustrates one of the major advantages of model-free factor analytical methods. The chemistry is deduced *after* the concentration profiles have been extracted. The profiles are not forced to fit theoretical speculations, as required by many classical techniques.

9.2   EMISSION SPECTRA

Factor analysis has proved to be a unique aid for determining the number of components responsible for an emission spectrum. In methods such as Raman and fluorescence spectroscopy, the emission intensities depend on the concentrations as well as the unique spectral properties of each emitting species. Consequently, factor analysis is readily applicable.

9.2.1 Raman

Factor analysis was successfully applied to the laser Raman spectra of aqueous indium (III) chloride solutions by Jarv et al.[31] A single, broad, asymmetric Raman band profile was observed, with a maximum that shifted from 311 to 279 cm^{-1} as the chloride-to-indium concentration ratio, R, was increased. This suggested that the single band might be a composite of several indium–chloride ion species. This problem is ideal for factor analysis, since the Raman intensity, I_{ik}, observed at the ith wavenumber for the kth solution, obeys the following expression:

$$I_{ik} = \sum_{j=1}^{n} J_{ij} C_{jk} \qquad (9.9)$$

where J_{ik} is the molar intensity of the jth species at the ith wavenumber and C_{ik} is the concentration of the jth species in the kth solution.

For each of 31 solutions of varying R, the Raman spectrum was recorded in digital format. Each spectrum was scanned between 170 and 410 cm^{-1}, corrected for background, and digitized into 481 points. Factor analysis of the resulting 481 $\times$ 31 data matrix gave evidence that four species were present. This conclusion was verified by examining the residual standard deviation, chi-squared, the number of misfits greater than 4 times the standard deviation, and the standard error in the eigenvalue.

Factor analysis was then used to estimate the range of R where less than four components were present. The method gave evidence that only two species were present when R was less than 2.36. This disagreed with other estimations based on the semi-half-band widths and the incomplete third moments, which indicated three species in this range. The discrepancy could be the result of an accidental linearity between the intensities of two of the three species in this region, making these species indistinguishable by factor analysis.

The following four species were postulated to be responsible for the spectra: $[\text{InCl}(\text{H}_2\text{O})_5]^{2+}$, $[\text{InCl}_2(\text{H}_2\text{O})_4]^+$, $[\text{InCl}_3(\text{H}_2\text{O})_3]$, and $[\text{InCl}_4(\text{H}_2\text{O})_2]^-$. Because of limited accuracy and very severe band overlap, the equilibrium constants between these species could not be calculated. This study shows that the observation of a single band in a Raman spectrum is not sufficient evidence for the existence of a single species.

Raman spectra of aqueous mixtures of ZnCl_2 and HCl were studied by Shurvell and Dunham.[32] Spectra were recorded in the Zn–Cl stretching regions using various concentration ratios of chloride and zinc ions. Factor analysis showed that only two light-scattering components existed. These were postulated to be ZnCl_2 and ZnCl_4^{2-}. Using band resolution techniques, the equilibrium constant for the reaction $\text{ZnCl}_2 + 2\text{Cl}^- \rightleftharpoons \text{ZnCl}_4^{2-}$ was estimated to be $0.22\,M^{-2}$.

Fluorescence presents a major problem in Raman spectra. Because fluorescence is generally much more intense than nonresonance Raman scattering, a trace amount of a fluorescent contaminant will often bury the Raman signal in the noise level. Hasegawa and co-workers[33] demonstrated how PFA could be used to isolate the weak Raman spectrum of indene from the strong fluorescence background of fluorescein. The method has been labeled FARMS, an acronym for factor analytical resolution of minute signals.

A data matrix, appropriate for PFA, was obtained by varying the fluorescence background. This was accomplished by irradiating the sample with a strong laser beam (500 mW). Fluorescein is bleached by the radiation while the Raman intensity of indene remains unaffected. The Raman spectra are displayed in Figure 9.8. Even after 50 replicate measurements were co-added to produce an enhanced spectrum, the Raman signal represented only one 1% of the signal. The signal of indene was indiscernible. Second-derivative techniques as well as other popular techniques were unsuccessful in recovering the minute signal of indene.

The first principal component extracted by PFA accounted for the strong fluorescence background. The second principal component, displayed in Figure 9.9, is the Raman signal of indene. In this case, it is important to realize that signal magnification and high-intensity background removal were accomplished without any prior knowledge of the spectra. Most interesting is the fact that abstract factors separated, directly, the weak Raman from the strong fluorescence signals, without rotational ambiguity.

FARMS has been applied to other types of spectroscopy that exhibit strong, broad background signals. For example, it has been used to study the molecule interaction mechanism between the alkyl-deuterated dipalmitoylphosphatidylcholine (DPPC-d_{62}) monolayer and sucrose based on infrared reflection–absorption spectra.[34]

9.2.2 Fluorescence

Nearly all the applications described thus far in this chapter require data matrices involving not one but a series of mixtures composed of varying amounts of the same

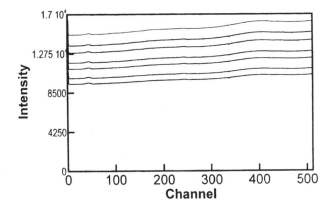

Figure 9.8 Raman spectra of seven mixtures containing a constant amount of indene and a variable amount of fluorescein. [Reprinted with permission from T. Hasegawa, J. Nishijo, and J. Umemura, *Chem. Phys. Lett.* **317**, 642 (2000). Copyright 2000 © Elsevier Science.]

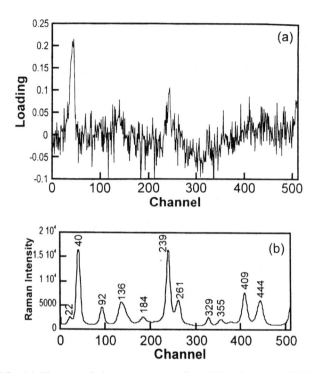

Figure 9.9 (*a*) The second abstract spectrum from PCA alone and (*b*) Raman spectrum of pure indene. [Reprinted with permission from T. Hasegawa, J. Nishijo, and J. Umemura, *Chem. Phys. Lett.*, **317**, 642 (2000). Copyright 2000 © Elsevier Science.]

components. The exceptions discussed earlier require that the components exist in chemical equilibrium so that a change in temperature or pH can produce a change in the composition. A method for determining the number of fluorescent components and their individual fluorescence spectra in a *single* mixture has aptly been demonstrated by Warner and co-workers.[35] Equilibrium between the components is not required for the factor analysis/fluorescence technique. However, the complete fluorescence spectrum of a single component in a mixture can be obtained when the data matrix includes wavelength regions at which only that component absorbs and emits. The analysis uses the fact that each fluorescent component is characterized by a unique dependence of its fluorescent intensities on two distinct parameters, the excitation wavelength λ_i, and the observed emission wavelength λ_j. The data matrix consists of an excitation–emission matrix, **M**, whose elements, M_{ij}, are the fluorescent intensities measured at λ_j when the excitation is at λ_i. For dilute mixtures, these intensities depend on the sum of product functions associated with each fluorescent component, k:

$$M_{ij} = \sum_{k=1}^{n} \alpha_k X_{ik} Y_{kj} \qquad (9.10)$$

where α_k is proportional to the concentration of component k, X_{ik} is proportional to the number of photons absorbed at wavelength λ_i per unit concentration of k, and Y_{kj}

is proportional to the fraction of fluorescence emitted by k at wavelength λ_j. Note that X_{ik} is independent of λ_j and Y_{kj} is independent of λ_i. This expression is ideal for factor analysis.

In a series of trial studies, Warner and co-workers[35] factor analyzed 10 two-component excitation–emission matrices involving 5 aromatic hydrocarbons: anthracene, pyrene, perylene, chrysene, and fluoranthene. Each data matrix was obtained in the following way. Initially, they held the excitation wavelength constant while the emission spectrum was scanned. The emitted spectrum was then digitized into 50 wavelengths. The intensities at these wavelengths were formatted and transmitted directly to a computer. The excitation wavelength was then changed and the scanning and digitizing procedures repeated at the same 50 wavelengths. This was repeated 50 times until 2500 data points were acquired. Each scan produced a row of the data matrix. Thus the data formed a 50 × 50 excitation–emission matrix.

A stray light component was found by recording the fluorescence of the pure solvent. The stray light contribution and the estimated dark current were first subtracted from each data matrix. Then a multiple of this stripped matrix was subtracted, to correct for the scattered light contribution. These pretreated data matrices were then subjected to AFA. The rank of each of the 10 different data matrices correctly equaled the number of fluorescent components. The fluorescence spectra of the pure components were deduced without recourse to any a priori knowledge of the pure components or their spectra.

The conventional procedure for determining the quantitative composition of a fluorescent mixture involves fitting the data to a set of simultaneous equations. This procedure requires a knowledge of the identity and the individual fluorescence spectra of all species in the mixture. Ho et al.[36] developed a rank annihilation method that yields the quantitative composition of a single fluorescent species in a mixture without requiring the identification of the other fluorescent components. The basis of the rank annihilation method is the following. The excitation–emission matrix, $\mathbf{M}$, of a multicomponent mixture has a rank that equals n_c, the number of components present. The rank of the corresponding excitation–emission matrix, $\mathbf{N}$, of a pure component ideally equals unity. If we subtract the correct amount of $\mathbf{N}$ from $\mathbf{M}$, we obtain a reduced matrix, $\mathbf{L}$, which has a rank equal to $n_c - 1$. The amount of $\mathbf{N}$ that must be subtracted from $\mathbf{M}$ to accomplish this task is equal to $(c_k/c_k^0)\mathbf{N}$, where c_k^0 is the concentration of pure component k, in the same solvent, used to obtain $\mathbf{N}$. In other words

$$\mathbf{L} = \mathbf{M} - \frac{c_k}{c_k^0}\mathbf{N} \tag{9.11}$$

Even when both data matrices, $\mathbf{M}$ and $\mathbf{N}$, are corrected for dark current and light scattering by the solvent, random noise tends to confuse the rank reduction process. An efficient way of determining the correct c_k/c_k^0 value is to examine the nth eigenvalue of $\mathbf{L}$ as a function of the ratio c_k/c_k^0. The n_cth eigenvalue will reach a minimum at the correct ratio value (see Figure 9.10). This method, called rank annihilation factor analysis (RAFA), was successfully applied to perylene and anthracene.[36]

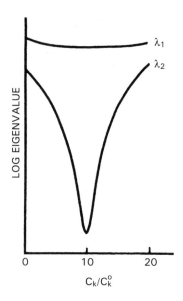

Figure 9.10   Results of rank annihilation of the fluorescence spectra of a binary mixture of perylene and anthracene. [Reprinted with permission from C. N. Ho, G. D. Christian, and E. R. Davidson, *Anal. Chem.*, **50**, 1108 (1978). Copyright 1978 © American Chemical Society.]

Important advances in the RAFA methodology are described in detail in Section 6.2.

9.2.3   Auger

The Fourier-transformed carbon Auger spectra of thin films of polyethylene and five poly(alkyl methacrylates) (alkyl = methyl, ethyl, isobutyl, n-butyl, and octadecyl) were recorded and factor analyzed by Gaarenstroom.[37] Two factors accounted for 99.8% of the variance. Semiquantitative analysis of the polymer films were obtained by treating the polymers as mixtures of polyethylene and poly(methyl methacrylate). The two eigenvectors were target transformed into axes corresponding to these two components. The factor (composition) loadings obtained from target factor analysis (TFA), as shown in Table 9.2, agree reasonably well with the known mixture fractions.

9.3   KINETICS

Determining the number of reacting species and their concentrations as a function of time is the basis of chemical kinetics. Factor analysis has been very valuable in such studies, yielding information that could not be obtained by any other means. Its use

TABLE 9.2 Composition of Polymeric Films Determined by Target Factor Analysis Compared to Known Fractions[a]

Polymer Films	TFA		Known	
	PE	PMMA	PE	PMMA
Poly(methyl methacrylate) (PMMA)	0.00	1.00	0.00	1.00
Poly(ethyl methacrylate)	0.19	0.82	0.17	0.83
Poly(isobutyl methacrylate)	0.12	0.88	0.37	0.63
Poly(n-butyl methacrylate)	0.20	0.80	0.37	0.63
Poly(octadecyl methacrylate)	0.65	0.36	0.77	0.23
Polyethylene (PE)	1.00	0.00	1.00	0.00

[a]Reprinted with permissioin from S. W. Gaarenstroom, *J. Vac. Sci. Technol.*, **16**(2), 600 (1979).

will undoubtedly increase because of the advent of sophisticated computer-interfaced data collection systems, which are being developed for rapid-scan wavelength-kinetic experiments.

The investigations of Ainsworth[38,39] paved the way to applying factor analysis to kinetics. In his pioneering work, Ainsworth[38] used the rank-analysis method of Wallace[1] to determine the number of absorbing species in reaction mixtures. A venom-solubilized preparation of cytochrome oxidase was catalytically reduced, and the reduced cytochrome oxidase was reacted with oxygen. Absorbances were measured at 11 wavelengths at 4 specific time intervals using a stopped-flow technique. The rank of the data matrix was found to be 3, implying that at least 3 components were present. This was interpreted as the result of 2 successive reactions of the type A $\rightarrow$ B $\rightarrow$ C.

Ainsworth[38] also studied reaction mixtures of oxyhemoglobin and reduced hemoglobin. Because the rate of reaction with oxygen was slow, no appreciable change in composition occurred during the time needed to obtain a set of absorbance readings of the reaction mixture. Rank analysis of the data gave evidence that four hemes of the hemoglobin molecule were present.

In another study, Ainsworth[39] recognized that it was possible, under certain conditions, to obtain the absorption spectrum of a component in the mixture without prior knowledge of the identities or spectra of any of the components. This may be accomplished when there exists a situation where the component does not contribute to the total absorption. Often this condition exists at the very beginning or at the very end of a chemical reaction, when the concentration of either a product or reactant is negligibly small. When this occurs, the rank of the absorbance matrix is diminished by one unit. By calculating the determinants of all the submatrices, it is possible to determine whether or not such a situation exists. If it does, then a further study, at times when the component contributes to the absorption, will reveal the spectrum of the component in question. Ainsworth successfully applied this technique to artificially computed data concerning mixtures of acridene orange, diiodo-(R)-fluorescein, and rhodamine B in alcohol.

The studies of Wallace[1] and Ainsworth[38,39] were based on determining the rank by examining submatrices for singularity using the standard deviation criterion. As described in Section 9.1.2, Katakis[5] deduced the rank by examining the residual matrices produced by the Gauss process of elimination. Using this procedure, he investigated the reaction between Cr^{2+} and maleic acid in 1 M perchloric acid solution. His absorbance data matrix consisted of wavelengths as rows and reaction times as columns. The fact that the first residual matrix was found to be less than the error matrix showed conclusively that only one absorbing component was present. Such information was a valuable aid in deducing the true mechanism of the reaction.

In order to study the kinetics of the hydrolytic depolymerization of certain cobalt (III) complexes, Hugus and El-Awady[13] developed and applied a variety of factor analytical techniques (see Section 4.3). The results and conclusions of their studies are described in detail in Section 9.1.3.

Cochran and Horne[40] discussed the problems of applying factor analysis to wavelength-kinetic experiments. Using mathematical models, they showed that experimental errors that vary with wavelength can lead to incorrect estimates of the number of species. They then showed how this error can be eliminated by statistically weighting the absorbance matrix. This method is not restricted to wavelength kinetics but is applicable in a more general sense. Unfortunately, they did not apply it to any real kinetic problem.

9.4   OPTICAL ROTATORY DISPERSION

Rank analysis can be used to determine the number of optically active species in a mixture when the angle of rotation of a plane-polarized beam of light of each component is proportional to the concentration of the component. Under these conditions, an equation analogous to (9.2) can be written:

$$\alpha_{\lambda k} = \sum_{j=1}^{n_c} \bar{\alpha}_{\lambda j} c_{jk} \qquad (9.12)$$

Here $\alpha_{\lambda k}$ is the angle of rotation of the kth mixture observed at wavelength λ, $\bar{\alpha}_{\lambda j}$ the specific rotation of the jth component observed at λ, c_{jk} the concentration of component j in mixture k, and n_c the number of optically active components in the mixtures. An optical rotatory dispersion (ORD) curve is obtained by measuring the angle of rotation as a function of wavelength.

McMullen et al.[41] applied the matrix rank-analysis method of Wallace and Katz[2] to investigate the optical rotatory dispersion of tobacco mosaic virus TMV RNA in solution. Their purpose was to determine the number of components present and to identify them.

The ORD of TMV RNA was measured in the wavelength region between 230 and 350 nm, over a wide range of temperatures and at four different ionic strengths. The resultant data for a specific ionic strength were digitized into a matrix in which the row designees were the different wavelengths and the column designees were the different temperatures. Four data matrices were examined, one for each ionic strength. This methodology was quite unique because temperature was used to alter the compositions. It was based on the premise that the equilibrium between different geometrical conformers would be temperature dependent. The method was fruitful; in each case, rank analysis revealed that two components were present. An attempt was made to fit all the experimental spectra with only two typical spectra chosen from the experimental data. The low-temperature spectrum of the $1\,M$ Na^+ case was chosen to represent the first vector. The second vector was chosen to be the high-temperature spectrum of the $0.004\,M$ Na^+ case. These typical vectors were selected because they represent two extremes, particularly so since increasing ionic strength had the same effect on the spectrum as decreasing the temperature. With these two typical vectors the entire data could be reproduced to within 4%.

Encouraged by these results, McMullen and co-workers attempted to transform the two factors into two different forms of TMV RNA molecule. They postulated that these two forms were the single-strand and double-strand helical conformations that coexist in chemical equilibrium. At high temperatures the equilibrium is shifted so that very little double strand remains. They proposed that the optical rotation of the single strand depends on temperature, whereas that of the double strand is independent of temperature. The optical rotations of both conformers are insensitive to ionic strength. However, a change in ionic strength shifts the equilibrium. The model led to a direct calculation of the percent composition of the double strand and the equilibrium constant within the range of the experimental conditions. At 2.5°C, TMV RNA appears to he approximately 50% double helix, whereas at 74.5°C it appears to be approximately 3% double helix.

9.5 CHROMATOGRAPHIC RESOLUTION

MacNaughtan et al.[42] were the first to report the successful application of factor analysis for deconvoluting two or more overlapping peaks in chromatography. The method requires several chromatograms of mixtures having different compositions but the same components. High precision, particularly on the time axis, is necesssary. The data matrix is constructed by digitizing each chromatogram at equal intervals of time. Each row of the data matrix corresponds to a given mixture and each column corresponds to a given elution time.

One of their studies was concerned with mixtures of benzene and perdeuterobenzene. The chromatograms of four mixtures were recorded. The areas were normalized to unity to compensate for any errors due to variations in sample size. The resulting data were factor analyzed and subjected to a deconvolution program. The quantitative results of the deconvolution study agreed within 2% of the results obtained by complete chromatographic separation.

The deconvolution program has one limitation: the chromatogram must have regions arising from each of the pure components. For a two-component system, this restriction is not too severe because the two extreme tails of the chromatogram meet this criterion.

Davis and co-workers[43] made use of mass spectra to determine the number of components under a single chromatographic peak. They recorded the mass spectra at fixed time intervals during the elution of a chromatographic peak. Thus each scan recorded the complete mass spectrum of a different composition of the same components. The data matrix consisted of the mass spectral intensities, wherein each column designated a time interval and each row designated a given mass-to-charge ratio. This data matrix was then factor analyzed.

The method was successfully applied to isotopic mixtures of carbon dioxide, $^{13}C^{16}O_2$ and $^{12}C^{16}O_2$, and to mixtures of n-hexane and n-heptane. The investigation also included a study of the effects of differences in chromatographic resolution, peak heights, peak widths, and peak tailing. Peak distortion from chemical or electronic sources, channel-to-channel carryover, and changes in baseline were found to have no significant effect on the ability of FA to detect the second component. Noise constituted the most serious problem, sometimes producing a fictitious component. However, this situation could he diagnosed quickly by visual inspection of the experimental graphs.

Since the entire PFA calculations can be carried out in 3 to 5 sec for a 200×5 matrix, this approach affords a quick and useful method for detecting the presence of more than one component in a chromatographic peak, which may appear to be due to a single species. In contrast to the usual deconvolution technique, the factor analysis method requires no prior assumption concerning the chromatographic peak shape of any component. The method can thus be used to confirm the purity of a peak or to give warning that the chromatographic separation has not been effective. A minicomputer interfaced with a gas chromatography–mass spectrometry (GC–MS) system is especially appropriate for such studies. The FA method is rapid, sensitive, and reliable.

A variety of techniques that have been successful in resolving multiple overlapping chromatographic bands, as shown in Figure 6.8, are described in Section 6.1.

9.6   MASS SPECTRA

9.6.1   Abstract Factor Analysis Studies

The use of FA–MS (factor analysis of mass spectra) for component analysis was explored by Ritter et al.[44] Mass spectral data of mixtures are factor analyzable because the signal intensity at a given m/e position is a linear sum of the corresponding intensities of the pure components weighted by their compositions. Ritter and co-workers factor analyzed the mass spectra of four sets of mixtures: cyclohexane/cyclohexene, hexane/cyclohexane, heptane/octane, and unknown xylenes. The mixture in a given set contained the same two components but differed in their composition. The data matrix consisted of the MS intensities measured at the same m/e positions for each mixture belonging to the given set. By studying the residual error in the covariance matrix (see Section 4.3), these investigators correctly identified the number of components in the mixtures and predicted that the unknown xylenes contained three components.

Not all m/e positions need be recorded. In fact, by deleting particular m/e positions from the data matrix, one can identify components that have unique mass positions. For example, deleting m/e 28 from the hexane/cyclohexane data matrix, Ritter and coworkers found that the factor space was reduced from 3 to 2, giving clear evidence that nitrogen was present as an impurity. Factor analysis of the heptane/octane mixtures indicated that three components instead of two were present. Careful examination of the spectra showed that the ion source was contaminated by nitrobenzene derivatives that had been run earlier.

This study showed that FA–MS can be used as a rapid and accurate method for determining the number of components in a series of mixtures.

9.6.2 Target Factor Analysis Studies

Malinowski and McCue[45] showed how target factor analysis of mass spectral data could be used for qualitative identification of substances suspected to be present in a series of related mixtures. They also showed how TFA can be used to obtain the chemical compositions of the mixtures as well. This unique approach to compound identification and subsequent quantitative analysis illustrates the power of TFA in analytical chemistry.

The basis of the methodology is as follows. The intensity (height) of each mass peak in the MS of a mixture is a linear sum of contributions due to each component:

$$H(i, \alpha) = \sum_{j=1}^{n} h^0(i, j) p(j, \alpha) \tag{9.13}$$

where $H(i, \alpha)$ is the height of the ith m/e peak in mixture α, $h^0(i, j)$ the height of the ith peak of the pure jth component per unit pressure, and $p(j, \alpha)$ the partial pressure of the component in the ionization chamber. Because of mass discrimination, the ratio of the partial pressures in the ionization chamber to that in the sample reservoir is different for each component. Because these pressures are extremely low, Dalton's law applies, so that

$$H(i, \alpha) = \sum_{j=1}^{n} H^0(i, j) F(j, \alpha) \tag{9.14}$$

where

$$F(j, \alpha) = X(j, \alpha) \frac{D(j)}{p^0(j)} P(\alpha) \tag{9.15}$$

Here $H^0(i, j)$ is the height of the ith m/e peak in the MS of pure j at pressure $p^0(j)$ in the ionization chamber, $P(\alpha)$ the total pressure in the ionization chamber, $D(j)$ the mass discrimination factor, and $X(j, \alpha)$ the mole fraction of j in the original sample mixture α.

Equation 9.14 shows that the spectral heights of the pure components are true factors and can be used as test vectors in TFA. Equation 9.15 shows how the corresponding factors, $F(j, \alpha)$, are related to the mole fractions. By employing a solution of known composition, one can obtain the compositions of the mixtures, independent of pressure measurements. This is possible because, for a given solution having components $1, 2, \ldots, n$, the ratios of the $F(j, \alpha)$ factors are independent of the total pressure $P(\alpha)$.

In order to use FA–MS for qualitative and quantitative analysis of mixtures, the following sequence of operations is carried out. First, the number of components is deduced by decomposing the covariance matrix. Second, the components are identified by target testing the MS of pure components suspected to be present. Finally, the compositions are obtained by adding the MS of a solution of known composition to the data matrix and then carrying out combination TFA using the MS of the pure components as real vectors.

In order to illustrate the steps in the TFA technique, Malinowski and McCue[45] subjected the MS data of Ritter and co-workers[44] to the foregoing sequence of operations. The data matrix consisted of the intensities of 18 m/e values for 7 cyclohexane/hexane mixtures. A previous AFA study[14] of this data gave the results shown in Table 9.3. Evidence for the presence of three components was given by the fact that the IE function showed little or no improvement on using four or more eigenvectors and the fact that the IND function reached a minimum at $n = 3$. The third component was suspected to be nitrogen gas as a contaminant.[44] When the intensities of the mass 28 peak, characteristic of nitrogen, were deleted from the data matrix, the results given in Table 9.3 were obtained. Both the IE and IND functions showed that only two components were responsible for the remaining spectral data,

TABLE 9.3   Factor Analysis Results Concerning Mass Spectra Intensities Used to Determine the Number of Components in a Series of Related Mixtures[a]

	Cyclohexane/Hexane				Cyclohexane/Hexane (Without m/e 28)			
n	RE	IE	IND $\times 10^3$	%SL	n	RE	IE	IND $\times 10^3$	%SL
1	1.810	0.684	50.27	0.5	1	1.812	0.685	50.35	0.5
2	0.465	0.249	18.62	0.2	2	0.134	0.071	5.36	0.0
3	0.128	0.084	8.03	0.7	3	0.106	0.070	6.65	26.1
4	0.111	0.084	12.30	37.9	4	0.092	0.070	10.25	38.7
5	0.098	0.073	24.56	46.1	5	0.072	0.061	18.08	37.3
6	0.074	0.068	73.51	47.2	6	0.058	0.054	58.18	50.7

[a]Reprinted with permission from E. Malinowski, *Anal. Chem.*, **49**, 612 (1977) and E. R. Malinowski, *J. Chemometrics*, **3**, 49 (1988).

thus confirming the presence of nitrogen. This was the same conclusion reached by Ritter and co-workers, who used the residual error in the covariance matrix as the error criterion.

From Table 9.3 we see that the real error (RE) corresponding to $n = 3$ for the total data matrix and corresponding to $n = 2$ for the reduced data matrix (without m/e 28) is 0.13 intensity unit. This is considerably greater than the error, ± 0.05, reported by the original investigators. The value 0.13 is much more reliable since it is a composite of all sources of error, whereas the value 0.05 was simply the error in reading the MS intensities from the experimental graphs. These conclusions were verified[16] by the F statistic. As shown in Table 9.3, for the full data set and the reduced data set, the significance levels are above the 5 to 10% cutoff for n greater than 3 and 2, respectively.

Using the reduced data matrix and two factors, target tests were carried out[45] using the MS intensities of pure cyclohexane and pure hexane as test vectors. In both cases the predicted intensities agreed with the test vectors within the expected error limit 0.13, as shown in Table 9.4 for cyclohexane. Intensities for those masses that were free-floated in the test vector were predicted correctly, thus providing further evidence for the presence of these components.

TABLE 9.4   Mass Spectral Intensities of Cyclohexane Obtained from Target Testing[a] and from Spectral Isolation[b]

m/e	Test[c]	TFA Prediction	Spectral[d] Isolation
27	(1.8)	1.9	1.8
29	1.3	1.3	1.1
39	2.5	2.3	2.3
40	0.7	0.6	0.7
41	(7.1)	7.3	6.9
42	(3.5)	3.5	3.2
43	2.2	2.1	1.8
44	0.2	0.2	0.1
54	(0.8)	0.7	0.8
55	4.6	4.9	4.7
56	13.5	13.6	13.5
57	(1.2)	1.2	0.7
69	3.8	4.1	4.0
83	0.8	0.7	0.7
84	10.7	10.4	10.7
85	0.9	0.9	0.9
86	(0.1)	0.1	0.0

[a]Reprinted with permission from E. R. Malinowski and M. McCue, *Anal. Chem.*, **49**, 284 (1977).
[b]Reprinted with permission from F. J. Knorr and J. H. Futrell, *Anal. Chem.*, **51**, 1236 (1979).
[c]Values in parentheses were free floated (i.e., left blank in the test vector).
[d]Adjusted so that the base peak is 13.5 rather than 100, as reported in the original study.

TABLE 9.5 Composition of Cyclohexane/Hexane Mixtures Obtained from Target Factor Analysis[a] and from Spectral Isolation[b]

	Mole Fraction Cyclohexane		
Mixture	Experiment	TFA (Prediction)	Spectral Isolation (Prediction)
1	1.00	1.00	0.96
2	0.92	0.88	0.84
3	0.83	0.81	0.78
4	0.55[c]	0.55[c]	0.54
5	0.23	0.30	0.30
6	0.12	0.16	0.17
7	0.00	0.00	0.01

[a]Reprinted with permission from E. R. Malinowski and M. McCue, *Anal. Chem.*, **49**, 284 (1977).
[b]Reprinted with permission from F. J. Knorr and J. H. Futrell, *Anal. Chem.*, **51**, 1236 (1979).
[c]Represents the standard solution.

When the two test vectors were used in combination, TFA yielded the complete set of $F(j, \alpha)$ factors. Mixture 4, containing 55 mol % cyclohexane, was considered to be the standard solution of known composition. Using (9.15), the values of the $F(j, \alpha)$, and the known composition of mixture 4, Malinowski and McCue[45] determined the compositions of the solutions given in Table 9.5. The agreement between the calculated and reported compositions is good considering the fact that the original solutions were prepared crudely.

In an elaborate study involving target testing, Rasmussen et al.[46] compiled a library file of approximately 17,000 mass spectra, which were used to correctly identify the components in several different mixtures and to estimate their relative concentrations. Their computational strategy was made efficient by employing a prefilter to remove library entries having masses greater than the highest mass of the mixture spectra. The validity of a target was judged by the following approach, which made use of Bessel's inequality test. The *coefficient of fit*, b, can be shown to be equal to the following:

$$b = \|\hat{\mathbf{x}}\|^2 / \|\mathbf{x}\|^2 \tag{9.16}$$

where $\|\mathbf{x}\|$ and $\|\hat{\mathbf{x}}\|$ are the norms of the target vector and the predicted vector. The coefficient of fit measures the extent to which the target vector lies inside the factor space. According to Bessel's inequality, a value of 0 indicates that the target is completely orthogonal to the factor space and lies entirely outside the space. A value of 1 indicates that the target lies completely inside the space. Hence target vectors associated with true components will have b values close to unity. This approach provides us with a method of searching through a huge library of spectra and sorting out the true components of mixtures.

Under certain conditions, the mass spectra of the components can be separated from the mass spectra of mixtures without recourse to library information. This can be achieved if there exists at least one mass peak in the mixed spectra that is unique to each component. Such pure mass peaks are most likely to occur when the number of components is small and the number of mass points is large and divergent. The pure mass points need not be specified by the chemist; instead, they are automatically selected by the algorithm involved.

A method for selecting pure mass points, similar to the separation technique of Lawton and Sylvestre,[24] was proposed by Knorr and Futrell,[47] which led to the development of key set factor analysis (KFSA)[48] and, eventually, to iterative key set factor analysis (IKSFA).[49] These methods are explained in detail in Section 3.4.5. Initially, the abstract row-factor matrix $\bar{\mathbf{R}}$ is normalized so that the sum of squares across each row is unity. This places each mass point on the surface of an n-dimensional sphere subtended by real, oblique axes that pass through the pure mass points. IKSFA searches for the most orthogonal set of these row vectors, which, in this case, corresponds to the unique masses.

The components of the normalized row vectors associated with the pure masses specify the relative amounts of the eigenvectors that comprise those mass points. In fact, the n-normalized rows, associated with the n pure masses, constitute the rows of the inverse of the transformation matrix, $\mathbf{T}^{-1}$. Having thus found the transformation matrix, we can convert the abstract factor matrices into real matrices—one that traces out the mass spectra of the pure components and one that reveals the relative concentrations of the components.

When this technique is applied to the mass spectral data of Ritter et al.,[44] the spectra of the pure components and their relative concentrations are obtained, independent of any other information. A typical example of such spectral isolation is shown in Table 9.4, giving clear evidence for the presence of cyclohexane. Prediction of the compositions of the cyclohexane/hexane mixtures is given in Table 9.5.

IKSFA has been used to identify as many as 10 unique mass points in complicated mixtures, as demonstrated at the bottom of Table 3.1.

9.6.3 The Variance Diagram

Pyrolysis mass spectra (Py-MS) of bio-organic mixtures are not amenable to key set factor analysis because these spectra are so complex that "pure mass" points for each of the components are unlikely to occur. The problem is aggravated by the fact that reference spectra of pure components are usually not available or even obtainable due to intermolecular interactions that occur during pyrolysis. Consequently, library search procedures are not applicable.

Windig and Meuzelaar[50] developed a graphical rotation technique, called the variance diagram (VARDIA), to extract pure component spectra from Py-MS data. With this method the data matrix is arranged so each row represents the spectra of a pyrolized sample. The starting point for VARDIA requires standardizing each mass in the set of spectra, so that correlation around the origin can be performed. The score (row) matrix concerns the sample mixtures and the loading (column) matrix concerns the mass spectra.

The underlying principles of VARDIA are best learned by an illustration taken from a simulated experiment. Figure 9.11 is a plot of the factor scores resulting from AFA of the mass spectra of 13 sample mixtures containing 3 components [bovine serum albumin (BSA), glycogcn (GLY), and peptidoglycan (PGL)], including the spectra of the 3 components. Because of the use of correlation around the origin, the scores and loadings lie in a common two-dimensional plane with a common origin. Prior to standardization, the data occupy three-dimensional factor space as illustrated in Figure 9.12. Standardization reduces the factor space to a two-dimensional subspace. Thus the analysis yields only 2 factors. Points 11, 12, and 13 in the figures represent the pure components, whose spectra were included in the original data matrix. All sample mixtures must lie inside the composition triangle defined by the three component points. The pure component axes that we seek are represented by lines drawn from the origin to the 3 corners of the composition triangle.

With real Py-MS data, however, because pure component samples are not members of the data set, the corners of the composition triangle are not known. VARDIA is designed to locate these corners. Spectral loadings characteristic of a component will tend to lie along the component axis, forming clusters in factor space directed toward the components. Such clusterings can be located from a variance diagram created by summing the squares of the lengths of the projected loadings that lie within 5° of an axis, and then rotating the axis in 10° steps for a full

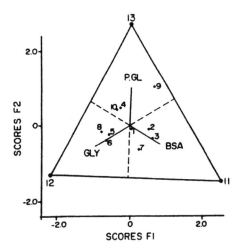

Figure 9.11   Plot of factor scores obtained from pyrolysis mass spectra of biopolymer mixtures, showing the pure component axes (PGL, BSA, and GLY). [Reprinted with permission from W. Windig and H. L. C. Meuzelaar, in H. L. C. Meuzelaar and T. L. Isenhour (Eds.), *Computer-Enhanced Analytical Spectroscopy*, Plenum Press, New York, 1987, Chap. 4.]

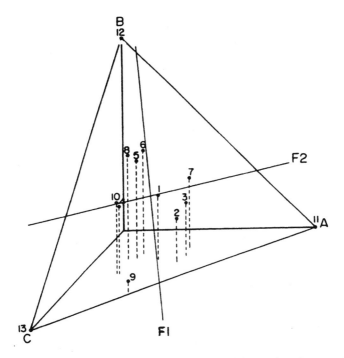

Figure 9.12 Three-dimensional plot of mixture compositions. The mixture points lie in a plane because the sum of the three components in the mixtures is 100%. Also, the mixture points lie within a triangle spanned by the scores of the three pure components (points 11, 12, and 13). [Reprinted with permission from W. Windig and H. L. C. Meuzelaar, in H. L. C. Meuzelaar and T. L. Isenour (Eds.), *Computer-Enhanced Analytical Spectroscopy*, Plenum Press, New York, 1987, Chap. 4.]

360°. This creates a polar plot of the variances within the angular windows. Figure 9.13 shows the clusters of the loadings (i.e., the normalized spectra) and the variance diagram obtained from the model data. The variance diagram clearly locates the component axes.

For more complex mixtures, the procedure requires examining variance diagrams of F1-F2, F2-F3 (where F1, F2, F3, etc. represent the abstract factor axes) and other two-dimensional projections of the factor space. The VARDIA plot of the first two principal factors (F1-F2) resulting from Py-MS analysis of grass leaves is given in Figure 9.14. The spectrum resulting from projections onto the axis marked with an asterisk, shown in Figure 9.15, is remarkably similar to the model spectrum of polyisoprenoids.

Unlike other methods, VARDIA does not require pure mass points. Instead, VARDIA relies on the proposition that the mass points have the greatest variance along the component axes resulting from factor analysis of standardized masses.

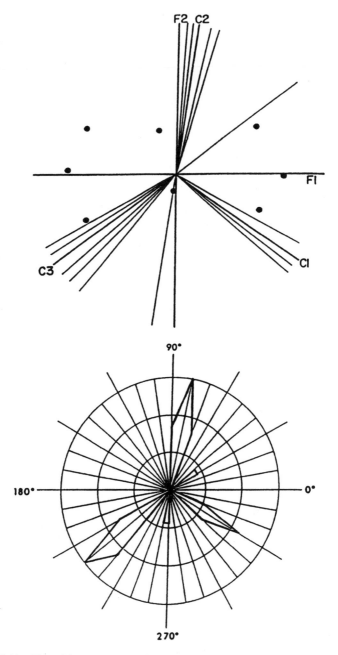

Figure 9.13 Plot of factor scores and loadings. Upper plot shows clear clusterings of the loadings. Lower plot is a quantification of the clusterings obtained by summing the squares of the lengths of the projected loadings that lie within 5° of an axis. (Reprinted with permission from W. Windig and H. L C. Meuzelaar, in H. L. C. Meuzelaar and T. L. Isenhour (Eds.), *Computer-Enhanced Analytical Spectroscopy*, Plenum Press, New York, 1987, Chap. 4.]

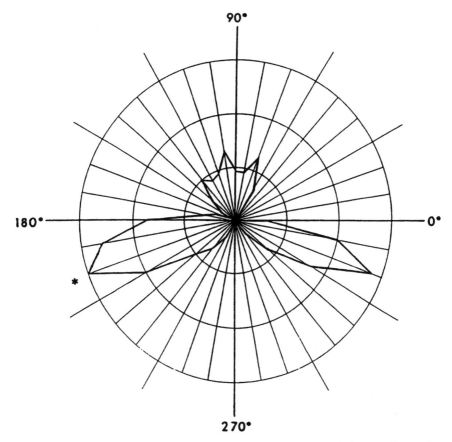

Figure 9.14 VARDIA plot of the first two discriminant functions of the grass leaves data. [Reprinted with permission from W. Windig and H. L. C. Meuzelaar, in H. L. C. Meuzelaar and T. L. Isenhour (Eds.). *Computer-Enhanced Analytical Spectrocopy.* Plenum Press, New York, 1987, Chap. 4.]

9.7 NUCLEAR MAGNETIC RESONANCE (NMR)

9.7.1 Solid State

The abstract factor analysis of ^{13}P-NMR spectra of octacalcium phosphate, OCP, has shown that this material exists as three district components in the solid state. The technique for such analysis was devised by Kormos and Waugh.[51] It involves differential cross polarization combined with magic angle sample spinning and high-power proton decoupling. Most important, only a single sample of powdered material is required.

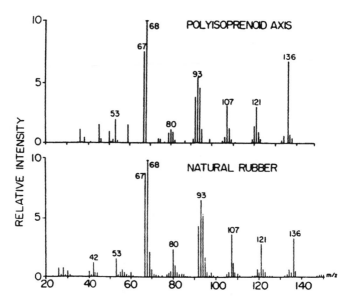

Figure 9.15   Spectrum (upper) derived from the VARDIA plot of Figure 9.14 by projections onto the 200° axis marked with an asterisk, compared to library spectrum (lower) of natural rubber. [Reprinted with permission from W. Windig and H. L. C. Meuzelaar, in H. L. C. Meuzelaar and T. L Isenhour (Eds.), *Computer-Enhanced Analytical Spectoscopy.* Plenum Press, New York, 1987, Chap. 4.]

Factor analysis is applicable for such experiments because the signal intensity, $D_{cp}(i, t)$, at frequency i and cross-polarization contact time t, is a linear sum of product terms:

$$D_{cp}(i, t) = \sum_{j=1}^{n} g(i,j)c(j)e(j, t) \qquad (9.17)$$

where the sum is taken over the n spin components; $g(i, j)$ is the line shape function for spin type j; $e(j)$ is the concentration of spin species j in the mixture; and $e(j, t)$ is the cross-polarization efficiency for type j spins with contact time t. A single, constant spinning speed for all spectra is required to produce a data matrix that will be factor analyzable.

According to this equation, if different spin species have significantly different cross-polarization efficiencies, their line intensities can be made to vary by altering the contact time. An appropriate data matrix was obtained by Kormos and Waugh[51] by recording the ^{13}P-NMR spectra of powdered OCP at nine contact times from 0.5 to 2.5 ms. The spectra were digitized into 195 points, covering the zero- and first-order sidebands. Factor analysis of the 195×9 matrix provided evidence for the existence of three components because the IND (see (4.63)] reached a minimum at this level.

This study shows the potential for factor analysis in solid-state NMR. Because the method is applicable to other nuclei, such as ^{13}C, it should be useful for the study of coal, wood, biological samples, and minerals, and especially useful for phase transition studies in the solid state.

9.7.2 Polymer Solutions

The direct exponential curve resolution algorithm (DECRA; see Section 6.3.5) has been used to resolve the NMR spectra of a polymer reaction product, completely separating the low-, mid-, and high-molecular-weight (MW) components.[52] DECRA takes advantage of the fact that pulse gradient spin echo (PGSE) techniques produce signals that decay exponentially with time. The decay signals are directly proportional to the diffusion coefficients of the components. In this case the polymer product consists of three distinct MW classes, low, mid, and high MW.

A sample of a copolymer incorporating low-molecular-weight polydimethylsiloxane (PDMS) was prepared by dissolution in dichloromethane-d_2. Stacked plots of 16 spectra of the sample, recorded at various magnetic field gradient pulse strengths, are displayed in Figure 9.16. The decay signals of water, dichloromethane, and PDMS are quite apparent. DECRA-resolved spectra, normalized to portray

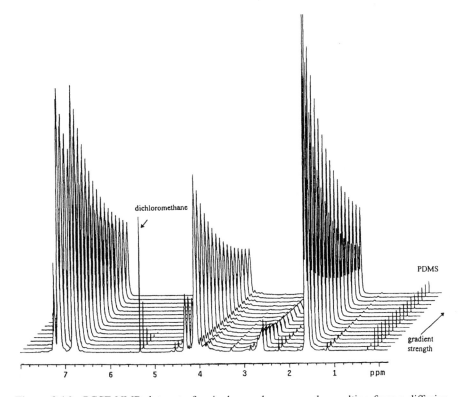

Figure 9.16 PGSE NMR data set of a single copolymer sample resulting from a diffusion experiment. [Reprinted with permission from W. Windig, B. Antelac, L. J. Sorriero, S. Bijlsma, D. J. Louwerse (AD), and A. K. Smilde, *J. Chemometrics*, **13**, 95 (1999). Copyright 1999 © John Wiley & Sons, Ltd.]

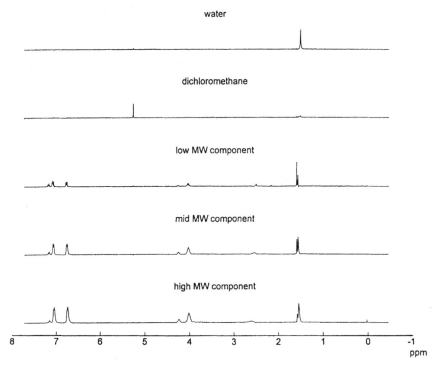

Figure 9.17   NMR spectra of five components obtained by DECRA analysis of the data portrayed in Figure 9.16. [Reprinted with permission from W. Windig, B. Antelac, L. J. Sorriero, S. Bijlsma, D. J. Louwerse (AD), and A. K. Smilde, *J. Chemometrics*, **13**, 95–110 (1999). Copyright 1999 © John Wiley & Sons, Ltd.]

accurate component compositions, are shown in Figure 9.17. The analysis also yielded the following associated diffusion coefficients (m^2s^{-1}): water, 6.06×10^{-9}; dichloromethane-d_2 (the solvent), 3.07×10^{-9}; low-MW polymer component, 0.734×10^{-9}; mid-MW polymeric component, 0.355×10^{-9}; high-MW polymeric component, 0.140×10^{-9}.

It is important to recognize that, unlike other chemometric methods, DECRA requires only a single sample, the spectra being resolved in a few seconds.

REFERENCES

1. R. M. Wallace, *J. Phys. Chem.*, **64**, 899 (1960).

2. R. M. Wallace and S. M. Katz, *J. Phys. Chem.*, **68**, 3890 (1964).

3. S. Perlis, *Theory of Matrices*, Addison-Wesley, Reading, MA, 1952, p. 45.

4. L. P. Varga and F. C. Veatch, *Anal. Chem.*, **39**, 1101 (1967).

5. D. Katakis, *Anal. Chem.*, **37**, 876 (1965).

6. E. Bodewig, *Matrix Calculus*, North-Holland, Amsterdam, 1959.

7. J. S. Coleman, L. P. Varga, and S. H. Mastin, *Inorg. Chem.*, **9**, 1015 (1970).

8. J. J. Kankare, *Anal. Chem.*, **42**, 1322 (1970).

9. J. L. Simmonds, *Photogr. Sci. Eng.*, **2**, 205 (1958); *J. Opt. Soc. Am.*, **53**, 968 (1963).

10. R. L. Reeves, *J. Am. Chem. Soc.*, **88**, 2240 (1966).

11. G. Wernimont, *Anal. Chem.*, **39**, 554 (1967).

12. P. J. Gemperline and J. C. Hamilton, *J. Chemometrics*, **3**, 455 (1989).

13. Z. Z. Hugus, Jr., and A. A. El-Awady, *J. Phys. Chem.*, **75**, 2954 (1971).

14. E. R. Malinowski, *Anal. Chem.*, **49**, 612 (1977).

15. E. R. Malinowski, in B. R. Kowalski (Ed.), *Chemometrics: Theory and Application*, ACS Symp. Ser. 52, American Chemical Society, Washington, DC, 1977, Chap. 3.

16. E. R. Malinowski, *J. Chemometrics*, **3**, 49 (1988).

17. G. T. Rasmussen, T. L. Isenhour, S. R. Lowry, and G. L. Ritter, *Anal. Chim. Acta*, **103**, 213 (1978).

18. M. McCue and E. R. Malinowski, *Appl. Spectrosc.*, **37**, 463 (1983).

19. M. K. Antoon, L. D'Esposito, and J. L. Koenig, *Appl. Spectrosc.*, **33**, 351 (1979).

20. J. T. Bulmer and H. F. Shurvell, *J. Phys. Chem.*, **77**, 256 (1973).

21. J. T. Bulmer and H. F. Shurvell, *Can. J. Chem.*, **53**, 1251 (1975).

22. J. T. Bulmer and H. F. Shurvell, *J. Phys. Chem.*, **77**, 2085 (1973).

23. J. Korppi-Tommola and H. F. Shurvell, *Can. J. Chem.*, **56**, 2959 (1978).

24. W. H. Lawton and E. A. Sylvestre, *Technometrics*, **13**, 617 (1971).

25. N. Ohta, *Anal. Chem.*, **45**, 553 (1973).

26. E. A. Sylvestre, W. H. Lawton, and M. S. Maggio, *Technometrics*, **16**, 353 (1973).

27. K. Yamaoka and M. Takatsuki, *Bull. Chem. Soc. Jpn.*, **51**, 3182 (1978).

28. C. H. Lin and S. C. Liu, *J. Chin. Chem. Soc.*, **25**, 167 (1978).

29. C. H. Lin and L. C. Lin, *Proc. Natl. Sci. Counc. Republic of China*, **3**, 1 (1979).

30. Z. Zhao and E. R. Malinowski, *J. Chemometrics*, **13**, 83 (1999).

31. T. Jarv, J. T. Bulmer, and D. E. Irish, *J. Phys. Chem.*, **81**, 649 (1977).

32. H. F. Shurvell and A. Dunham, *Can. J. Spectrosc.*, **23**, 160 (1978).

33. T. Hasegawa, J. Nishijo, and J. Umemura, *Chem. Phys. Lett.*, **317**, 642 (2000).

34. T. Hasegawa, *Anal. Chem.*, **71**, 3085 (1991); *Trends in Anal. Chem.*, **20**, (2001).

35. I. M. Warner, G. D. Christian, E. R. Davidson, and J. B. Callis, *Anal. Chem.*, **49**, 56 (1977).

36. C. N. Ho, G. D. Christian, and E. R. Davidson, *Anal. Chem.*, **50**, 1108 (1978).

37. S. W. Gaarenstroom, *J. Vac. Sci. Technol.*, **16** (2), 600 (1979).

38. S. Ainsworth, *J. Phys. Chem.*, **65**, 1968 (1961).

39. S. Ainsworth, *J. Phys. Chem.*, **67**, 1613 (1963).

40. R. N. Cochran and F. H. Horne, *Anal. Chem.*, **49**, 846 (1977).

41. D. W. McMullen, S. R. Jaskunas, and I. Tinoco, Jr., *Biopolymers*, **5**, 589 (1967).

42. D. MacNaughtan, Jr., L. B. Rogers, and G. Wernimont, *Anal. Chem.*, **44**, 1421 (1972).

43. J. E. Davis, A. Shepard, N. Stanford, and L. B. Rogers, *Anal. Chem.*, **46**, 821 (1974).

44. G. L. Ritter, S. R. Lowry, T. L. Isenhour, and C. L. Wilkins, *Anal. Chem.*, **48**, 591 (1976).

45. E. R. Malinowski and M. McCue, *Anal. Chem.*, **49**, 284 (1977).

46. G. T. Rasmussen, B. A. Horne, R. C. Wieboldt, and T. L. Isenhour, *Anal. Chim. Acta*, **112**, 151 (1979).

47. F. J. Knorr and J. H. Futrell, *Anal. Chem.*, **51**, 1236 (1979).

48. E. R. Malinowski, *Anal. Chim Acta*, **134**, 129 (1982).

49. K. J. Schostack and E. R. Malinowski, *Chemometrics Intell. Lab. Syst.*, **6**, 21 (1989).

50. W. Windig and H. L. C. Meuzelaar, in H. L. C. Meuzzelaar and T. L. Isenhour (Eds.), *Computer-Enhanced Analytical Spectroscopy*, Plenum, New York, 1987, Chap. 4; *Anal. Chem.*, **56**, 2297 (1987).

51. D. W. Kormos and J. S. Waugh, *Anal. Chem.*, **55**, 633 (1983).

52. W. Windig, B. Antelac, L. J. Sorriero, S. Bijlsma, D. J. Louwerse (AD), and A. K. Smilde, *J. Chemometrics*, **13**, 95 (1999).

No problem is completely solvable.

NUCLEAR MAGNETIC RESONANCE

This chapter concerns factor analytical studies of the effects of substituents and solvents on nuclear magnetic resonance (NMR) chemical shifts. Other applications, based on DECRA, are described in Sections 6.3.5 and 12.6.

10.1   PROTON SOLVENT SHIFTS

The NMR spectral features (chemical shifts and coupling constants) are strongly influenced by the solvent. This is unfortunate because it makes spectral interpretation difficult, but fortunate because it provides the chemist with a clue for probing the liquid solution state and for studying intermolecular interactions.

Homer[1] has reviewed the factors that are believed to contribute to the solvent shift. Theoretical expressions for many of these factors have been derived and attempts have been made to isolate one factor from another by the judicious choice of solvent, solute, and other experimental variables. All such attempts have had only limited success, chiefly because it is impossible to find experimental conditions such that the effects of all but one factor are held constant.

The series of target factor analysis (TFA) studies by Weiner and Malinowski[2-5] led to a complete solution of the proton solvent shifts of some simple nonpolar solutes. In this section we systematically trace through the intricate steps that eventually lead to the complete solution. Details of the development are presented in this section from a chronological viewpoint in order to emphasize the systematic and deductive reasoning involved.

10.1.1 Key Solvents

The following is a brief description of the first investigation of proton-NMR solvent shifts using TFA, reported by Weiner et al.[2] The goal of the investigation was to develop a procedure for predicting the shifts of simple solutes in a large variety of solvents using a minimum of shift data.

Proton shifts of a series of simple substituted (polar and nonpolar) methanes, measured in a variety of solvents with tetramethylsilane (TMS) as an internal standard, were chosen as an ideal system to study. The raw data that were subjected to factor analysis consisted of the proton shifts of nine halogenated methanes in nine solvents. The shifts of CH_4, CH_3CN, CH_2Cl_2, CH_2ClCN, and $CHBrCl_2$ were not included in the data matrix but were purposely set aside for later testing purposes. The factor analytical reproduction step showed that three factors were sufficient to span the factor space and to reproduce the data within experimental error, ± 0.5 Hz. This implied that only three solvent–solute interaction terms were involved.

Since the nature of the interaction terms was unknown, attempts were made to find a set of "typical" factors. A typical factor is a column of the data matrix. Although any column of data can be used as an axis of the factor space, care must be exercised in choosing a set of three data columns, since an arbitrary combination of three columns may not span the factor space. A given combination may contain data that do not involve a particular solute–solvent interaction. For example, if hydrogen bonding is an important interaction, it is imperative that at least one of the three axes account for this factor.

Acetonitrile, carbon tetrachloride, and methylene bromide were chosen as a key set of typical solvent vectors for the following reasons. Acetonitrile possesses a large dipole moment and has π electrons. Methylene bromide has a large polarizability and a sizable quadrupole moment. Carbon tetrachloride is nonpolar and contains bulky chlorine atoms. It is believed that these features would adequately account for all possible solute–solvent interactions involved in the factor space. In fact, after all combinations of typical column vectors were target tested, these three solvent factors were found to give the best data reproduction. Although there is nothing unique about this choice, other combinations did not span the factor space. For example, the set composed of methylene chloride, chloroform, and carbon tetrachloride did not satisfactorily reproduce the data. Evidently, at least one of the important solute–solvent factors of the space was not sufficiently represented by this group.

A simultaneous combination transformation onto the three data columns yielded equations of the following kind:

$$\delta(u, CH_3CN) = 1.002f_1 - 0.004f_2 + 0.002f_3$$
$$\delta(u, CH_2Cl_2) = 0.081f_1 + 0.715f_2 - 0.207f_3$$
$$\delta(u, CHCl_3) = -0.046f_1 + 0.817f_2 + 0.230f_3$$
$$\delta(u, CCl_4) = -0.002f_1 + 1.004f_2 - 0.002f_3 \tag{10.1}$$
$$\delta(u, CS_2) = 0.006f_1 + 1.128f_2 - 0.139f_3$$
$$\delta(u, CH_3I) = 0.561f_1 - 0.224f_2 + 0.653f_3$$

Here $\delta(u, v)$ is the chemical shift of solute u in solvent v and $f_1 = \delta(u, CH_3CN)$, $f_2 = \delta(u, CCl_4)$, and $f_3 = \delta(u, CH_2Br_2)$. These equations predict the chemical shift of a solute in a given solvent in terms of its measured shift in the three key solvents. Because of experimental error and computer roundoff, the equations given above for acetonitrile, methylene bromide, and carbon tetrachloride each exhibit three finite coefficients, two near zero and one near unity, rather than one.

The utility, as well as validity, of these equations was further substantiated by examining shifts predicted for the solutes that were purposely omitted from the factor analysis scheme. For example, by measuring the shifts of CH_2Cl_2 in the three key solvents, we can use (10.1) to predict the shift of methylene chloride in the other solvents. The shifts of CH_2Cl_2 in the three key solvents (CH_3CN, CCl_4, and CH_2Br_2) are, respectively, 326.9, 317.1, and 321.2 Hz. Placing these values for f_1, f_2, and f_3 into the 10.1 expression corresponding to CH_3I solvent, we make the following calculation:

$$\delta(CH_2Cl_2, CH_3I) = (0.561)(326.9) - (0.224)(317.1) + (0.653)(321.2)$$
$$= 322.1 \text{ Hz}$$

This value, representing the proton shift of CH_2Cl_2 in CH_3I solvent, agrees with the measured value, 322.5 Hz, within experimental error, ± 0.5 Hz.

Predictions for the methanes not included in the analysis were also found to be in excellent agreement with their measured shifts (see Table 10.1). Equations 10.1 have also been used to predict the shifts of substituted ethanes.[2] This was accomplished by measuring the shifts of various ethanes in the three key solvents. Typical predicted shifts for CH_2ClCCl_3 are shown in Table 10.1. We see here that even though the real nature of the factors is unknown, it is still possible to predict solvent shifts by means of factor analysis.

10.1.2 Theoretical Considerations

Encouraged by the success of the initial factor analytical studies, Weiner et al.[2] pursued the problem in a more fundamental way, reasoning as follows: Without

TABLE 10.1   Experimental and Predicted Shifts[a] of Some New Solutes Using (10.1)[b]

	Solute					
	CH_2ClCN		$CHClBr_2$		CH_2ClCCl_3	
Solvent	Experimental	Predicted	Experimental	Predicted	Experimental	Predicted
CS_2	242.8	241.9	427.9	427.5	254.1	253.8
$CHCl_3$	246.1	245.7	432.1	431.0	257.0	256.8
$CHBr_3$	253.0	252.3	433.0	433.0	—	261.1
CH_3I	255.3	254.2	439.6	440.4	263.0	264.6
CH_2I_2	257.3	258.5	434.9	434.9	—	241.8

[a]In hertz at 60 MHz, relative to internal TMS.
[b]Reprinted with permission from P. H. Weiner, E. R. Malinowski, and A. R. Levinstone, *J. Phys. Chem.*, **74**, 4537 (1970).

proof, Buckingham et al.[6] postulated that the solvent shift can be expressed as a linear sum of terms:

$$\delta(u, v) = \delta(u, \text{gas}) + \sigma_b(v) + \sigma_a(v) + \sigma_w(u, v) + \sigma_E(u, v) \\ + \sigma_H(u, v) + \cdots \tag{10.2}$$

where $\delta(u, v)$ is the chemical shift of solute u in solvent v, $\delta(u, \text{gas})$ is the gas-phase shift of the solute, $\sigma_b(v)$ is due to the bulk susceptibility of the solvent, $\sigma_a(v)$ is the solvent shift caused by the anisotropy of the solvent, $\sigma_w(u, v)$ is the van der Waals dispersion interaction between the solute and solvent, $\sigma_E(u, v)$ is the reaction field interaction between the solute and solvent, and $\sigma_H(u, v)$ is due to hydrogen bonding.

Factor analysis gave evidence that the solvent shift must be a linear sum and that each of the terms must be a product function of solute and solvent parameters. Furthermore, the hydrogen-bonding contribution should be negligible, since none of the solutes and solvents studied form strong hydrogen bonds. Hence (10.2) can be written as

$$\delta(u, v) = \delta(u, \text{gas}) \cdot 1 + 1 \cdot \sigma_b(v) + 1 \cdot \sigma_a(v) + \sigma_w(u) \cdot \sigma_w(v) \\ + \sigma_E(u) \cdot \sigma_E(v) \tag{10.3}$$

Equation (10.3) involves five factors, whereas factor analysis clearly indicated that only three factors are operative. The answer to this dilemma lies in the fact that the chemical shifts were referenced with respect to a trace of TMS that was dissolved in the same solvent. The internal standard, TMS, also comes in contact with the solvent and experiences a solvent shift:

$$\sigma(\text{TMS}, v) = \delta(\text{TMS}, \text{gas}) \cdot 1 + 1 \cdot \sigma_b(v) + 1 \cdot \sigma_a(v) \\ + \sigma_w(\text{TMS}) \cdot \sigma_w(v) + \sigma_E(\text{TMS}) \cdot \sigma_E(v) \tag{10.4}$$

The experimental shift, $\delta^{TMS}(u, v)$, represents, in reality, the difference between (10.3) and (10.4):

$$\delta^{TMS}(u, v) = \delta^{TMS,g}(u, \text{gas}) \cdot 1 + [\sigma_w(u) - \sigma_w(\text{TMS})]\sigma_w(v) \\ + [\sigma_E(u) - \sigma_E(\text{TMS})]\sigma_E(v) \tag{10.5}$$

where $\delta^{TMS,g}(u, \text{gas}) = \delta(u, \text{gas}) - \sigma(\text{TMS}, \text{gas})$ is the gas-phase shift of the solute relative to the gas-phase shift of TMS. Equation (10.5) involves a sum of three terms. Each term is a product of a solute and solvent contribution. This expression predicts that the data space is three dimensional, in complete accord with the results of factor analysis.

10.1.3   Solute Gas-Phase Shift

According to (10.5) the gas-phase shift of the solute should be a basic factor. Because each suspected factor can be examined independently, via target transformation, the gas-phase shift can be tested solely on its own merits, without our having to invoke any model or specify any details concerning the other two factors. Transformation into the solute gas-phase shifts, relative to gaseous TMS, was successful (see Table 10.2), giving clear evidence that the gas-phase shift is a true fundamental factor.

In Chapter 2 we learned that not all the data points are required for target transformation. A fringe benefit of TFA is its ability to predict those points not inserted (free floated) in the target test vector. In Table 10.2 the third column shows predicted gas-phase shifts for CH_3Cl, CH_2I_2, CHI_3, and CH_2ClBr that were free-floated. In spite of the fact that two other unknown factors are simultaneously active,

TABLE 10.2   Test of Gas-Phase Chemical Shifts[a] as a Solute Factor Using Three Factors[b]

Solute	Test	Predicted
CH_3Cl	—	168.2
$CHCl_3$	427.3	427.1
CH_3Br	146.9	147.1
CH_2Br_2	285.0	285.5
$CHBr_3$	406.9	406.8
CH_3I	119.0	118.5
CH_2I_2	—	227.6
CHI_3	—	301.5
CH_2ClBr	—	297.7

[a]In hertz at 60 MHz, relative to gaseous TMS.
[b]Reprinted with permission from P. H. Weiner, E. R. Malinowski, and A. R. Levinstone, *J. Phys. Chem.*, **74**, 4537 (1970).

we see how valuable information can be obtained by means of a single successful target transformation.

Target transformation into gas-phase shifts relative to methane gas failed. The shifts involved in the test vector must be referenced with respect to gaseous TMS as dictated by the theoretical equation (10.5). The choice of a reference standard in the test vector is not arbitrary. This illustrates the sensitivity of the target test to the exact details of the basic factor being investigated.

10.1.4   External Standards

Subsequent discussions require an understanding of NMR referencing procedures. There are two different methods for measuring and reporting solute chemical shifts: internal and external reference standards.

The internal standard method, the more common procedure, involves dissolving a trace of a reference compound in the solute–solvent solution and observing the shift between the reference and the solute. There are several advantages to this procedure. The main experimental advantage is the ease of measurement. This procedure also has a theoretical advantage in that the bulk magnetic susceptibility term, $\sigma_b(v)$, and the solvent anisotropy term, $\sigma_a(v)$, are eliminated [see (10.5)]. These two terms depend primarily on the solvent and produce the same shift effect on both the solute and the internal reference. The primary disadvantage of using an internal standard is that one measures not the effect of the solvent on the solute but the difference between the effect of the solvent on the solute and the effect of the solvent on the reference compound. Also, anomalous intermolecular interactions between the reference and solute molecules may occur, seriously complicating the situation.

The second major experimental approach for measuring solvent effects involves the use of an external standard. In this case the reference material and solution are placed in separate compartments of a coaxial cell consisting of two concentric glass cylinders. An external reference is a reliable standard because it does not come in contact with the solute solution. However, since the reference standard and the solute do not experience the same bulk magnetic susceptibility shielding effect, the observed chemical shift must be appropriately corrected. Theoretical equations involved in this correction are well established. In fact, an NMR technique[7] can be used to determine the bulk susceptibility corrections. In addition to making the problem more complex from an experimental point of view, the use of external standards allows the solvent anisotropy effect to be active. Although this increases the complexity of the theoretical problem, more useful information can be obtained.

10.1.5   Solvent Anisotropy

The success of the earlier investigation[2] encouraged Weiner and Malinowski[3] to probe deeper into the theoretical aspects of the problem. They decided to simplify the problem by studying a subspace composed of nonpolar solutes. In this situation the reaction field, $\sigma_E(u, v)$, is absent because nonpolar solutes lack a permanent electric dipole moment necessary to produce a reaction field. Hexamethyldisiloxane

(HMD) was used as an external standard and corrections were made for bulk susceptibility, $\sigma_b(v)$. The proton shifts of six nonpolar solutes (methane, ethane, neopentane, cyclohexane, cyclooctane, and tetramethylsilane) dissolved in 22 solvents were measured. The experimental accuracy of the data, knowledge crucial to proper factor analysis, was estimated to be ± 1.5 Hz.

When the shift data were initially factor analyzed, two factors almost seemed sufficient to reproduce the data matrix within experimental error. Only a small improvement was observed when three factors were invoked. With a two-factor model, one might erroneously conclude that the solvent anisotropy effect was negligible. However, the two-factor model was rejected in favor of a three-factor model when serious inconsistencies arose. The nature of these inconsistencies will become apparent at a later point in our discussion.

To study the three fundamental factors, we again examine the equation of Buckingham, Schaefer, and Schneider, which can appropriately be expressed as follows:

$$\delta^{HMD,X}(u, v) = \delta^{HMD,X}(u, \text{gas}) \cdot 1 + \sigma_w(u) \cdot \sigma_w(v) + 1 \cdot \sigma_a(v) \tag{10.6}$$

where $\delta^{HMD,X}(u, v)$ is the chemical shift of solute u in solvent v, relative to external HMD, and $\delta^{HMD,X}(u, \text{gas})$ is the gas-phase shift of solute u relative to external HMD (the shifts having been corrected for bulk susceptibility).

According to factor analysis, each term in this equation must be a product function of solute and solvent parameters. For the gas-phase term, this constraint is satisfied since the gas-phase chemical shift is only a function of the solute being studied, independent of the solvent. Similarly, the solvent anisotropy term is dependent entirely on the solvent and is independent of the solute. These two terms can therefore be expressed as product functions with the solvent part of the former term being unity and the solute part of the latter term equaling unity. Several investigators[8,9] have suggested that the van der Waals shielding can be expressed as a product function, in accord with (10.6).

To test this equation in detail, we need accurate estimations for all three terms on the right side of this equation. Such information is not readily available since the van der Waals term and solvent anisotropy are theoretical quantities. For these quantities, only crude estimates for a very limited number of solvents exist. Theoretical estimates of solvent anisotropy have been made for only two solvents, benzene and carbon disulfide. Schug[10] estimated the anisotropies of these solvents to be -30 and $+18.1$ Hz, respectively, whereas Homer[11] suggested -35 and $+7$ Hz, respectively.

This perplexing problem was solved by Weiner and Malinowski[3] in the following manner. For methane as a solute, (10.6) takes the form

$$\delta^{HMD,X}(CH_4, v) = \delta^{HMD,X}(CH_4, \text{gas}) \cdot 1 + \sigma_2(CH_4) \cdot \sigma_w(v) + 1 \cdot \sigma_a(v) \tag{10.7}$$

This equation was solved for $\sigma_w(\alpha)$ and substituted into (10.6), yielding

$$\delta^{\text{HMD},X}(u, v) = \delta^{\text{HMD},X}(u, \text{gas}) \cdot 1 + \frac{\sigma_w(u)}{\sigma_w(\text{CH}_4)} \cdot \delta^{\text{CH}_4,\text{g}}(\text{CH}_4, v)$$
$$+ \left(1 - \frac{\sigma_w(u)}{\sigma_w(\text{CH}_4)}\right) \cdot \sigma_a(v)$$

(10.8)

In this expression, $\delta^{\text{CH}_4,\text{g}}(\text{CH}_4, v)$ represents the gas-to-solution shift of methane in solvent v. The advantage of this expression in comparison to (10.6) lies in the fact that it is independent of the van der Waals model, circumventing the need to specify $\sigma_w(v)$. Instead, this solvent factor is replaced by a new factor, the methane gas-to-solution shift. This procedure illustrates the intricate interdependency between fundamental factors and the fact that there are many different ways of expressing the factor space, each way being an equally valid representation.

From the viewpoint of solvent tests, the unity and methane-gas-to-solution vectors are easily constructed since all necessary data are available. On the other hand, the solvent anisotropy test vector poses a serious problem, as described earlier. To solve this problem, Weiner and Malinowski[3] proceeded as follows. They ascribed a zero value for the solvent anisotropy of carbon tetrachloride because it is nonpolar and symmetric. They then systematically varied the solvent anisotropies of benzene and carbon disulfide in accord with various theoretical estimates. Because there were three factors and only three solvents for which only crude estimates of solvent anisotropies were available, it was not possible to conclude which of these estimates were the best from the results of the target tests alone. This anomaly arose because, with a three-factor space, any test factor containing only three defined numbers would yield a perfect fit. Since, in this case, any three random numbers would fit perfectly, other criteria had to be devised.

Equation (10.8) provides the necessary criteria for selecting the proper anisotropy test vector. According to this equation, the coefficients of the solvent unity test vector should correspond to the gas-phase chemical shifts of the solutes. Furthermore, the sum of the last two solute coefficients in (10.8) should equal unity. If a simultaneous target transformation into the three-solvent test vectors (unity, methane gas-to-solution shifts, and solvent anisotropy) is performed, the best anisotropy test vector can be deduced by comparing the predicted loadings [i.e., the solute coefficients in 10.8 with the criteria given above.

The results of the target tests involving these three solvent factors are shown in Table 10.3. In order to obtain the best solvent anisotropy vector, a variety of test vectors were generated and examined in the following manner. The solvent anisotropy of carbon tetrachloride was set equal to zero, and all combinations of the estimated values of benzene and carbon disulfide were employed in the combination step along with the unity and methane gas-to-solution shifts test vectors. The solute coefficients generated by this procedure did not agree satisfactorily with the specified criteria.

TABLE 10.3 Solvent Factor Target Tests[a]

	Unity		$\delta^{CH_{4,g}}_{(CH_{4,l})}$		$\sigma_a(v)$		$\sigma_w(v)$	
Solvent	Test	Predicted	Test	Predicted	Test	Predicted	Test	Predicted
CCl_4	1.0	1.014	25.2	25.4	0.0	0.0	0.245	0.273
$CHCl_3$	1.0	1.009	23.0	23.1	—	0.8	0.238	0.242
$CHBr_3$	1.0	1.002	34.4	34.5	—	4.9	0.323	0.310
CH_2I_2	1.0	0.974	43.4	43.2	—	14.8	0.301	0.286
CH_2ClCCl_3	1.0	1.007	24.4	24.5	—	−0.1	—	0.264
C_6H_{12}	1.0	1.001	16.9	16.9	—	−1.3	—	0.203
CS_2	1.0	0.992	32.7	32.6	9.0	9.0	—	0.246
$(CH_3)_2CO$	1.0	1.003	9.5	9.5	—	−5.9	0.175	0.180
C_6H_6	1.0	0.977	−9.5	−9.7	−24.0	−24.0	0.199	0.186
C_6F_6	1.0	1.002	−12.5	−12.4	—	−27.6	—	0.200

[a]Reprinted with permission from P. H. Weiner and E. R. Malinowski, *J. Phys. Chem.*, **75**, 1207, 3160 (1971).

A more detailed search was undertaken. The solvent anisotropy of carbon tetrachloride was again set equal to zero while the anisotropy value of benzene was systematically varied from −40 to −20 Hz and the value for carbon disulfide was varied from +5 to +20 Hz. For each combination, a comparison was made of the predicted solute coefficients with the proposed criteria. The best fit was obtained when $\sigma_a(\text{benzene}) = -24.0$ Hz and $\sigma_a(CS_2) = +9.0$ Hz, yielding the following equations:

$$\delta^{HMD,X}(CH_4, v) = -8.3f_1 + 0.99f_2 + 0.01f_3$$

$$\delta^{HMD,X}(CH_3CH_3, v) = 35.9f_1 + 0.72f_2 + 0.28f_3$$

$$\delta^{HMD,X}(\text{neo-}C_5H_{12}, v) = 41.6f_1 + 0.64f_2 + 0.32f_3$$

$$\delta^{HMD,X}(C_6H_{12}, v) = 75.8f_1 + 0.49f_2 + 0.55f_3$$

$$\delta^{HMD,X}(C_8H_{16}, v) = 83.8f_1 + 0.38f_2 + 0.63f_3$$

$$\delta^{HMD,X}(\text{TMS}, v) = 12.8f_1 + 0.63f_2 + 0.35f_3$$

Here f_1 = unity, f_2 = methane gas-to-solution shift, and f_3 = solvent anisotropy. The coefficients of the unity terms can be compared to the actual experimental gas-phase shifts[12] (relative to external HMD): methane = −8.4 Hz, ethane = 35.5 Hz, neopentane = 42.1 Hz, cyclohexane = 75.6 Hz, and tetramethylsilane = 16.4 Hz. Except for tetramethylsilane, the predicted gas-phase shifts shown in the equations above are in excellent agreement with the experimental values. Furthermore, for each equation shown, the sum of the solute coefficients for the last two terms on the right is close to unity. Since the two criteria concerning the solute coefficients have been satisfied, the solvent anisotropy test vector should be reliable.

The "best" anisotropy test vector is shown in Table 10.3. Solvent anisotropies predicted for the free-floated solvents are shown in the table. These predictions represent the first empirical estimates of solvent anisotropy. Halogenated solvents such as methylene iodide and bromoform are predicted to have large solvent anisotropies. These values serve as guides in the theoretical development of solvent anisotropies for the halogenated methanes.

Recall that the data matrix could be reproduced with only two abstract factors, but the two-factor solution was rejected in favor of a three-factor solution. This conclusion can now be understood in light of the present discussion. With the two-factor model, the assumption is made that the anisotropy contribution to the chemical shift is negligible for all solvents in the scheme. This is a particularly poor assumption for solvents such as benzene and carbon disulfide. If this assumption is made, (10.8) reduces to

$$\delta^{HMD,X}(u, v) = \delta^{HMD,X}(u, \text{gas}) \cdot 1 + \frac{\sigma_w(u)}{\delta_w(CH_4)} \cdot \delta^{CH_4,g}(CH_4, v) \qquad (10.9)$$

When an equation of this form was tested, using a two-dimensional transformation matrix, the solute coefficients of the unity test differed from the true gas-phase values by more than 10 Hz. These large deviations were far beyond experimental error, approximately ± 1.5 Hz. The two-factor model was rejected in favor of the three-factor model, which, in marked contrast, did predict the gas-phase shifts within experimental error.

10.1.6 Van der Waals Effect

In Section 10.1.5 the van der Waals shift contribution was circumvented by using the experimentally determined methane gas-to-solution shifts as a measure of this term. This procedure yielded the "best" set of solvent anisotropies, independent of any particular theoretical model that might be proposed for the van der Waals contribution. Having obtained a satisfactory set of solvent anisotropies, Weiner and Malinowski[4] then attempted to unravel the van der Waals factor.

Any model of the van der Waals shift must meet four criteria. The first criterion concerns the overall ability of the three solvent test factors to predict the data matrix within experimental error. The last three criteria concern the solute coefficients in (10.6). If the van der Waals model is correct, the solute coefficients of the unity test factor should correspond to the solute gas-phase chemical shifts; the solute coefficients of the anisotropic shift term should be unity; and the solute coefficient of the van der Waals term should agree with the model.

Several theoretical models have been proposed to explain the van der Waals effect involving nonpolar solutes. Linder and co-workers[12,13] treated the solute as an oscillating dipole and the solvent as a dielectric continuum. Rummens and co-workers[14] used a virial expansion of solute–solvent interactions. Both theories have much in common but differ in detail. Because of its relative simplicity, Weiner and Malinowski[3] tested the continuum approach of Linder. Following the theory of

Linder, and after removing an unnecessary approximation, Weiner and Malinowski obtained the following expression for the van der Waals contribution:

$$\sigma_w(u, v) = \left(\frac{k}{V_u}\right)_u \left(\frac{n^2 + 2}{2n^2 + 1} \times \frac{\Sigma \langle r_j^2 \rangle}{V}\right)_v \tag{10.10}$$

Here k is a constant; V_u and V are the molar volumes of the solute and solvent, respectively; n is the refractive index of the solvent; and r_j is the electron cloud distribution of an electron in the solvent molecule, the sum being taken over all electrons of the molecule. Based on this theoretical expression, values for the solvent van der Waals term were estimated from information taken from the literature. Using target transformation, the results shown on the extreme right in Table 10.3 were obtained. Considering the crudeness of the model, the fit is surprisingly good.

10.1.7 Combination of the Basic Solvent Factors

In target factor analysis each basic vector can be tested individually; therefore, we can focus attention on any one fundamental factor. The ultimate test of the theoretical model requires a simultaneous transformation into the three fundamental solvent factors: unity, $\sigma_w(v)$, and $\sigma_a(v)$, in accord with (10.6). Simultaneous transformation into these three test vectors yielded the following equations:

$$\delta^{HMD,X}(CH_4, v) = -10.4(1) + 103.8\sigma_2(v) + 1.10\sigma_a(v)$$
$$\delta^{HMD,X}(CH_3CH_3, v) = 34.3(1) + 72.5\sigma_w(v) + 1.07\sigma_a(v)$$
$$\delta^{HMD,X}(neo\text{-}C_5H_{12}, v) = 40.2(1) + 65.0\sigma_w(v) + 1.03\sigma_a(v)$$
$$\delta^{HMD,X}(C_6H_{12}, v) = 74.7(1) + 45.4\sigma_w(v) + 1.04\sigma_a(v) \tag{10.11}$$
$$\delta^{HMD,X}(C_8H_{16}, v) = 82.9(1) + 39.0\sigma_w(v) + 1.06\sigma_a(v)$$
$$\delta^{HMD,X}(TMS, v) = 14.2(1) + 63.5\sigma_w(v) + 1.04\sigma_w(v)$$

These equations reproduced the measured shifts within experimental error, ±1.5 Hz.

The validity of this analysis is also substantiated by the following observations concerning the loadings (i.e., the solute coefficients) in the equations above. The coefficients of the unity factor compare quite favorably with the measured gas-phase shifts. Also, the solute coefficients of solvent $\sigma_a(v)$ terms compare quite closely to unity. Finally, a plot of the predicted solute coefficients of the van der Waals terms in the equations above versus the solute parts of (10.10) exhibits a linear trend, with the points being somewhat scattered, however. This scatter is probably due to the fact that hydrogen atoms in different solute molecules occupy different positions relative to the center of the molecule. The van der Waals model used in the factor analytical studies does not take such site factors into account.

10.1.8 Summary

The ultimate objective of factor analysis is to convert a matrix of experimental points into a set of equations that reveal the true origin of the observations. These studies clearly demonstrate that such an objective can be reached. We have seen here how factor analysis can be used to unravel the important interactions that are responsible for proton-NMR solvent shifts. Until the advent of TFA, it was impossible to isolate the important interaction effects so that each basic factor could be tested individually against prevailing theories.

The main accomplishments that resulted from applying TFA to proton solvent shifts are summarized below:

1. Solvent shifts, obtained from internal standard data, were predicted empirically without recourse to any particular theory.
2. Solute gas-phase shifts were isolated from solvent contributions by a mathematical route rather than by an experimental route.
3. The additivity model of Buckingham was shown to describe the nature of the effects of solvent on the chemical shifts of nonpolar solutes.
4. Within the accuracy of the data, it was shown that each interaction term could be expressed as a product function.
5. The solvent anisotropy effect was successfully isolated from the van der Waals shift.
6. Quantitative values of the solvent anisotropy were predicted for solvents, for which no theory exists.
7. The continuum dispersion model of Linder was redeveloped to better represent the nature of the van der Waals shift term and was target tested successfully.
8. The combination of the three identified factors, (1) solute gas-phase shift, (2) solvent anisotropy, and (3) van der Waals contribution, predicted the solvent shifts within experimental error.

These studies focused attention on simple nonpolar solutes. The total solution to the solvent shift phenomenon, encompassing every conceivable solvent and solute, still remains to be unraveled.

10.2 ^{1}H, ^{13}C, AND ^{29}Si SOLVENT SHIFTS OF TMS AND CYCLOHEXANE

Bacon and Maciel[15] applied abstract factor analysis to the "intrinsic" solvent shifts of TMS and cyclohexane. The intrinsic shift is defined as the shift of the solute in some solvent relative to the shift of the solute in pure TMS, corrected for bulk susceptibility differences. The study involved the ^{1}H, ^{13}C, and ^{29}Si shifts of the two solutes in 38 solvents. Because of experimental difficulties in recording ^{13}C signals in natural abundance, Bacon and Maciel did not employ pure solvents. Instead, their solvents contained 20 vol % TMS. In the case of the TMS nuclides, the shifts were referenced with respect to pure TMS. For the cyclohexane nuclides the solutions contained 2 vol % cyclohexane in addition to the 20 vol % TMS, and the shifts were referenced relative to cyclohexane, 2 vol % in TMS.

Principal factor analysis was performed on four sets of solvents. Set 1 consisted of cyclohexane, benzene, *o*-dichlorobenzene, and 1,2,4-trichlorobenzene; set 2 consisted of 14 solvents; set 3 consisted of 15 solvents, hexafluorobenzene being added to set 2; and set 4 involved all 38 solvents but only the 4 nuclides for which all data were available, thus excluding ^{29}Si.

The first two major eigenvectors reproduced the data of solvent sets 1, 2, and 3 within experimental error. For set 4, three factors were required, suggesting that an additional effect such as a dipole-induced dipole interaction between the solvent and solute is present in the larger scheme. In order to ascertain which solvent or solvents were principally responsible for the additional factor, a "suspicious" solvent was added to set 3 and a factor analysis was performed. A series of such tests revealed that hexafluorobenzene and carbon disulfide were the major contributors to the mysterious third factor. Solvent polarity was ruled out as a cause of the third factor because acetone and chloroform, which are highly polar, had no appreciable effect on the third eigenvalue. It was postulated that perhaps the partitioning into solvent–solute pairs was beginning to break down, thus producing an additional factor.

For data sets 1, 2, and 3, a two-dimensional abstract transformation matrix, involving a single rotation angle θ, is applicable:

$$\mathbf{T} = \begin{bmatrix} \cos\theta & -\sin\theta \\ \sin\theta & \cos\theta \end{bmatrix} \qquad (10.12)$$

The entire range of solvent factors was examined by rotating the eigenvectors at 5° increments. The results of this procedure are illustrated in Figure 10.1 for the monohalogenated benzenes and cyclohexanes. To interpret this graph in terms of two pairs of solvent–solute vectors that correspond to dispersion interaction and anisotropy, the following observations were made. An eigenvector rotation angle of 90° produces a common factor for the cyclohexanes and a different common factor for the benzenes, as illustrated by the intersecting lines in Figure 10.1. This factor was interpreted to be due to solvent anisotropy. The solute coefficients of the anisotropy factor showed a variation from one nucleotide to another: 1.04 for ^{1}H (TMS), 1.43 for ^{13}C (TMS), 1.26 for ^{1}H (CHX), and 1.11 for ^{13}C (CHX). These coefficients evidently are a quantitative measure of the various site factors of the different nuclei of the solute molecules.

A rotation angle of 160° produces a situation where the halogen substituent is important. At this angle the solvent factor associated with a given halobenzene is identical with the solvent factor associated with the corresponding halocyclohexane (see Figure 10.1). This situation, according to Bacon and Maciel, is consistent with their dispersion interaction guidelines but inconsistent with all other dispersion models.

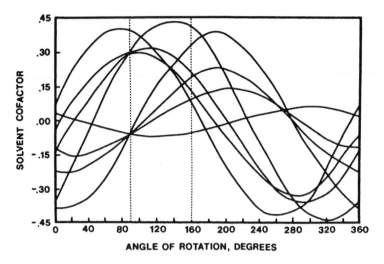

Figure 10.1   Solvent factors of data set 3 as a function of rotation angle. In order of increasing value at $0°$, the solvents are iodocyclohexane, iodobenzene, bromocyclohexane, bromobenzene, chlorocylohexane, chlorobenzene, cyclohexane, and benzene. [Reprinted with permission from M. R. Bacon and G. Maciel, *J. Am. Chem. Soc.*, **95**, 2413 (1973). Copyright 1973 © American Chemical Society.]

10.3   FLUORINE SOLVENT SHIFTS

Abraham et al.[16] measured the ^{19}F shifts of 19 rigid, nonpolar solutes in 8 solvents. They suggested that 2 factors, the solute gas-phase shift and the van der Waals effect, were responsible for the solvent shifts. When these data were subjected to factor analysis, Malinowski[17] found evidence, based on the IE and IND functions (see Section 4.3.2), that not 2 but 3 factors were active. Although small, the third factor was suspected to be due to solvent anisotropy even though benzene and carbon disulfide, which are known to exhibit large anisotropic effects, were not among the solvents employed. There was no way of testing this effect because the anisotropies of the solvents involved were not known.

In another study, using the same data, Malinowski[18] confirmed the solute gas-phase shift to be a basic factor by target testing the gas-phase shifts. The test vector and the predicted vector are given in Table 4.9. The theoretical target errors, also listed in the table, give evidence that the gas-phase shift is a basic factor. This study also shows that the predicted gas-phase shifts are more accurate than the measured gas-phase shifts (REP = 0.05 ppm, whereas RET = 0.19 ppm). This occurs because the solution shifts are more accurate than the gas shifts.

10.4   SUBSTITUENT EFFECTS ON ^{13}C SHIFTS

In order to probe into the origins of ^{13}C shifts, Wiberg et al.[19] focused attention on halogen-substituted hydrocarbons. They factor analyzed the "differential" chemical shift (i.e., the difference in shift between the halogen-substituted compound and the unsubstituted parent compound). Their data matrix consisted of 62 differential shifts as column designees and four halogens (F, Cl, Br, I) as row designees.

Two major factors emerged from the analysis. Target tests yielded three successful halogen property vectors: a unity vector (1 1 1 1), an arithmetic progression vector (1 2 3 4), and a bimodal vector (1 0 0 1). On using these three vectors in combination, three associated loadings were obtained for each carbon nucleus involved. The three different types of loadings were believed to be due, respectively, to a "hypothetical" halogen having the same quantum number, the "freeness" of the valence electrons, and a "conformation" effect.

A factor analysis study of the ^{13}C shifts of 75 3- and 4-substituted styrenes involving 15 common substituents [H,N(CH_3)$_2$, NH_2, OCH_3, SCH_3, CH_3, Si(CH_3)$_3$, F, Cl, Br, $COCH_3$, CO_2CH_3, CF_3, CN, and NO_2] has been reported by Reynolds and co-workers.[20] Three abstract factors accounted for the *actual* chemical shifts (measured relative to internal TMS) within experimental uncertainty, whereas only two factors were required to account for the *substituent* shifts (measured relative to the parent compound). Of the various substituent parameters, target testing showed that Taft's σ_F and σ_R^0 constants (exhibiting SPOILs of 2.70 and 1.69, respectively, see Section 4.6.2) are the two most appropriate substituent factors. The unity test also proved to he a valid factor for the actual shift data. This was expected because the TMS reference is arbitrary. To account for the *ipso*, *ortho*, and *meta* carbon shifts, an additional factor was required for each carbon.

In a similar study involving the ^{13}C substituent shifts of 4-substituted phenols and 2-nitrophenols in dimethyl sulfoxide solution, Hutton and co-workers[21] also found that the *para* carbon shifts require two factors whereas the *ipso*, *ortho*, and *meta* carbon shifts require three factors. Their analysis agrees with conclusions of Reynolds and co-workers[20] that the Taft inductive and field-effect parameters are the probable factors.

A 26 × 4 data matrix containing the ^{13}C substituent shifts of 13 axial and 13 equatorial substituted 4-*tert*-butylcyclohexanes for the 4 different carbon atoms was investigated by Urbaniak and Zalewski.[22] The 13 substituents consisted of Cl, Br, I, OH, OCH_3, $OOCCH_3$, $OOCCF_3$, OTs, NH_2, $NHCH_3$, N(CH_3)$_2$, NO_2, and CH_3. Two abstract factors accounted for more than 99.60% of the total variation in the data. By target transformation these 2 factors were identified as electronegativity and substituent field effects.

These investigations complement each other. They attempt to determine the substituent factor space, which, in principle, is responsible for other chemical phenomena as well.

REFERENCES

1. J. Homer, *Appl. Spectrosc. Rev.*, **9**, 132 (1975).
2. P. H. Weiner, E. R. Malinowski, and A. R. Levinstone, *J. Phys. Chem.*, **74**, 4537 (1970).

3. P. H. Weiner and E. R. Malinowski, *J. Phys. Chem.*, **75**, 1207 (1971).

4. P. H. Weiner and E. R. Malinowski, *J. Phys. Chem.*, **75**, 3160 (1971).

5. P. H. Weiner, Ph.D. thesis, Stevens Institute of Technology, Hoboken, NJ: *Diss. Abstr.*, **32**(6), Publ. No. 3290B (1971).

6. A. D. Buckingham, T. Schaefer, and W. G. Schneider, *J. Chem. Phys.*, **32**, 1227 (1960).

7. E. R. Malinowski and A. R. Pierpaoli, *J. Magn. Reson.*, **1**, 509 (1969).

8. A. A. Bothner-By, *J. Mol. Spectrosc.*, **5**, 52 (1960).

9. W. T. Raynes, A. D. Buckingham, and H. J. Bernstein, *J. Chem. Phys.*, **36**, 3481 (1962).

10. J. C. Schug, *J. Phys. Chem.*, **70**, 1816 (1970).

11. J. Homer, *Tetrahedron*, **23**, 4065 (1967).

12. B. Linder, *J. Chem. Phys.*, **33**, 668 (1960).

13. B. B. Howard, B. Linder, and M. T. Emerson, *J. Chem. Phys.*, **36**, 485 (1962).

14. F. H. A. Rummens, W. T. Raynes, and H. J. Bernstein, *J. Phys. Chem.*, **72**, 2111 (1968).

15. M. R. Bacon and G. Maciel, *J. Am. Chem. Soc.*, **95**, 2413 (1973).

16. R. J. Abraham, D. F. Wileman, and G. R. Bedford, *J. Chem. Soc. Perkin Trans. II*, **1973**, 1027.

17. E. R. Malinowski, *Anal. Chem.*, **49**, 612 (1977).

18. E. R. Malinowski, *Anal. Chim. Acta.*, **103**, 339 (1978).

19. K. B. Wiberg, W. F. Pratt, and W. F. Bailey, *Tetrahedron Lett.*, **1978** (49), 4861, 4865 (1978); *J. Org. Chem.*, **45**, 4936 (1980).

20. W. F. Reynolds, A. Gomes, A. Maron, D. W. MacIntyre, R. G. Maunder, A. Tanin, and H. E. Wong. *Can. J. Chem.*, **61**, 2367 (1983).

21. H. M. Hutton, K. R. Kunz, I. D. Bozek, and B. J. Blackburn, *Can. J Chem.*, **65**, 1316 (1987).

22. Z. Urbaniak and R. I. Zalewski, *Polish J. Chem.*, **61**, 773 (1987).

11

The harder we try to find the solution, the more complicated the problem becomes.

CHROMATOGRAPHY

11.1 INTRODUCTION

In this chapter we show how factor analysis has been used to broaden our understanding of chromatographic processes. In particular, factor analysis has been employed to identify basic factors that influence solute–solvent interactions and to classify solutes and solvents into groups having similar characteristics. Results from nearly 20 factor analyses of chromatographic data are discussed.

Factor analytical solutions to chromatographic retention problems have the general form

$$d_{ik} = \sum_j u_{ij} v_{jk} \tag{11.1}$$

311

where d_{ik} is the measured retention for the ith solute chromatographed on the kth stationary-phase solvent, and u_{ij} and v_{jk} are the jth factors for the solute and solvent, respectively. In areas such as gas–liquid chromatography (GLC) and liquid–liquid chromatography, the stationary phase is an immobile liquid that acts as a solvent for solutes in the mobile phase. The matrix expression corresponding to (11.1) is

$$\mathbf{D} = \mathbf{UV} \tag{11.2}$$

where $\mathbf{D}$ is the matrix of chromatographic retention data, and $\mathbf{U}$ and $\mathbf{V}$ are the solute- and solvent-factor matrices, respectively.

The measured retention represents a composite of all interactions that govern the movement of a solute through the chromatographic column. The energetics of solute–solvent interactions in chromatography are described in Section 2.2 of the Karger et al. monograph.[1] The ubiquitous interaction in chromatography is due to London dispersion forces arising from induced dipole–induced dipole interactions. Dispersion interactions play the dominant role in selective retardation of nonpolar solutes. Polar interactions involving dipole-induced dipole forces and dipole–dipole forces become major factors if either the solute or the solvent are polar. In some separations, interactions involving electron donor–acceptor forces such as hydrogen bonding play a crucial role. In special circumstances, steric factors can assume significance.

Nearly all applications of factor analysis to chromatography have involved GLC data. Two methods for reporting GLC data, based on the specific retention volume and the retention index system, are appropriate for factor analysis. The logarithm of the corrected specific retention volume, V_r^0, is proportional to the standard molar free energy of solution for chromatographic distribution phenomenon. Since careful control is required to determine accurate values of V_r^0, the retention index system has become the preferred method for presenting GLC results. Retention volumes of solutes in a homologous series are used as reference points in the retention index approach. A solute's position on this scale is measured relative to those of the marker solutes. The retention index I_{ik} of solute i on stationary-phase solvent k is defined as

$$I_{ik} = 100n + 100\frac{\log V_i - \log V_n}{\log V_{n+1} - \log V_n} \tag{11.3}$$

Here, n and $n + 1$ are the numbers of carbon atoms in the marker solutes eluted directly before and directly after solute i, and V_i, V_n, and V_{n+1} are the retention volumes of the ith solute and the two marker solutes, respectively. Based on thermodynamic arguments of Rohrschneider,[2] the retention index can be expressed as a linear sum of products and therefore matrices of retention indices should have factor analysis (FA) solutions.

11.2 SUMMARY OF PROBLEMS

In this section we present an overview of the applications of factor analysis to chromatography. A total of 22 problems, involving 19 publications, are summarized.

Scope. Factor analytical contributions in chromatography have been of two main types. The majority of the applications use target factor analysis to acquire knowledge about the factors that govern GLC separations. In a second type of application, both target factor analysis and abstract factor analysis have been employed to classify solutes and stationary-phase solvents into groups having similar behavior.

A concise summary of the factor analysis/chromatography studies is presented in Table 11.1. Information concerning the nature of the data matrices is given in the first three columns of the table. With the exception of two problems based on liquid–liquid chromatographic data,[10,13] the problems used GLC data. The extensive

TABLE 11.1   Summary of Applications of Factor Analysis to Chromatography

Data[a]	Solutes[b]	Solvents	Type of FA[c]	Reference
log *a*	39	7 polymeric	A	3
I	30	23 polymeric	T	4
I	30	23 polymeric	T	5
I	26 alcohols	25 polymeric	T	6
I	15 esters	6 polymeric	T	7
I	15 esters	6 polymeric	T	8
I	26 alcohols	25 polymeric	T	8
I	10	210 polymeric	A	9
log *K*	26 steroids	6 polymeric	A	10
I	10	5 mixed	T	11
I	25 hydrocarbons	12 polymeric	T	12
log *V*	15 carboranes	8 polymeric	T	13
I	44	25 polymeric	T	14
I	39	12 polymeric	T	14
I	18 ethers	25 polymeric	T	15
I	10	225 polymeric	A	16
I	10	226 polymeric	A	17
I	30	23 polymeric	A	18
I	10	226 polymeric	A	18
I	49	7 monomeric	T	19
I	53	18 monomeric	T	20
I	39 carbonyls	18 monomeric	T	21

[a]*a*, activity coefficient; *I*, retention index; *K*, liquid-chromatographic partition coefficient; *V*, liquid-chromatographic retention volume.
[b]Incorporates a variety of organic solutes unless otherwise indicated.
[c]A, abstract factor analysis; T, target factor analysis.

compilations of McReynolds[23,24] have been especially useful sources of GLC data. The retention index, with but one exception,[3] has been the type of GLC data that were factor analyzed. The number of rows and columns in the data matrices has generally been much greater than the number of factors (see Section 11.4). A variety of organic solutes and solvents have been incorporated in the studies. Some problems[6–8,10,12,13,15,21] were restricted to a single class of solutes; the remaining problems involved solutes from several different families. Stationary-phase solvents in most cases spanned a wide range of polarities. In all but four problems,[11,19–21] the solvents were primarily polymeric. The type of factor analysis employed is given in the fourth column of Table 11.1. Fifteen problems involved target factor analysis; the other problems involved abstract factor analysis. The last column in the table lists the references of each study. The factor analytical applications were due mainly to two researchers, Howery and Weiner. In a series of pioneering collaborations,[4–6,14] Howery and Weiner demonstrated that target factor analysis furnished insights into the basic factors affecting GLC retention.

Information. A summary of information from GLC applications using abstract factor analysis only is given in Table 11.2. These studies are discussed in Section 11.3.

Information from target factor analyses (TFA) is summarized in Table 11.3. These studies are discussed in Section 11.4 and following. Ranks of the data matrices, estimated from the reproduction step, are shown in the second column of Table 11.3 (see Section 11.4 for discussion). The next five columns in the table indicate which of the specified steps of TFA were carried out in each problem. The symbols u and v signify that the designated TFA procedure was performed for solutes and solvents, respectively. For example, for the third problem[6] in the table, uniqueness values for both solutes and solvents were determined: key, typical vectors were not determined for either solutes or solvents; and target tests for basic vectors, combinations of basic vectors, and predictions based on basic vectors were published for solutes only.

General conclusions and representative examples from each step of factor analysis are discussed in the following sections. To keep the discourse in perspective, the reader will find Tables 11.1, 11.2, and 11.3 useful.

TABLE 11.2 Summary of Abstract Factor Analyses

Reference	Factors	Main Objective
3	5	Prediction of new data
9	3	Classification of solvents
10	3	Calculation of abstract factors
16	2, 3	Classification of solvents
17	2	Classification of solvents
18	2–5	Classification of solutes and solvents

TABLE 11.3   Summary of Information from Target Factor Analyses

Reference	Factors[a]	Uniqueness Tests[b]	Typical Designees[b,c]	Target Transformations[b]	Key Basic Factors[b,d]	Predictions[e]
4	8	—	—	u	—	—
5	8	u, v	u, v	u, v	—	t
6	5, 6[f]	u, v	—	u	u	b
7	4	—	v	—	—	—
8	4	u	—	u	u	e
8	5	—	v	—	—	e
11	1, 2[f]	—	—	v	v	—
12	6	u	u	u	u	—
13	3	—	—	u, v	u	e
14	1–8[f]	u	u	u	—	—
14	1–3[f]	—	—	—	—	—
15	6	u, v	u, v	u	u	b^g
19	3	u, v	u, v	u, v	u, v	b
20	4–7[f]	u, v	u, v	u, v	u, v	—
21	4–7[f]	u, v	u, v	u, v	u, v	—

[a]Based on number of abstract factors to reproduce data near experimental error.
[b]u and v denote results published for solutes and solvents, respectively; a blank signifies that the step was not carried out.
[c]Based on combination step using typical vectors.
[d]Based on combination step using basic vectors.
[e]t, based on solvent-associated typical vectors; b, based on basic vectors; e, based on matrix expanded with a basic vector (see Section 11.7).
[f]Depends on submatrix studied (see original articles).
[g]Results given in Ref. 22.

11.3   ABSTRACT FACTOR ANALYSES

Applications of abstract factor analysis (AFA) to chromatography are listed in Table 11.2. Funke et al.[3] were the first to apply factor analysis to chromatography. They used the principal factor solution to predict activity coefficients. Key sets of five typical solutes and of five typical solvents were chosen from chemical criteria. The factor coefficients needed for the predictions were taken from the principal factor model rather than from a combination TFA solution. Activity coefficients for a new solute and for five new solvents were predicted quite favorably. Although target testing was not employed, the study foretold the essential features of target factor analysis.

The main goal in the remainder of the abstract factor analyses listed in Table 11.2 was to classify solvents. There exists a profusion of stationary-phase solvents. Hence mathematical methods for classifying solvents and for detecting redundancies among GLC packings are of practical interest. The ultimate goal is to identify a small set of key, preferred solvents that can accomplish the vast majority of GLC separations.

Abstract factor analysis has been used to identify major similarities and differences in stationary phases. Clusterings were based on the similarities of scores (or loadings) in the principal factor solution. Unfortunately, rotational techniques such as varimax (see Section 3.4.2) have not been applied to GLC data. Users of AFA have employed the percent variance (see Section 4.3.2) to estimate the number of factors. Typically, an AFA–GLC solution has been considered adequate if, say, 98% of the variance is accounted for. Two-factor or three-factor models have been used for classification purposes. (For target factor analysis, the variance can be deceiving since, even if 99% of the variance is accounted for, GLC data often are not reproduced near experimental error.)

In the ground-breaking use of AFA to categorize solvents, Wold and Andersson[9] selected a three-factor model as a reasonable compromise between generality and practicality. Three principal components reproduced the data within about 30 retention index units. The primary abstract factor was attributed to the polarity of the solvent, the second factor was accounted for by an unspecified but relatively constant solute parameter, and the third factor was attributed to hydrogen-bonding interactions involving alcoholic solutes.

The approaches employed by McCloskey and Hawkes,[16] Lowry et al.,[17] and Chastrette[18] were similar. In order to describe the solvents in an easily visualized manner, two-factor solutions were used. Plots of solvent loadings on factor 1 against solvent loading on factor 2 indicated clusterings of solvents. Each point in such a plot represents a single solvent. The utility of this approach is illustrated in Figure 11.1. The clusterings of 225 stationary-phase solvents depicted in the figure are

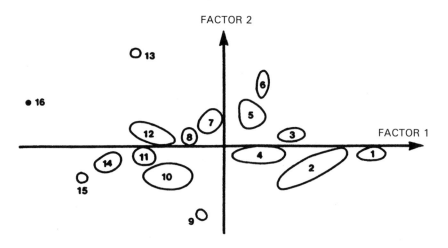

Figure 11.1 Two-dimensional plot of principal factor loadings for 225 solvents. Types of solvents in each cluster: 1, apolars; 2, silicons; 3, alkyl adipates; 4, phthalates; 5, ucon-related; 6, amides; 7, ethofat-related; 8, igepal-related; 9, fluorinated; 10, cyano-containing; 11, esters of polyols; 12, carbowax-related; 13, siponat and stepan; 14, succinates; 15, cyanoethoxypropane-related; and 16, diglycerol. [Reprinted with permission from M. Chastrette, *J. Chromatogr. Sci.* **14**, 357 (1976).]

consistent with chemical insight. Preferred stationary phases are likely to be those that have loadings far apart on both axes of the two-dimensional plots.[17] Redundant solvents form clusters in the two-factor space. In another use of the two-dimensional plots, Chastrette[18] suggested that the plot for solute factors be superposed on the plot for solvent factors. Solute–solvent pairs that interact the most with each other tended to have points near each other on the superposition.

11.4 NUMBER OF FACTORS

The number of factors found for 15 different target factor analyses varied between one and eight, as shown in Table 11.3. As the complexity of the problems increase, the number of factors generally increase. Only one or two factors were required for problems that were considered to be chemically simple.[11,14] In the most complicated problems,[4,14,20,21] as many as seven or eight factors were needed. In some cases,[15,19–21] the ranks were substantiated using the methods discussed in Chapter 4.

The average error or the root-mean-squared (RMS) error of data reproduction was usually employed to estimate the number of factors. For example, Selzer and Howery[15] reported that, with five factors, the average error of reproduction was 2.8 r.i. units, the largest error was 14 r.i. units, and about 8% of the data were reproduced with errors exceeding 5 r.i. units. With six factors, the average error dropped to 2.1, the largest error decreased to 8.2, and only 3.6% of the data had errors greater than 5. Since the experimental error had an upper limit of 5 r.i. units. six factors were deemed necessary. Malinowski[25] reached the same conclusion using the factor indicator function (IND; see Section 4.3), which exhibited a minimum at six factors. In other GLC problems, Howery and Soroka[19–21] also found that the number of factors based on the experimental error criterion was the same as the number of factors deduced from the factor indicator function.

Three other uses of the reproduction step are of interest. Weiner et al.[11] employed the dimensions of the factor space to verify retention mechanisms. They found that one factor was required if only nonpolar solutes were considered. whereas two factors were needed if both polar and nonpolar solutes were considered. The number of factors estimated from factor analysis was consistent with a chemical model for the retention mechanism. Howery and co-workers[12,14,15,19–22] used the reproduction step to furnish insights into the effects of solutes on the complexity of data spaces. They studied the change in data reproduction when columns of data were added to or deleted from the data matrix. In one approach, Howery et al.,[14] starting with a data matrix containing only alkane solutes, determined the increase in the factor size when data columns for one or two new solutes were added to the data matrix. The increases in rank were related to the complexities of the functional groups of the new solutes. For example, a solute with an alkyl side chain had almost no effect on the factor size, while an alcohol solute increased the rank by nearly 1 unit. Using the opposite approach, Howery and Soroka[19] studied the effects of removing one or more solutes from the data matrix. Usually, the greatest reduction in rank was observed for polar solutes and for chemically unique solutes.

11.5 UNIQUENESS AND UNITY TESTS

The uniqueness test (p. 60) is a useful, preliminary target test that enables the chromatographer to ferret out designees that are responsible for special interactions or that contain gross errors. Uniqueness tests can also be employed to identify clusters between similar solvents.

A summary of solutes and solvents that yielded high uniqueness values is tabulated in Table 11.4. In most cases the unique molecules have either polar functional groups or small molecular masses. For example, the pronounced uniqueness of the solvent diglycerol[6,15,20,21] is indicative of the hydrogen-bonding capability of diglycerol. High uniqueness values, not anticipated from chemical insight, might indicate gross errors. With this observation in mind, Weiner and Parcher[8] employed uniqueness tests to single out suspicious data points. Two esters were found to be quite unique (see Table 11.4), yet chemically the esters were not considered to be atypical. Suspected of having gross experimental error, these two esters were eliminated from the data matrix.

A concise table can be employed to summarize the results of uniqueness tests. Werner and Howery[5] published such a table for a number of solute uniqueness tests. Representative results are shown in Table 11.5. Pyridine and t-butanol, both relatively unique chemically, had the largest uniqueness values. Solutes with fairly

TABLE 11.4 Summary of Unique Solutes and Solvents

	Reference
Solutes[a]	
Pyridine (0.73), t-butanol (0.56)	5
2-Propynl-l-ol (0.81), methanol (0.71)	6
Isobutylbutyrate[b] (1.00), isopropylisobutyrate[b] (0.81)	8
Ethylbenzene (0.90), 2-octyne (0.74), cyclohexane (0.70)	12
2-Propynl-l-ol (0.75), octadecane (0.55)	14
Dimethyl ether (0.84), isobutylmethyl ether[b] (~0.8), t-butylmethyl ether (0.54)	15
Acetaldehyde (0.51)	21
Solvents[a]	
Fluorosilicone QF-1 (0.85), diethyleneglycol succinate (0.60), silicone fluid nitrile XF 1150 (0.60)	5
Diglycerol (0.93), polyphenyl ether—5 rings (0.56)	6
Diglycerol (0.90), Hyprose (0.60), Zonyl E7 (0.60)	15
Hexadecyl iodide (0.90), heptadecane (0.65)	19
Diglycerol (0.99), Zonyl E7 (0.94), sucrose octaacetate (0.61), Quadrol (0.55)	20
Diglycerol (0.99), Zonyl E7 (0.98), sucrose octaacetate (0.83), Hyprose (0.66)	21

[a]Predicted uniqueness value given in parentheses.
[b]Designee not included in final TFA (see original study for discussion).

TABLE 11.5   Uniqueness Table for Solutes[a]

Solutes Tested	Uniqueness Value	Other Solutes Clustering with Tested Solute
2,4-Dimethyl pentane	0.34	Cyclohexane
Cyclohexane	0.25	2,4-Dimethyl pentane
Benzene	0.16	Toluene, styrene
Toluene	0.18	Benzene, styrene
Styrene	0.26	Toluene
Methyl iodide	0.27	Cyclohexane
Ethyl bromide	0.07	Methyl iodide
Acetone	0.17	Methyl ethyl ketone
Methyl ethyl ketone	0.19	Acetone
Propionaldehyde	0.11	Acetone, methyl ethyl ketone
Nitromethane	0.35	Nitroethane
Nitroethane	0.23	Nitromethane
Pyridine	0.73	—
Ethanol	0.19	—
2-Propanol	0.18	Ethanol, t-butanol
t-Butanol	0.56	2-Propanol

[a]Reprinted with permission from P. H. Weiner and D. G. Howery, *Anal. Chem.*, **44**, 1189 (1972).

high predicted values on a given uniqueness test were assumed to form a cluster of similar solutes. For example, although benzene, toluene, and styrene were not unique individually, these chemically similar solutes were clustered as indicated in the table. Uniqueness tests do not necessarily cluster designees that are chemically alike, as shown by the alcohols in Table 11.5.

In GLC studies the results of unity tests (3.131) have been consistent with chemical expectations. When all the solutes contain at least one real factor in common, such as the same functional group, unity should be a factor. For problems involving solutes from one family,[6,8,13,16,21] unity transformed quite well. In most of the other studies referred to in Table 11.2 the unity vector was not a factor.

11.6   KEY SETS OF TYPICAL VECTORS

To search for a key set of typical vectors, columns of the data matrix are employed in combination TFA (see Section 3.4.5). For GLC, a summary of the best key sets is presented in Table 11.6. In two problems,[7,15] every possible combination of typical vectors was tested.

The results demonstrate that retention index data can be represented adequately by small sets of typical vectors. As shown in the right-hand column of Table 11.6, every acceptable key set reproduced the data matrix near or within experimental error. Molecules represented in the key sets were usually similar to those that might be selected on the basis of chemical insight.[7] Many combinations gave almost

TABLE 11.6 Summary of Key Combinations of Typical Vectors

| | | Error, TFA Solution[a] | | |
Reference	Scope[b]	Key Set Solutes	Key Set Solvents	Experimental Error[c]
5	s	4	3	5
7	a	—	2.5	2
8	s	—	2.5	5
12	s	0.8	—	1
14	s	2.1	—	5
15	a	2.8	2.9	5
19	a	0.7	0.6	1
20	s	2.4	2.0	5
21	s	2.3	2.2	5

[a]Average or RMS error (r.i. units) for key combination.
[b]s, selected combinations; a, all combinations.
[c]Estimated upper limit of experimental error (r.i. units).

equivalent results, indicating considerable redundancy of both solutes and solvents. Testing all combinations has the advantage of identifying the best solution. Selzer and Howery[15] designed a pattern table that listed the percentage of times each typical vector occurred in the better combinations for a range of factors. By examining the trends in the pattern table, they were able to estimate the relative importance of each vector.

11.7 PREDICTIONS BASED ON KEY SETS OF TYPICAL VECTORS

Predictions of new retention data based on key combinations of typical vectors (see Section 3.4.5) have been accomplished in three GLC studies.[5,8,13]

Weiner and Howery[5] used typical vectors to predict retention indices for four new solutes on 15 nonkey solvents. Retention data for each of the new solutes on eight key solvents were required because the factor space was found to be eight-dimensional. Retention indices were predicted with an average error of 4.4 r.i. units, within the upper limit of the experimental error.

Weiner and Parcher[8] introduced a novel procedure from which new values of basic factors, rather than new retention indices, were predicted. They expanded the data matrix by adding a complete basic vector to the data matrix as the $(c + 1)$th data column. The key set of typical solvent vectors for the original data matrix was determined by combination in the usual manner. The enlarged matrix was then combination reproduced with the same set of key typical vectors to obtain a target-transformed column matrix. Then data were predicted for new row-designee solvents. For nonkey columns up to the cth column, new retention data resulted. More intriguingly, for the $(c + 1)$th column, values of the basic parameter corresponding to the added column were predicted for the new solutes. Predictions based on the expanded-matrix method were considered valid only if the dimensionalities of the original data matrix and the expanded matrix were identical. The method was used by Weiner and Parcher[8] to predict values for 10 chemical parameters, such as boiling points, carbon numbers, and van der Waals constants. Agreement between predicted and known values was quite satisfying. Kindsvater et al.[13] used the unity vector as the added column and found that the predicted values were close to unity only if the new solute was chemically similar to the original solutes.

11.8 TARGET TESTS

Testing parameters of solutes and solvents via target transformation has been the most fundamental aspect of the factor analytical investigations of chromatographic data.

11.8.1 Overview of Successful Tests

Properties of solutes and, to a lesser extent, of solvents have been target tested in several GLC studies, as shown in Table 11.3. An overview of possible basic factors based on successful target tests is presented in Table 11.7. Parameters that occur more than once in the table are abbreviated in the tabulation and identified in a footnote.

In the pioneering application of TFA to gas chromatography, Weiner and Howery[4,5] studied a complex data set that contained a diversity of solute and solvent types. They found that target testing isolated many chemically reasonable factors. As GLC–TFA researchers have gained experience in developing test vectors, the scope of target testing has increased. For example, Selzer and Howery[15] tested over 50 solute vectors and, in addition, tested the square, reciprocal, and logarithm of each vector.

Several conclusions can be drawn from an examination of the solute factors listed in Table 11.7. Significantly, a large number of test parameters have been found to be factors in more than one problem. Three real vectors—carbon number (CN), molar refraction (MR), and molecular weigh (MW)—were judged to be solute factors in at least eight problems. Chemically, each of these factors might be related to dispersion interactions. Other factors being constant, elution according to molecular weight is a well-known empirical rule of GLC.

Parameters related to the vaporization of the solutes have frequently tested well. Rohrschneider[2] postulated that the retention index is proportional to the free-energy change for solute retention and therefore to the enthalpy of solution. As pointed out by Weiner and Parcher,[8] the enthalpy of solution can be divided into the heat of vaporization for the solute and the heat of mixing. From Table 11.7 we see that the enthalpy of vaporization (EV) is a solute factor in eight problems. On the basis of chemical thermodynamics, parameters proportional to the enthalpy of vaporization, such as the boiling point (in kelvins) and the logarithm of the vapor pressure, should

TABLE 11.7 Summary of Basic Factors Identified by Target Testing

	Reference
A. Basic Solute Factors[a]	
HU,[b] nitrogen uniqueness,[b] VW[b]	2
DM, square DM,[b] EV, MR,[b] VW	3
BP,[b] CN,[b] DC, DM,[b] EV,[b] hydroxyl position, log VP, MR,[b] MW,[b] UN[b]	4
BP, CN, EV,[b] log VP,[b] MR,[b] MW, UN,[b] VW[b]	6
AU,[b] carbon-bond scale,[b] CN,[b] enthalpy of formation, EV, heat capacity,[b] heat content, HC,[b] MR, NV,[b] MW, multiple-bond scale,[b] multiple-bond uniqueness,[b] triple-bond uniqueness[b]	11
CN,[b] DM, square DM, log GLC, retention volume[b]	12
CN,[b], DE, DM, EV, MR,[b] MW, log VP, VW[b]	13
AN, BP, square BP,[b] CN, chain-related vectors, CN difference, freezing point, methyl uniqueness,[b] methyl ether uniqueness,[b] MR, MW, UN, unsaturation uniqueness,[b] vinyl uniqueness	15[c]
AN, BP, square BP,[b] CN, DE, DM, EV, GD,[b] IU, MR, MW, RD,[b] UN, XU	20
AU, BP, square BP, CN, cyclic uniqueness, DC, EV, HC, HU, ketone uniqueness, MS,[b] melting point, MR, MV,[b], MW, ON, UN	21
AU, BP, square BP, CN, DM, group DM, EV, HC, HU, MS, MW, surface tension, VW	22
B. Basic Solvent Factors[a]	
Etherlike uniqueness[b]	2
Interfacial area,[b] bulk volume[b]	10
DC,[b] UN[b]	12
DM,[b] GD,[b] halogen-ether uniqueness, IU, MR, MW, RD,[b] UN	20[c]
AN, EU, HU, MC, MR, MW, SU, UN	21[c]
AN, AU, EU, HU, MC,[b] MR,[b] ON,[b] SU, UN, XU	22[c]

[a]Vectors appearing more than once in the table are abbreviated as follows: AN, atom number; AU, aromatic uniqueness; BP, boiling point; CN, carbon number; DC, dielectric constant; DE, density; DM, dipole moment; EU, ether uniqueness; EV, enthalpy of vaporization; GD, group delta; HC, heat of combustion; HU, hydroxyl uniqueness; IU, iodine uniqueness; MC, McReynolds constant; MR, molar refraction; MS, magnetic susceptibility; MV, molecular volume; MW, molecular weight; ON, oxygen number; RD, r.i. dispersion; UN, unity; VP, vapor pressure; VW, van der Waals constant; XU, halogen uniqueness.
[b]Details of target test are given in the reference.
[c]Additional vectors are discussed in the reference.

be factors. Satisfyingly, the boiling point has been reported as a solute factor in six studies, and log VP was identified in three studies.

Another type of factor accounts for polar interactions. Dipole moment (DM) tested adequately in five problems, and the square of the dipole moment was a basic factor in three problems. Both forms of the dipole moment appear in thermodynamic interaction terms.[1]

Other types of factors have been suggested from TFA results. Weiner and Howery[4] showed that gas-phase nonideality can be relegated to the van der Waals constants. These constants tested well in five GLC problems. That target factor analysis could isolate such a minor factor attests to the sensitivity of target testing. The unity vector, which represents a constant solute property, passed the target test in five problems involving solutes with a common functional group.

Most of the other factors in Table 11.7 are specific kinds of structural parameters. Clustering uniqueness vectors for solutes having similar chemical properties usually tested successfully. For instance, uniqueness tests for aromatic solutes indicated that the benzene ring was responsible for a factor in three problems. The square of the boiling point tested very well and seemed to be a major factor in several problems, in agreement with Bach et al.[26]

Parameters that chromatographers would not expect to be factors, such as melting point, refractive index, surface tension, and viscosity, generally did not transform satisfactorily. Other examples of unsuccessful tests can be found in the studies referenced in Table 11.3.

In summary, applications of TFA to retention indices indicate an intuitively acceptable model for the solute part of the solute–solvent interaction space. Target FA has been used to identify solute factors related to dispersion interactions, enthalpy of vaporization, polar interactions, gas-phase imperfections, and specific structural parameters. Overall, the high correlation between TFA results and chemical intuition is quite encouraging.

Only a few detailed studies of solvent factors have been conducted. Basic solvent parameters identified by Howery and Soroka[20–22] are shown in the lower portion of Table 11.7. The parameters listed are quite similar to the kinds of basic solute parameters discussed above. Additional target factor analytical studies of solvent factors are needed to develop better models for solute–solvent interactions in gas chromatography.

11.8.2   Solute Factors

In this section we examine results of 10 representative target tests of solute parameters. The results, each involving 13 typical solutes, are shown in Table 11.8. The solutes in these tests were selected to span a wide range of chemical characteristics.

Examples A through D in Table 11.8, taken from four different investigations, show results of testing four physical parameters: molar refraction, molecular weight, van der Waals *b*, constant, and dipole moment squared. Examples E through J illustrate the versatility of structural test vectors. The table shows results concerning a uniqueness test for alcohols, a uniqueness test for unsaturation, a carbon number test, a test for a multiple-bond scale, a test for hydroxyl position, and a unity test. The advantages of free-floating test points are illustrated in examples A, E, F, and I. In each example the validity of the test was substantiated because known values of points that were deliberately free-floated were predicted quite well.

Most of the examples in Table 11.8 are self-explanatory. Examples E and I have special historical significance. The uniqueness test for alcohols in example E was the first use of target testing to single out a subgroup of designees. In this test Weiner and Howery[4] assigned values of 1 to two of the alcohols, free floated the test values

TABLE 11.8 Examples of Target Tests of Solute Vectors[a]

Solute	Test	Prediction
A. Molar Refraction[b]		
Hexane	29.9	28.8
Octadecane	86.2	86.5
Benzene	26.2	26.3
Ethyl benzene	(35.8)	34.5
Ethyl ether	22.5	23.0
Propylmethyl ether	22.0	23.0
Methyl acetate	(17.5)	18.7
Ethyl propionate	26.8	26.4
Acetone	16.2	15.8
2-Hexanone	30.0	30.1
Propanol	17.5	17.7
Octanol	40.7	40.8
Cyclohexanol	(28.7)	29.6
B. Molecular Weight[c]		
Acetaldehyde	44	41
Butanal	72	74
Isopentanal	86	84
2,2-Dimethyl propanal	86	89
Hexanal	101	101
2-Ethyl hexanal	128	128
Acetone	58	59
3-Pentanone	86	87
3,3-Dimethyl-2-butanone	100	101
2-Octanone	128	127
Cyclopentanone	84	85
2,3-Butadione	86	87
2,4-Pendadione	100	94
C. van der Waals b Constant[d]		
Methyl butyrate	0.166	0.167
Ethyl butyrate	0.194	0.194
Propyl butyrate	0.223	0.223
Butyl butyrate	—	0.250
Pentyl butyrate	—	0.279
Isopropyl butyrate	—	0.213
Isopentyl butyrate	0.274	0.274
Ethyl isobutyrate	0.189	0.186
Propyl isobutyrate	0.216	0.220
Butyl isobutyrate	—	0.246
Pentyl isobutyrate	—	0.273
Isobutyl isobutyrate	0.240	0.237
Isopentyl isobutyrate	—	0.269
D. Dipole Moment Squared[c]		
Cyclohexane	0.0	0.7
Toluene	0.1	0.1

TABLE 11.8 (*Continued*)

Solute	Test	Prediction
Acetone	8.4	8.9
Crotonaldehyde	13.5	10.8
Butyl acetate	3.4	5.0
Acetonitrile	14.8	13.9
Nitroethane	12.6	13.0
Dibutyl ether	1.5	−0.3
Chloroform	1.0	1.6
Methyl iodide	2.6	2.8
Ethyl bromide	4.3	3.6
2-Propanol	2.6	2.3
Pyridine	4.6	4.4
E. Hydroxyl Uniqueness[f]		
2,4-Dimethyl pentane	0	0.00
2-Ethyl hexane-1	0	0.00
Toluene	0	0.00
Acetone	—	−0.13
Dibutyl ether	—	0.03
Chloroform	—	0.37
Pyridine	—	0.69
Ethanol	1	1.00
Propanol	1	1.00
Isopropanol	(1)	0.94
t-Butanol	(1)	0.74
Cyclopentanol	(1)	0.95
Allyl alcohol	(1)	1.02
F. Unsaturation Uniqueness[g]		
Dimethyl ether	0	0.01
Propylmethyl ether	0	0.04
Butylmethyl ether	0	0.19
t-Butylmethyl ether	0	−0.05
Butylethyl ether	(0)	0.04
Dipropyl ether	(0)	−0.03
Isopropylpropyl ether	0	−0.04
Diphenyl ether	0	0.07
Diisopentyl ether	0	−0.06
Ethylvinyl ether	1	1.07
Isobutylvinyl ether	1	1.03
2-Ethyl-1-hexylvinyl ether	1	1.01
Allylethyl ether	1	0.74
G. Carbon Number[h]		
o-Carborane (θ)	0	0.07
HθCH$_3$	1	1.03
HθC$_2$H$_5$	2	1.90

(*continued*)

TABLE 11.8 (*Continued*)

Solute	Test	Prediction
$H\theta C_4H_9$	4	3.86
$H\theta C_6H_{13}$	6	6.09
$H\theta CHCH_2$	—	1.50
$H\theta C_2H_4CHCH_2$	—	3.03
$H\theta CH_3CCH_2$	—	2.27
$CH_3\theta CH_3$	—	1.83
$H\theta CH_2Br$	—	1.65
$CH_3\theta CH_2Br$	—	2.29
$H\theta C_6H_5$	—	3.10
$H\theta CH_2C_6H_5$	—	3.06
H. Multiple Bond Scale[i]		
Ethane	0	0.00
Hexane	0	0.02
2-Methane heptane	0	0.02
Decane	0	0.04
1-Octene	1	0.96
2-Ethyl hexene	1	0.95
1-Octyne	—	3.31
2-Octyne	—	3.01
Benzene	3	3.02
Toluene	3	2.98
Mesitylene	3	2.96
Styrene	4	3.98
Phenylacetylene	—	5.71
I. Hydroxyl Position[j]		
Methanol	(0)	0.01
Ethanol	1	1.00
Propanol	1	0.99
sec-Butanol	(2)	2.03
t-Butanol	3	3.00
Pentanol	1	1.00
3-Pentanol	2	2.03
Hexanol	1	0.98
2-Hexanol	(2)	2.08
2-Methyl-2-pentanol	3	2.99
3-Methyl-3-pentanol	3	2.98
Octanol	1	1.01
2-Propyn-1-ol	—	−2.06
J. Unity[j]		
Methanol	1	0.98
Ethanol	1	1.01
Propanol	1	0.99
sec-Butanol	1	1.00
t-Butanol	1	1.05

TABLE 11.8 (*Continued*)

Solute	Test	Prediction
3-Pentanol	1	0.97
Hexanol	1	1.00
2-Methyl-2-pentanol	1	1.01
4-Heptanol	1	0.99
Octanol	1	1.03
Cyclohexanol	1	1.04
2-Propen-l-ol	1	0.99
2-Propyn-l-ol	1	1.04

[a]Results are given for representative solutes. For complete results, consult the original articles. Test points in parentheses were known values free floated on the test vectors; dashes indicate that points were free floated.

[b]Reprinted with permission from D. G. Howery, P. H. Weiner, and J. S. Blinder, *J. Chromatogr. Sci.*, **12**, 366 (1974).

[c]Reprinted with permission from J. M. Soroka and D. G. Howery, unpublished work.

[d]Reprinted with permission from P. H. Weiner and J. F. Parcher, *Anal. Chem.*, **45**, 302 (1973).

[e]Reprinted with permission from P. H. Weiner and D. G. Howery, *Anal. Chem.*, **44**, 1189 (1972).

[f]Reprinted with permission from P. H. Weiner and D. G. Howery, *Can. J. Chem.*, **50**, 448 (1972).

[g]Reprinted with permission from R. B. Selzer and D. G. Howery, *J. Chromatogr.*, **115**, 139 (1975).

[h]Reprinted with permission from J. H. Kindsvater, P. H. Weiner, and T. J. Klingen, *Anal. Chem.*, **46**, 982 (1974).

[i]Reprinted with permission from D. G. Howery, *Anal. Chem.*, **46**, 829 (1974).

[j]Reprinted with permission from P. H. Weiner, C. Dack, and D. G. Howery, *J. Chromatogr.*, **69**, 249 (1972).

for most of the remaining solutes, and assigned a 0 to each of the hydrocarbons. The predicted values for all the alcohols were near unity except for *t*-butanol. The somewhat lower predicted value for *t*-butanol and the rather high predicted value for pyridine were attributed to the high uniqueness of the two solutes (see Table 11.3). Example I shows an early attempt to test detailed structural parameters with TFA. In order to learn whether the position of the hydroxyl group in a series of alcohols was a factor, Weiner et al.[6] constructed a structural vector that indicated the number of carbon atoms attached to the carbon atom bonded to the hydroxyl group. The vector was essentially a test for primary, secondary, and tertiary character, with the proviso that methanol should have a value of zero on the vector. Free floating methanol and all the secondary alcohols was a severe test of the idea. As can be seen from the results for example I, the vector tested quite well and the free-floated points were predicted within chemical expectations. The negative values predicted for the free-floated unsaturated alcohols were believed[6] to be caused by the inductive effects of the multiple bond on the hydroxyl group.

Chemically meaningful scales have been developed using the free-floating capability of TFA. For example, Kindsvater et al.[13] (see example G of Table 11.8) assigned carbon numbers to five straight-chain solutes and free floated all other solutes. Predicted values for the free-floated designees were used to devise an effective carbon number scale for solutes having unsaturation, branching, and aromatic character. A scale for interrelating the effects of various kinds of carbon–carbon multiple bonds was developed by Howery[12] (see example H). On this scale saturated solutes were assigned a value of 0, solutes containing a double bond were assigned a value of 1, solutes containing a benzene ring were given a value of 3, and hydrocarbons containing a triple bond were free floated. Styrene was assigned a test value of 4 since the contributions of aromaticity and double bonding were considered to be additive. The results of the test confirmed the general form of the vector and predicted an average value slightly greater than three for the triple bond on this scale. The additivity assumption was substantiated by free floating phenylacetylene. That solute had a predicted value of 5.7, only slightly less than the value of six postulated for a molecule having a triple bond plus benzene ring.

11.8.3   Solvent Factors

Most GLC stationary-phase solvents are polymeric and often are poorly character-ized. Physical property data even for the common solvents are not available. Because it is extremely difficult to formulate structural vectors for complex polymers, nearly all insight into solvent factors has been obtained from studies involving relatively pure, well-defined monomeric solvents.

Four examples of solvent tests are given in Table 11.9. Examples A and B concern aqueous solvents. In example A, Kindsvater et al.[13] showed that the unity vector tested adequately for a group of mixed, liquid chromatographic solvents if only the first six, most polar solvents were included in the factor analysis (results given in the third column). Unity did not transform as a factor when the last two mixed solvents, which have considerably lower dielectric constants, were incorpo-rated in the data matrix (results in the fourth column). In example B, the volume of liquid per gram of packing was felt by Weiner et al.[11] to be a measure of the second-most important solvent factor. In confirmation of theory, the vector tested well in a two-factor space (fourth column), whereas the transformation was unsuccessful for a one-factor model (third column).

Target testing has been used to corroborate empirical scales for solvents. For instance, Howery and Soroka,[19] in a study of well-characterized monomeric solvents, tested a scale designed to measure the van der Waals interaction strengths of solvent functional groups. This parameter was calculated by Zielinski and Martire[27] from a least-squares analysis of retention indices. The results in example C confirmed that the interaction parameter is a basic solvent factor. In another problem involving monomeric solvents, the well-known McReynolds constants[24] were shown by TFA to be solvent factors. According to McReynolds approach, a solvent is characterized by a set of 10 constants associated with 10 simple solute "probes." In example D, Howery and Soroka[20] verified that the McReynolds constant for iodobutane solute

TABLE 11.9 Examples of Target Tests of Solvent Vectors

		Test	Prediction	
A. Unity[a]				
			6 Solvents	8 Solvents
$CH_3CN:H_2O$ (11:9)		1.0	0.94	1.12
$CH_3CN:H_2O$(6:4)		1.0	0.93	0.81
$CH_3CN:H_2O$ (4:6)		1.0	1.04	1.39
$CH_3CN:H_2O$ (3:1)		1.0	1.19	0.99
$CH_3CN:H_2O$ (4:1)		1.0	0.98	0.90
$CH_3COCH_3:H_2O$ (1:1)		1.0	0.86	0.75
$CH_3OH:H_2O$ (12:1)		1.0	—	0.95
Dioxane: H_2O (3:1)		1.0	—	0.25
B. Bulk Volume per Gram[b]				
			(1 factor)	(2 factors)
Packing 1[c]		0.29	0.13	0.30
Packing 2[c]		0.25	0.17	0.26
Packing 3[c]		0.20	0.18	0.19
Packing 4[c]		0.17	0.22	0.17
Packing 5[c]		0.14	0.25	0.15
C. Interaction Parameter[d]				
Heptadecane		0.0	−0.1	
1-Hexadecane		1.6	1.8	
1-Hexadecyl chloride		5.8	6.0	
1-Hexadecyl bromide		6.5	6.5	
1-Hexadecyl iodide		7.1	7.1	
Dioctyl ether		4.3	4.1	
Dioctyl thioether		6.2	6.1	
D. McReynolds Iodobutane Constant[e]				
Diglycerol		245	248	
Diisodecyl phthalate		83	86	
Dioctyl sebacate		68	70	
Flexol 8N8		98	90	
Hallcomid M18		82	77	
Hyprose SP-80		310	287	
Isooctyldecyl adipate		(72)[f]	69	
Quadrol		208	233	
Sucrose octaacetate		292	300	
TMP Tripelargonate		77	85	
Tricresol phosphate		169	159	
Zonyl E7		146	144	

[a]Reprinted with permission from J. H. Kindsvater, P. H. Weiner, and T. J. Klingen, *Anal. Chem.*, **46**, 982 (1974).
[b]Reprinted with permission from P. H. Weiner, H. L. Liao, and B. L. Karger, *Anal. Chem.*, **46**, 2182 (1974).
[c]Packings contain different loadings of aqueous tetraethylammonium bromide.
[d]Reprinted with permission from D. G. Howery and J. M. Soroka, *Anal. Chem.*, **58**, 3091 (1986).
[e]Reprinted with permission from D. G. Howery and J. M. Soroka, *J. Chemometrics*, **1**, 91 (1987).
[f]Known value in parentheses was free floated.

transformed as a solvent factor. The predicted constant for isooctyldecyl adipate, free floated on the test vector, substantiated the test.

11.9 KEY SETS OF BASIC VECTORS

The best models resulting from combination TFA of basic vectors are summarized in Table 11.10. Each vector that occurs in more than one solution is abbreviated with

TABLE 11.10 Summary of Key Combinations of Basic Vectors

Reference	Error, TFA Solution[a]	Experimental Error[b]	Factors in Best Solution[c]
		Solutes	
6	8.4	5	BP, CN, MR, MW, UN
8	5	3	BP, either CN or MR, UN, van der Waals a constant
12	8	2	Aromatic uniqueness, carbon bond scale, CN, either enthalpy of vaporization or molecular volume, multiple-bond scale, triple-bond scale
13	0.04	0.03	CN, log GLC retention volume, UN
15	5.3	5	AN, CN, chain difference, chain ratio, square BP, either square CN or square AN or square (C + O) number
19	2.7	1	AN, CN, DM
20	9.8	5	CN, enthalpy of combustion, ketone uniqueness, MS, vapor pressure
21	11.0	5	Alcohol uniqueness, DM, MS, surface tension, total number of carbons, van der Waals b constant
		Solvents	
11	0.5[d]	2[d]	Interfacial area/g, bulk volume/g
19	0.1	1	Log refractive index, retention index dispersion, UN
21	6.5	5	Correlation vector, DG, hydroxide uniqueness, McReynolds constant for nitropropane, log sum McReynolds constants, Zonyl uniqueness
20	6.4	5	Correlation vector, DG, McReynolds constants for butanol and iodobutane McReynolds b constant, UN

[a]Average or RMS error in r.i. units.
[b]Estimated upper limit of experimental error in r.i. units.
[c]Vectors appearing more than once in the table are abbreviated as follows: AN, atom number; BP, boiling point; CN, carbon number; DM, dipole moment; DG, diglycerol; MR, molar refraction; MS, magnetic susceptibility; MW, molecular weight; UN, unity.
[d]Percent error.

symbols defined in the footnote. Considering the complexity of chromatographic processes, several of the TFA models are strikingly good. On comparing the RMS errors resulting from the key combinations with the experimental errors (see Table 11.10), we find that especially satisfactory solutions have been formulated for solutes in two problems.[13,15] and for solvents in two problems.[11,19] In the earlier combination studies, only a few selected sets of basic vectors were tested. By contrast, in several of the more recent studies, all possible combinations of 25 or more vectors involving as many as seven vectors per combination set have been examined.

Even in the earliest use of the combination procedure, Weiner et al.[6] found a set of solute vectors that accounted for the major factors, Kindsvater et al.[13] obtained a TFA model that reproduced liquid-chromatographic data within experimental error. As an example of thoroughness, Selzer and Howery[15] studied all combinations of 33 basic vectors in a six-factor space. Over 1 million combinations were required. They proposed the use of pattern tables (see Section 11.6) for identifying the more important factors and for pinpointing similar factors.

In the key combinations for solutes in Table 11.10, carbon number appears seven times and unity appears three times. Structural vectors, including a number of clustering uniqueness tests, were incorporated in a majority of the key combinations. In most of the combination studies, several different models gave nearly equivalent data reproduction, implying the equivalence of some of the basic factors.

The most satisfying tie-ins with theory have been obtained for tests of solvent factors. Using TFA, Weiner et al.[11] verified, in detail, a theoretical model for retention mechanisms. For saturated alkane solutes, they substantiated a model involving only one factor, gas–liquid interfacial surface adsorption. When polar and unsaturated solutes were added to the data matrix, a second solvent factor, related to bulk, gas–liquid partition (see Table 11.9, example B), was identified by target transformation, again consistent with the theoretical expectations. As a second example, Howery and Soroka[19] verified and extended a three-factor model for solute–solvent interactions proposed by Zielinski and Martire.[27] Several empirical parameters for both solutes and solvents were devised by Zielinski and Martire. Howery and Soroka corroborated these scales using target testing (see Table 11.9, example C). A thorough search for solvent factors led to a solution that reproduced the data within experimental error, as shown in Table 11.10. A moderately good solution was also formulated for the solute part of the same problem. The combination TFA solution consisting of three solute factors and three solvent factors yielded a complete model that was theoretically reasonable. The three factors in the model consisted of a dispersion term, a dipole-induced dipole term, and a dipole–dipole term.

11.10 PREDICTIONS BASED ON KEY SETS OF BASIC VECTORS

For three GLC problems,[6,19,22] new rows for a data matrix have been predicted using sets of basic vectors (see Section 3.4.5). Since retentions for even simple alkane–

alkane GLC systems have not been predicted theoretically, these results demonstrate the power of target factor analysis.

In the earliest application using basic vectors, Weiner et al.[6] predicted the retention indices of four new alcohols on 20 solvents. The five basic factors employed in the calculation are listed in Table 11.10. Data points were predicted with an average error of about 20 r.i. units, an encouraging result for an exploratory effort. In a second application, Selzer and Howery[22] predicted the retention indices of two new ethers on 20 solvents with an average error of only 3.5 r.i. units, well within the experimental error of 5 r.i. units.

A different approach was explored by Howery and Soroka.[19] Key sets of three basic factors were found for both the solutes and the solvents (see Section 11.9). Accordingly, (11.1) was rewritten as

$$d_{ik} = k_1 u_{i1} v_{1k} + k_2 u_{i2} v_{2k} + k_3 u_{i3} v_{3k} \tag{11.4}$$

Here k_m is the overall proportionality constant for the mth factor, and u_{im} and v_{mk} are the mth key basic factors for solute i and solvent k, respectively. Constant k_m is the product of the proportionality constants for the mth solute term and the mth solvent term. The basic factors used in the model are listed in Table 11.10. In view of (11.4), if values for three datum and for the corresponding three solute factors and three solvent factors are known, values for the three unknown k_m's could be calculated algebraically from three simultaneous equations. Using this approach, Howery and Soroka predicted r.i. data for new solute–solvent pairs for which values of the key factors were available from other sources. The errors in prediction for 11 new r.i. data averaged 10.3 r.i. units, thus confirming the validity of the overall TFA model.

11.11 EVALUATION OF POLARITY SCALES

The main goal of chromatography is to separate the components. Selecting the proper stationary phase is one of the most important steps in this endeavor. A variety of scales have been proposed to characterize the liquid phase. In spite of the differences, all of these methods characterize the liquid in terms of "polarity/ selectivity," a nebulous term that has somewhat different meanings to different investigators.

To contrast different polarity scales Herberger[28] subjected Castello's data matrix[29] to principal component analysis. The 8 × 30 matrix concerned 30 stationary phases covering a wide range of polarity, and the following 8 polarity scales:

MR	McReynolds polarity[30]
K_C	Kováts coefficient[31]
RP	Retention polarity of Tarján et al.[32]
DC	Castello's coefficients[33]
x_b	Snyder's selectivity parameter for butanol[34]

x_n	Snyder's selectivity parameter for nitroproane[35]
x_d	Snyder's selectivity parameter for dioxane[36]
y_b	Castello's parameter,[33] $y_b = x_b/(x_b + x_d)$

Three abstract factors accounted for more than 99% of the total variance. The first abstract factor was highly correlated with MR, DC, K_C, and RP; the second with x_b, x_d, and y_b; and the third with x_n. Based on the correlations, the physical meanings of the abstract factors were interpreted[28] to be the following: factor 1, polarity (according to McReynold's, Kováts, and Costello); factor 2, hydrogen accepting and donating ability; and factor 3, dipole interactions. Rotating the factor axis did not alter the correlations. Although no single polarity scale is sufficient to characterize the stationary phases, there is no need to apply more than three scales.

11.12 SUMMARY OF RESULTS

Factor analytical studies of gas-chromatographic retention data have served as a major testing ground for applications of factor analysis in chemistry. Some preliminary conclusions from FA–GLC are listed below.

1. The number of factors predicted from factor analysis generally has been consistent with chemical insight.
2. Solvents have been classified using principal factors.
3. The uniqueness test has identified atypical molecules and has grouped molecules in clusters consistent with chemical insight.
4. Data matrices have been reproduced adequately using small sets of typical vectors.
5. New retention data have been predicted accurately using key sets of typical vectors.
6. A large number of physical and structural factors for solutes and solvents have been target tested successfully.
7. Molecular weight, heat of vaporization, and dipole moment appear to be solute factors in several problems.
8. Gas-phase imperfection, various structural vectors, and unity (for solutes having a common functional group) appear to be solute factors.
9. Complete TFA models involving basic vectors have been developed separately for solutes and for solvents.
10. An overall model involving pairs of solute and solvent factors has been formulated using combination TFA.
11. New retention data have been predicted adequately using key sets of basic vectors.
12. Eight polarity scales are interrelated. Only three of the polarity scales are required to characterize Castello's data.

REFERENCES

1. B. L. Karger, L. R. Snyder, and C. Horvath, *An Introduction to Separation Science*, Wiley-Interscience, New York, 1973.

2. L. Rohrschneider, *J. Chromatogr.*, **22**, 6 (1966).

3. P. T. Funke, E. R. Malinowski, D. E. Martire, and L. Z. Pollara, *Sep. Sci.*, **1**, 661 (1966).

4. P. H. Weiner and D. G. Howery, *Can. J. Chem.*, **50**, 448 (1972).

5. P. H. Weiner and D. G. Howery, *Anal. Chem.*, **44**, 1189 (1972).

6. P. H. Weiner, C. Dack, and D. G. Howery, *J. Chromatogr.*, **69**, 249 (1972).

7. P. H. Weiner and J. F. Parcher, *J. Chromatogr. Sci.*, **10**, 612 (1972).

8. P. H. Weiner and J. F. Parcher, *Anal. Chem.*, **45**, 302 (1973).

9. S. Wold and K. Andersson, *J. Chromatogr.*, **80**, 43 (1973).

10. J. F. K. Huber, E. T. Alderlieste, H. Harren, and H. Poppe, *Anal. Chem.*, **45**, 1337 (1973).

11. P. H. Weiner, H. L. Liao, and B. L. Karger, *Anal. Chem.*, **46**, 2182 (1974).

12. D. G. Howery, *Anal. Chem.*, **46**, 829 (1974).

13. J. H. Kindsvater, P. H. Weiner, and T. J. Klingen, *Anal. Chem.*, **46**, 982 (1974).

14. D. G. Howery, P. H. Weiner, and J. S. Blinder, *J. Chromatogr. Sci.*, **12**, 366 (1974).

15. R. B. Selzer and D. G. Howery, *J. Chromatogr.*, **15**, 139 (1975).

16. D. H. McCloskey and S. J. Hawkes, *J. Chromatogr. Sci.*, **13**, 1 (1975).

17. S. R. Lowry, G. L. Ritter, H. S. Woodruff, and T. L. Isenhour, *J. Chromatogr. Sci.*, **14**, 126 (1976).

18. M. Chastrette, *J. Chromatogr. Sci.*, **14**, 357 (1976).

19. D. G. Howery and J. M. Soroka, *Anal. Chem.*, **58**, 3091 (1986).

20. D. G. Howery and J. M. Soroka, *J. Chemometrics*, **1**, 91 (1987).

21. D. G. Howery and J. M. Soroka, private communication.

22. D. G. Howery, in B. R. Kowalski (Ed.), *Chemometrics: Theory and Applications*, ACS Symp. Ser. 52, American Chemical Society, Washington, DC, 1977, p. 73.

23. W. O. McReynolds, *Gas Chromatographic Retention Data*, Preston Technical Abstract Co., Niles, IL, 1966.

24. W. O. McReynolds, *J. Chromatogr. Sci.*, **8**, 685 (1970).

25. E. R. Malinowski, *Anal. Chem.*, **49**, 612 (1977).

26. R. W. Bach, E. Dotsch, H. A. Friedrichs, and L. Marx. *Chromatographia*, **4**, 459 (1971).

27. W. L. Zielinski and D. E. Martire, *Anal. Chem.*, **48**, 1111 (1976).

28. K. Heberger, *Chemometrics Intell. Lab. Syst.*, **47**, 41 (1999).

29. G. Castello, G. D. D'Amato, and S. Vezzani, *J. Chromatogr.*, **646**, 361 (1993).

30. W. O. McReynolds, *J. Chromatographic Science*, **8**, 685 (1970).

31. E. Fernández-Sánchez, A. Fernández-Torres, J. A. Garcia-Dominquez, and J. M. Santiuste, *Chromatographia*, **31**, 75 (1991).

32. G. Tarján, A. Kiss, G. Kocsis, S. Mészáros, and J. M. Takács, *J. Chromatogr.*, **118**, 327 (1976).

33. G. Castello, G. D'Amato, and S. Vezzani, *J. Chromatogr.*, **646**, 361 (1993).

34. L. R. Snyder, *J. Chromatogr.*, **92**, 223 (1974).

35. L. R. Snyder, *J. Chromatographic Science*, **16**, 223 (1978).

36. M. S. Klee, M. A. Kaiser, and K. B. Laughlin, *J. Chromatogr.*, **279**, 681 (1983).

12

There is no such thing as a constant variable or a variable constant.

ADDITIONAL APPLICATIONS

The variety of applications discussed in this chapter further demonstrate the broad applicability of factor analysis. A summary of the kinds of problems covered in the chapter is tabulated in Table 12.1. In Sections 12.1 through 12.5 we focus attention on fundamental data that might be expected, theoretically or intuitively, to have factor analytical solutions. In Sections 12.6 through 12.10 we discuss some very complicated problems of a practical vein for which factor analysis also has furnished useful insights.

12.1 LINEAR FREE-ENERGY RELATIONSHIPS

Linear free-energy relationships (LFER) have received much attention during the past 50 years.[1] The LFER approach attempts to express reaction rates and equilibrium constants for a series of structurally related compounds in terms of properties of the substituent groups and the reaction media. The empirical Hammett equation,

$$\log K' = \log K^0 + \rho\sigma \tag{12.1}$$

335

TABLE 12.1   Summary of Problem Types Discussed in This Chapter

Field	Data Factor Analyzed	Kinds of Designee
Linear free-energy relationships	Log acidity constant	Solutes, solvents
Mass spectrometry[a]	Spectral intensity	Molecules, m/e values
Instrument comparison	Absorbance	Solutions, wavelengths
Method comparison	Concentration	Methods, blood sera
Medium comparison	Absorbance	Compounds, wavenumbers
Solubility	Log solubility	Solutes, solvents
Solution properties	Thermodynamic parameter	Solutes, solvents
Polarography	Half-wave potential	Metal ions, solvents
Chelation	Log stability constant	Metal ions, ligands
Bond energy	Bond dissociation energy	Radicals, radicals
Biomedical chemistry	Biological parameters	Drugs, drug properties
Environmental chemistry	Concentration	Components, sites

[a]See also Chapter 9.

constitutes the origin of LFER. In (12.1), K^0 is the rate constant or equilibrium constant for the parent molecule; K' is the same quantity measured under identical experimental conditions for the parent molecule, which has been modified by the presence of a substituent group; ρ is a reaction medium constant that is independent of the substituents; and σ is a constituent constant that is independent of the reaction medium.

Originally, the Hammett equation applied only to benzene compounds containing substituents in the meta and para positions, failing to account for ortho substituents. Higman[2] applied abstract factor analysis (AFA) to LFER in an attempt to account for the ortho substituents. Speculating that the Hammett equation might contain additional terms due to a variety of effects caused by substituent groups, Higman postulated that a more appropriate expression might be

$$\log K' = \log K^0 + \sum_i \rho_i \sigma_i \tag{12.2}$$

where the sum is taken over all possible effects of the substituents, such as reaction field, polarization, inductive, and steric. Equation (12.2) is ideally suited for factor analysis. Unfortunately, Higman was unable to find a satisfactory solution due to the inadequacy of the graphical techniques he employed.

In 1961, Malinowski[3] investigated the acidity constants of 13 ortho-, meta-, and para-substituted phenyl compounds in four reaction media. When the data were subjected to factor analysis, three primary factors emerged. Two solute factors were target tested: unity, to account for the unity coefficient multiplier of the $\log K^0$ term, and Hammett's σ constant, obtained from previous studies involving only meta and para groups. The success of the two tests is seen in Table 12.2. Since σ constants for ortho groups did not exist, these points were free floated, allowing the target test to yield predictions for these points. As seen in the table, σ values predicted for a substituent in the ortho position are much greater than those of the same substituent in the meta or para positions. At that time, these results were surprising since the electronic effect of a substituent in the ortho position was generally assumed to differ little from that in the meta or para position. The third solute factor was suspected to be caused by steric hindrance of the substituent in the ortho position. To account for this factor, Malinowski developed a test vector related to the size of the substituent group, consisting of the covalent radii of the ortho substituents and of zeros for the meta and para substituents. This test also gave excellent correlation, as shown in Table 12.2.

The three solute factors, used in combination, reproduced the data within experimental error and therefore accounted for all the important factors of the solute space. Furthermore, the loading coefficients obtained from combination target factor analysis (TFA), which correspond to the solvent factors, ρ, compared favorably with the reaction medium constants calculated by Jaffe[4] from data involving meta- and para-substituted compounds only.

Weiner[5] factor analyzed the acidities of 19 substituted benzoic acids in 7 solvents. Four factors were required to reproduce the data within the reported experimental error. Uniqueness tests for the solutes showed that none of the solutes were atypical in behavior. However, when applied to the solvents, the uniqueness tests indicated that ethylene glycol possessed a unique characteristic. This characteristic could not be attributed to hydrogen bonding because the other alcohol solvents exhibited low

TABLE 12.2 Target Tests of Solute Test Vectors for Acidity Data[a]

Solute	Unity		Hammett σ Constant		Covalent Radius	
	Test	Prediction	Test[b]	Prediction	Test	Prediction
—H	1	1.00	0.000	0.005	0.00	−0.18
o-F	1	1.04	—	0.690	0.72	0.87
m-F	1	0.93	0.337	0.339	0.00	0.23
p-F	1	1.00	0.062	0.059	0.00	0.07
o-Cl	1	1.07	—	0.823	0.99	0.97
m-Cl	1	1.03	0.373	0.373	0.00	−0.01
p-Cl	1	1.02	0.226	0.219	0.00	−0.03
o-Br	1	1.04	—	0.862	1.14	1.08
m-Br	1	1.00	0.391	0.374	0.00	0.02
p-Br	1	0.99	0.232	0.234	0.00	−0.05
o-I	1	0.90	—	0.920	1.33	1.23
m-I	1	0.98	0.352	0.371	0.00	0.04
p-I	1	0.98	0.276	0.276	0.00	−0.04

[a]Reprinted with permission from E. R. Malinowski, Ph.D. thesis, Stevens Institute of Technology, Hoboken, NJ, 1961.
[b]Dashes indicate free-floated test points.

predicted values on the test for ethylene glycol uniqueness. To account for the 4 factors, Weiner proposed the following model:

$$\log \frac{K(i,k)}{K^0} = \left(\log \frac{K(i, \text{gas})}{K^0}\right) \cdot 1 + U_E(i,j) \cdot V_E(j,k) + U_W(i,j)$$
$$\cdot V_W(j,k) + U_G(i,j) \cdot V_G(j, \text{ethylene glycol}) \qquad (12.3)$$

The first term on the right-hand side of this equation accounts for the acidity of the proton transfer reaction in the absence of solvent (i.e., in the gas phase) with the solvent factor for the gas-phase term being unity. The second term is the electrostatic contribution, the third term is the van der Waals (dispersion) effect, and the fourth term accounts for the unique effect produced by ethylene glycol solvent. The U's refer to solute factors and the V's refer to solvent factors.

Weiner used target testing to identify all four solute factors. Two of the solvent factors—unity (the solvent part of the gas-phase acidity term) and the uniqueness term for ethylene glycol—were easily target tested because a theoretical model was not required. For the electrostatic contribution, Weiner used the theoretical model proposed by Kirkwood and Westheimer,[6] which expressed this term as a product function of solute and solvent terms, as required for factor analysis. The solvent electrostatic factor was simply the reciprocal of the dielectric constant of the medium. For the van der Waals contribution, Weiner took advantage of the success of a similar term developed in nuclear magnetic resonance–FA (NMR–FA) studies (see Section 10.1.6), where the solvent part of the term was found to be a function of the polarizability and the electron cloud distribution of the solvent molecule. Target transformation into each of these four solvent factors was successful, as shown in Table 12.3.

TABLE 12.3   Target Tests of Solvent Vectors for Acidity Data[a]

Solvent	Unity		Reciprocal of Dielectric Constant		van der Waals Effect		Uniqueness for Ethylene Glycol	
	Test	Prediction	Test	Prediction	Test	Prediction	Test	Prediction
Methanol	1	1.02	3.17	2.89	7.44	7.41	0	0.00
Ethanol	1	1.00	4.14	4.98	6.93	6.78	0	−0.10
Ethylene glycol	1	0.96	2.66	2.95	7.59	7.59	1	0.93
Butanol	1	0.97	5.73	5.31	6.78	6.72	0	0.19
Propanol	1	1.03	4.98	4.66	6.85	7.08	0	0.00
Dioxane–water ($\epsilon = 55$)	1	1.00	1.82	1.29	7.84	7.84	0	0.00
Dioxane–water ($\epsilon = 40$)	1	1.00	2.50	3.36	7.72	7.68	0	0.00
Dioxane–water ($\epsilon = 15$)	1	1.00	6.68	6.23	7.46	7.49	0	0.00

[a]Reprinted with permission from P. H. Weiner, *J. Am. Chem. Soc.*, **95**, 5845 (1973).

Simultaneous target transformation into the combination of the four solvent test factors yielded a root-mean-square (RMS) error about three times the experimental error in the acidity data. According to (12.3), the coefficients resulting from combination TFA for the solvent unity term should correspond to the solute gas-phase acidities. Thus gas-phase acidities were predicted solely from solution measurements. Unfortunately, no experimental data for such acidities existed at that time for comparison. Weiner checked the validity of these results by repeating the factor analysis after deleting ethylene glycol from the data matrix. If the fourth factor were truly unique to ethylene glycol, the factor space would be reduced to three and the first three terms in (12.3) would suffice. When the unique solvent was removed, the same solute loadings were obtained for the first three terms.

Wold and Sjostrom[7] used principal factor analysis (PFA) to determine which reaction series follow the simple Hammett relationship (12.1) and which reaction series require an expanded Hammett equation (12.2). As a secondary goal, they also used PFA to obtain the best set of σ values from all the series that were shown by PFA to obey the simple Hammett relationship.

In one calculation, the constant in the Hammett equation was set equal to the logarithm of the equilibrium constant of the unsubstituted parent molecule of the series, giving a "restricted" constant. In a second calculation, the constant was set equal to the mean value of the logarithm of the equilibrium constants for the series, giving an unrestricted constant. A complete data set and an incomplete data set involving only meta substituents with free electron pairs (donors) and p-NO_2 groups were factor analyzed. The latter set was chosen to ensure the absence of resonance effects. Two conclusions were reached. First, the overall fit with the unrestricted constant was slightly better in all cases. Second, σ values generated by PFA for the unrestricted cases led to data reproduction that was far better than that obtained using previously published σ values.

12.2 FRAGMENTATION IN MASS SPECTROMETRY

Mass spectra (MS) of large organic molecules exhibit hundreds of fragmentation lines containing detailed structural and mechanistic information. Because of the complexity of the ionization process and the myriad of uncontrollable variables, MS patterns are poorly understood.

Justice and Isenhour[8] believed that factor analysis could show how the functional groups of the parent molecule influence the mass spectral pattern. They carried out an ambitious study involving 630 low-resolution mass spectra of compounds having an elemental composition of the form $C_{2-10}H_{2-22}O_{0-4}N_{0-2}$. Their data matrix incorporated the intensities of 119 m/e positions ranging from 12 to 141 mass units. The experimental data were standardized by subtracting the mean intensity of the mass position and then dividing by the standard deviation of the mass position, a modified form of correlation about the mean. Retaining only those factors having eigenvalues greater than 1 (see Section 4.3.2), 42 factors, accounting for only 73% of the total variance, were deemed important. Attempts were made to relate the loading coefficients to the presence of functional groups such as carbon chain, phenyl, carbonyl, ether, hydroxyl, and amine. Ranking procedures were developed for this purpose and the factors were apportioned among the various functional groups.

A more selective approach was taken by Rozett and Petersen.[9-11] A series of detailed factor analytical studies of the mass spectral intensities of 22 benzenoid isomers having the formula $C_{10}H_{14}$ were carried out. They reasoned that since all these hydrocarbon isomers had the same molecular weight and contained a benzene ring, the factor space should be relatively simple, while the side-chain structures of the isomers should provide sufficient variety to study phenomena such as ring opening, ring expansion, and ion-neutral complexing, phenomena known to influence mass spectra.

Of the various factor analytical pretreatment methods studied, Rozett and Petersen[9] found that the best technique for MS used the absolute intensities in conjunction with covariance about the origin (see Section 3.2.3). With three principal factors, 99.1% of the variation in the data was accounted for, and the data were reproduced within 1%, the average repeatability in measuring the height of a mass spectral line. For such a complicated problem, the factor size was surprisingly small. Combination TFA of various sets of isomers (typical rows) and fragments (typical columns) showed that there were a number of key sets of isomers and fragments that reproduced the data adequately.[10] Several key sets were found because some of the isomers and fragments were equally dependent on the same real factor. Conversely, sets of isomers and sets of fragments that did not represent all three factors gave poor reproductions in combinations. Such studies helped classify the isomers and fragments into clusters.

In order to examine the clusters that evolved from FA–MS, Rozett and Petersen[11] used triangular plots to represent the three-factor space. These plots were constructed so that the three triangular coordinates represented the three abstract factors. The sum of the squares of the three loadings for any specified isomer was standardized to equal 100, a mathematical requirement for any three-dimensional plot made on a two-dimensional plane. For the principal factor solution, the triangular plot exhibited three clusters of isomers. Sixteen, four, and two isomers clustered near the corners associated with the first, second, and third principal factors, respectively.

Because it is impossible to have negative intensities in mass spectra, the factor loadings obtained from PFA lacked theoretical interpretation. To give real significance to the loadings, Rozett and Petersen[10] used varimax rotation (see Section 3.4.2) to obtain a new set of abstract factors, called partial factors, all of which had positive loadings. Figure 12.1 is a triangular plot of the square of the loadings of the three factors obtained from the varimax rotation.[11] The clusters obtained after rotation were tighter and closer to the corners than the clusters in the principal factor plot. Furthermore, the three diethyl isomers formed a cluster midway along edge 1–2, and *t*-butylbenzene gave a unique point lying on edge 1–3. Clusters along the edges arose from linear combinations of the two connecting corner factors. Since no clusters appeared in the interior of the triangle, there were no isomers that depended

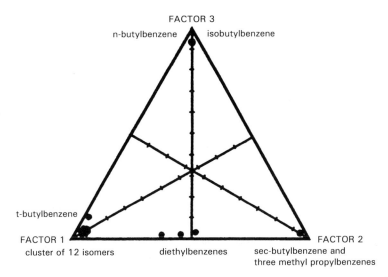

Figure 12.1 Triangular plot of the square of the loadings obtained from varimax rotation of mass spectral data. [Reprinted with permission from R. W. Rozett and E. M. Petersen, *Anal. Chem.*, **48** 817 (1976).

on all three rotated factors. In addition, such plots gave insights into the nature of the reaction mechanisms responsible for the mass spectral fragmentations. By studying the structures of the isomers in each cluster, Rozett and Petersen[11] concluded that corners 1, 2, and 3 represented fragmentation patterns indicative of a loss of one carbon atom, two carbon atoms, and a propyl group, respectively, from the parent ion. Edges 1–2 and 1–3 were attributed to a loss of both methyl and ethyl groups and to a loss of both methyl and propyl groups, respectively.

Encoded within the three abstract factors were three basic factors truly responsible for the fragmentation. In a search for the real factors, Rozett and Petersen[11] target tested a variety of geometrical, physical, and thermodynamic properties. The geometric tests included structural features such as the Weiner path number (the sum of the number of bonds between all pairs of carbon atoms), the total number of side chains, the number of specific terminal groups (such as methyl, ethyl, and propyl), and the number of ways of obtaining certain designated masses. The physical tests included melting point, boiling point, density, and refractive index. Thermodynamic tests involved properties such as the heat of formation, entropy of formation, heat capacity, and ionization potential. Although some of the individual tests were encouraging, no combination set of three basic vectors was found to be satisfactory. While all the tests involved parameters associated with the neutral parent molecule, it was deemed more likely that the true factors were associated with properties of the parent ion.

The success of their initial studies led Rozett and Petersen[12] to expand their data base to include 70 aromatic hydrocarbons, ranging from C_7H_8 to $C_{12}H_{18}$, and 151 mass peaks, ranging from 12 to 162. The intensities of each compound were normalized by dividing each intensity by the square root of the sum of the squares of the intensities of the compound. To conserve on computer time, they developed a direct factor analysis procedure[12] for extracting the principal eigenvectors up to a specified number, without requiring the complete solution. This method is valuable when dealing with large data matrices that might have hundreds of eigenvectors.

Five factors, accounting for 91% of the variation of the data, were deemed acceptable.[12] An abstract rotation called *direct quartimin* was used to group the molecules into clusters. Quartimin involves rotating the eigenvectors so as to minimize the sum of the fourth powers of the loading cofactors. By studying the dependence of each compound on the quartimin vectors, a physical interpretation of the underlying factors emerged. Factor 1 was associated with the presence of the $C_9H_{11}^+$ ion, factor 2 with $C_8H_9^+$, factor 3 with $C_7H_7^+$, factor 4 with $C_{10}H_{13}^+$, and factor 5 with $C_{11}H_{15}^+$.

Burgard and co-workers[13] factor analyzed the mass spectra of oligodeoxyribo-nucleotides in order to determine the presence or absence of a given nucleoside, the relative number of different nucleosides, and the sequence of the nucleosides in the compound. The raw data matrix consisted of the intensities of 32 selected ions emanating from 32 nucleotides. The nucleotides were composed of various amounts and various linkages of four nucleosides: adenosine, guanosine, thymidine, and cytidine.

Each data column, associated with a specific nucleotide, was normalized by dividing each ion intensity by the sum of the intensities of the 32 ions in the column. Principal factor analysis of the normalized data matrix, using covariance about the origin, yielded four eigenvectors that accounted for 99% of the variance. Although varimax rotation indicated that the varimax factors closely resembled the four nucleosides, the loadings could not be used to establish the presence or absence of a particular nucleoside in a compound.

To establish the existence of a particular nucleoside, the following procedure was adopted. The initial set of 32 ions was chosen on the basis of previous mass spectral studies so that a set of eight selected ions represented each nucleoside. To test for the presence of a specific nucleoside, each column of the data matrix was normalized by dividing each ion intensity by the sum of the intensities of the ions belonging to the subset associated with the nucleoside under consideration and principal factor analysis was carried out. The factor loadings on the first two principal axes obtained from testing for the presence of adenosine are plotted in Figure 12.2. The loadings of the 32 compounds were divided into two clusters separating the compounds that contained adenosine from those that did not. Similar results were achieved for the other three nucleosides.

To determine the relative number of different nucleosides in a compound, ratios of the intensities of selected ions for each pair of nucleosides were considered. The loadings from PFA gave a surprisingly good estimate of the relative amount of the two nucleosides present in a compound. Attempts to use factor analysis to obtain sequence information were unsuccessful.

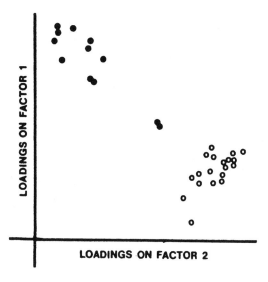

Figure 12.2   Plot of the loadings in the first two principal axes resulting from analysis of the normalized-to-sum adenosine ions for compounds containing adenosine (filled circles) and compounds not containing adenosine (open circles). [Reprinted with permission from D. R. Burgard, S. P. Perone, and J. L. Weibers, *Anal. Chem.*, **49**, 1444 (1977).

12.3   SPECTRAL LIBRARY COMPRESSION AND NOISE REDUCTION

Because of automatic data collection features of modern chemical instruments, the storage and retrieval of spectral data have grown prohibitively large. Researchers have attempted to compress the information by various techniques, such as binary encoding, which, unfortunately, lead to a loss of information and discriminating power. Isenhour and co-workers[14–16] have shown that abstract factor analysis is capable of not only compressing spectral data but also removing redundant information as well as reducing noise, without any loss in search performance.

Abstract factor analysis (AFA) is ideally suited to library compression because it defines the orthogonal coordinate system that spans the factor space of the spectral data most efficiently. A library of data with size $r \times c$ can be stored as two much smaller matrices, one with size $r \times n$ and the other with size $n \times c$. If each row of the library data consists of a normalized spectrum and each column represents a specific spectral interval (e.g., wavelength, wavenumber, or mass/charge), the two resulting matrices are abstract representations of the compounds and the spectra, respectively. Multiplication of the abstract vector representation of any compound by the abstract spectral matrix reproduces the spectra of the compound.

This technique has been applied to mass and infrared spectra.[14–16] Figure 12.3 illustrates the nature of the compression achieved with a library consisting of the absorbances of 2300 vapor-phase infrared spectra digitized into 198 discrete frequencies at 16-cm^{-1} intervals from 3840 to 800 cm^{-1}.[15] Approximately 95%

of the variance in the data was accounted for by 38 abstract factors. Thus all essential information can be stored in two abstract matrices, which together amount to a 79% reduction in required storage capacity.

Based on normalized spectra, four data pretreatment techniques were investigated: (1) covariance about the origin, (2) covariance about the mean, (3) correlation about the origin, and (4) correlation about the mean (see Section 3.2.3). Covariance about the origin (no standardization) was reported to have the best search performance, while correlation about the origin was reported to be the worst. Libraries containing poor-quality spectra reduce the efficiency and accuracy of AFA compression. Each outlying spectrum produces a unique factor, thus increasing the factor space undesirably. The removal of all spectra that exhibit a variance in AFA reproduction, which falls below a certain fraction of the total variance of the library, after compression, is recommended.[16]

One of the fringe benefits of AFA is the fact that data reproduced with a small number of factors will be inherently more accurate than the original data, provided, of course, that the factor space is adequately spanned. A discussion and theoretical derivation of this important feature are found in Section 4.2. Library compression by AFA, if done properly, should lead to data improvement. Theoretically, the uncertainty will be reduced from real error (RE) to imbedded error (IE) [see (4.40) and (4.41)], representing a reduction in uncertainty by a factor of $\sqrt{n/c}$. Koenig and co-workers[17,18] have verified this noise reduction feature of AFA in their investigations of polymer mixtures by fast Fourier–infrared (FT–IR).

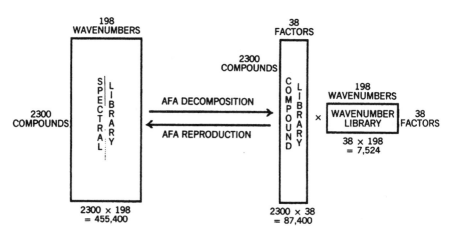

Figure 12.3 Illustrating library storage requirements before and after AFA. An infrared library containing the spectra of 2300 compounds encoded at 198 discrete wavenumbers (455,400 points) is compressed into a library of compound factors (87,400 points) and a library of wavenumber factors (7524 points), totaling 94,924 points (a 79% reduction in size). [Based on the work of P. B. Harrington and T. L. Isenhour, *Anal. Chem.*, **41**, 449 (1987).]

12.4 COMPARISONS IN ANALYTICAL CHEMISTRY

Three types of factor analytical applications involving comparison methods in analytical chemistry are presented in this section. These methods take advantage of the inherent statistics that is automatically built into the principal component feature of factor analysis.

Instrument Comparisons. Analytical chemists require methods to evaluate the performance of spectrophotometers. With this need in mind, Wernimont[19] examined the absorbance curves of a group of 16 spectrometers, all of the same model but spanning more than a 20-year manufacturing period. The absorbances of three solutions of potassium dichromate were measured at 20 wavelengths on two different days, yielding a 6×20 data matrix for each instrument. The rank of such a data matrix will be unity if the instrument is performing properly, whereas the rank will be two or more if the instrument is improperly adjusted, improperly calibrated, or improperly read. The first factor arises from Beer's law for one component behaving ideally (see Chapter 9); additional factors arise from unique error factors such as wavelength shifts in the instrument or large reading errors by the chemist. The AFA methodology is simple to carry out on a routine basis. Regular application assures the spectroscopist that the instruments are performing properly and gives a warning when performance is unsatisfactory.

Using covariance about the origin, Wernimont calculated the residual standard deviations [see (4.44)] for each of the 16 data matrices, assuming rank one. Several instruments showed residuals that were much greater than 0.004 absorbance unit, the error estimated from individual measurements. In some cases close examination of the data revealed gross reading mistakes, which were easily corrected. More interestingly, six instruments still had unsatisfactory residual errors. After careful mechanical, electrical, and optical adjustments were made, all these erratic instruments produced acceptable spectra.

Method Comparison. New methods of analysis are a main pursuit of analytical chemists. Before replacing an old method with a new one, the chemist must evaluate the new technique for systematic and random errors, and for possible interfering substances that may be specific to the new method. Recognizing this problem, Carey et al.[20] applied principal factor analysis to method comparisons. PFA offers an advantage over older methods such as regression analysis by eliminating the need to choose a reference method that must be assumed to be free from errors and interferences. Instead, PFA allows the chemist to use a composite of different methods as a reference, thus involving multivariate statistics rather than univariate statistics.

Carey and co-workers[20] investigated six methods for measuring glucose, listed in Table 12.4, involving 130 blood sera that were divided into uremic and nonuremic sera. Three PFA models were used to evaluate the data. Model I assumed that a

TABLE 12.4   Standard Errors from PFA of Six Methods for Measuring Glucose in Uremic Sera[a,b]

	Model I		Model II		Model III	
Method	E_1	E_2	E_1	E_2	E_1	E_2
Ferricyanide	10.11	3.47	11.63	3.42	22.48	4.05
Neocuprine	7.06	5.37	6.96	5.41	11.53	5.35
o-Tuluidine	4.21	4.16	4.93	4.24	4.67	4.57
Oxidase ABTS	6.76	2.64	99.50	2.69	22.56	3.06
Oxidase MBTH–DMA	3.11	2.15	4.26	2.14	4.14	2.92
Hexokinase	4.77	3.71	5.22	4.16	6.02	4.19

[a]E_1, RMS error assuming no interfering species; E_2, RMS error assuming one interfering species.
[b]Reprinted with permission from R. N. Carey, S. Wold, and J. O. Westgard, *Anal. Chem.*, **47**, 1824 (1975).

systematic error C_i, as well as a random error E_{ik} existed. The glucose value for the ith method applied to the kth sample, Y_{ik}, was expressed as

$$Y_{ik} = C_i + E_{ik} + \sum_{j=1}^{m} U_{ij}X_{jk} \qquad (12.4)$$

where U_{ij} was the response of the jth species and X_{jk} was the concentration of the jth species. The sum was taken over all m species that contribute to the measurement. If there were no interfering species, m would equal 1; if there were a single interfering species, m would equal 2. Based on (12.4), PFAs were carried out and RMS errors in the reproduced glucose values were calculated for each method. These errors for uremic sera, labeled E_1, and E_2 for m equaling 1 and 2, respectively, are shown in Table 12.4. Model II assumed that the systematic error C_i was zero. Since the PFA results in Table 12.4 for uremic sera based on this model show that the E_2 values for models I and II do not differ significantly, the absence of systematic error was indicated. Model III assumed that the responses due to the first factor, U_{i1}, were equal to 1.00 for those methods calibrated by aqueous standards (ferricyanide, o-toluidine, and hexokinase), thus allowing for calibration constants that depended on the method employed.

By comparing E_1 to E_2 for a given method, we can conclude quickly which methods are sensitive to interfering substances. If, for a given method, E_2 approximately equals E_1, no interfering species are present. If E_2 is significantly smaller than E_1, an interfering species is present. From all three models the same conclusions were reached. For both uremic sera and nonuremic sera, the ferricyanide and oxidase ABTS methods were found to be extremely sensitive to interfering substances; neocuprine and oxidase MBTH–DMA methods showed a moderate dependence on foreign species; and the hexokinase and o-toluidine methods were relatively free from interferences. Thus factor analysis can reveal which glucose measurements are subject to interference from foreign substances or to imprecision.

Such information is a valuable aid in selecting the most valid analytical procedure to be used in routine analysis.

Medium Comparisons. Correcting analytical measurements for medium effects has been a vexing problem for the chemist. In many situations factor analysis may afford the most powerful method for identifying and compensating for medium effects efficiently and inexpensively.

An example of the use of factor analysis for this purpose is found in the work of Edward and Wong[21] concerning measurement of the ionization of carbonyl compounds in sulfuric acid. The ultraviolet spectra of 16 carbonyl compounds (involving aldehydes, ketones, and amides) in various concentrations of sulfuric acid did not exhibit isosbestic points characteristic of protonation equilibria. Figure 12.4 concerning anthraquinone, shows a typical result. The spectra (top set of curves) were digitized and the absorbance matrix was subjected to PFA. The first principal eigenvector, accounting for 96% of the variance, was associated with the effect of protonation. The second eigenvector, accounting for 3% of the variance, was associated with the medium effect. The remaining 1% variability was due to experimental error. Using only the first principal eigenvector, the PFA-reconstituted curves exhibit clear isosbestic points as shown in the lower set of curves in Figure 12.4. The ionization ratios and equilibrium constants for the 16 carbonyl compounds were then obtained from the reconstructed spectra.

12.5   ADDITIONAL FUNDAMENTAL APPLICATIONS

Factor analyses of five kinds of physicochemical data are summarized in this section.

Solubility. Understanding the factors that influence the solubility of gaseous solutes in liquid solvents is one of the most fundamental problems in physical chemistry. Furthermore, the prediction of gas–liquid solubilities is of practical concern in chemical engineering. Factor analysis of solubility data has been employed for both of these purposes.

Howery and Chan[22] target-factor-analyzed a data matrix involving 8 nonpolar solutes (ranging from helium through oxygen to ethane) and 11 polar and nonpolar solvents (8 alkyl and aromatic hydrocarbons, chlorobenzene, ethanol, and dimethyl-sulfoxide). They analyzed the logarithm of the mole fraction solubility because, from thermodynamics, the free energy of transfer of a gaseous solute into solution was known to be proportional to the logarithm of solubility, rather than to the solubility itself. Theoretically, the free energy was suspected to be a linear sum of terms related to the solute–solvent interactions responsible for the data. Howery and Chan were interested in identifying those solute and solvent basic factors that best model the solute–solvent interactions.

Four factors reproduced the solubility matrix within 5%, the upper limit of the experimental error. The factor indicator function (see Section 4.3.2) reached a minimum at four factors, further confirming the factor size. Uniqueness tests showed

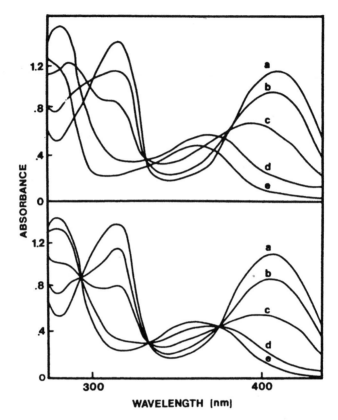

Figure 12.4 Ultraviolet spectra of anthraquinone in aqueous sulfuric acid: (a) 99.0%, (b) 91.6%, (c) 86.3%, (d) 80.7%, and (e) 73.0%. Raw data curves are shown in the upper block. Reconstituted curves, obtained from the mean curve and the first principal factor, are shown in the lower block. [Reprinted with permission from J. T. Edward and S. C. Wong, *J. Am. Chem. Soc.*, **99**, 4229 (1977).

that ethane (uniqueness value = 0.97) and methane were the most unique solutes and that dimethylsulfoxide (uniqueness value = 0.86) and hexane were the most unique solvents. Combination TFA was used to determine which sets of typical factors (data vectors from the data matrix) best represented the data. For both key combination sets, the RMS error for the reproduction was within experimental error. The key set of typical solute vectors was associated with neon, argon, methane, and ethane, while the key set of typical solvent vectors involved hexane, heptane, benzene, and dimethylsulfoxide. The most unique designees (see above) were included in the key sets, a result usually noted in combination FA.

To identify parameters associated with the solute–solvent interactions, Howery and Chan[22] target tested a large number of solute and solvent parameters, several of which had been suggested from previous theoretical and empirical studies of solubility. In the past, plots of solubility versus some property of the substances have been used to identify factors. Since such plots are expected to be linear only in one-factor problems, target testing affords a safer method for finding factors in multidimensional solubility problems. Solute factors that tested adequately included molecular mass, entropy of vaporization, entropy of solution, Henry's law constant, polarizability, hard-sphere diameter (results shown in Table 12.5), and Lennard-Jones force constant. Among the solvent parameters identified as basic factors were surface tension, log vapor pressure, rigid-sphere diameter, and a nonpolar interaction term. The well-known Hildebrand solubility parameter did not test very well. By including the functional forms of the test vectors, over 40 solute factors and nearly 35 solvent vectors tested moderately well or better.

In order to find which sets of parameters represented the best complete model for the solute and solvent parts of the interaction space, combination TFA of the basic factors was carried out. For both the solutes and the solvents, the key combination sets gave RMS errors less than twice the experimental error. Henry's law constant and polarizability were represented in nearly all the best combinations of basic solute vectors. Surface tension seemed to be an especially important solvent factor.

A simple target factor analytical procedure for predicting new solubility data was demonstrated by Howery and Chan.[22] Their method involved target testing incomplete vectors of solubility data, thereby predicting solubilities for the free-floated molecules. A typical result is shown in Table 12.5. The accuracy of such predictions depended on the extent to which the test points spanned the factor space. Predictions were more reliable if test points for unique molecules and for molecules represented in the key sets of typical vectors were incorporated in the test vector.

In another effort to develop a practical method for predicting solubility data, deLigny and co-workers[23] carried out a combined abstract factor analysis/multiple regression analysis on the free energy of solution and the entropy of solution for 20 gases in 39 solvents. The original study should be consulted for details concerning the data sets. An iterative regression procedure was developed to predict data

TABLE 12.5 Target Tests of Solute Test Vectors for Solubility Data (Howery and Chan[22])

Solute	Hard-Sphere Diameter		Log Solubility in Cyclohexanol	
	Test	Prediction	Test[a]	Prediction
He	2.63	2.62	−0.325	−0.327
Ne	2.78	2.79	—	−0.168
Ar	3.40	3.73	−0.724	−0.717
H_2	2.87	2.73	0.228	0.238
N_2	3.70	3.85	0.424	0.424
O_2	3.46	3.36	—	0.712
CH_4	3.70	3.65	1.099	1.100
C_2H_6	4.38	4.35	1.914	1.914

[a]Dashes indicate free-floated test points.

missing from the data matrices. Abstract factor analysis was then applied to the completed data matrices. Two factors accounted for the main features of both the free energy and the entropy data. The method was tested by seeing how well measured values were predicted from the abstract factors. The standard deviations for log solubility and entropy of solution were 0.13 and 0.84 cal mol^{-1} deg^{-1}, respectively. For missing data, the accuracy of predicted values for log solubility was estimated to be 0.20. The approach predicted adequately several data points that were published in the literature after the predictive equations were developed.

Solution Properties. In order to better understand physicochemical properties of solutions, Fawcett and Krygowski[24] applied factor analysis to several sets of thermodynamic data. Properties such as enthalpies of solution, free energies of solution, enthalpies of transfer, and free energies of transfer were factor analyzed. Included in the solute–solvent data matrices were both polar organic molecules and inorganic electrolytes. For each property, two principal eigenvectors, resulting from covariance about the mean, explained more than 90% of the variance in the property. In each problem, the two principal factors correlated linearly with empirical estimates of solvent acidity and solvent basicity.

Polarography. A study by Howery[25] of the polarographic half-wave potentials of five alkaline and four alkaline-earth ions in five solvents illustrates the application of target factor analysis in an area lacking a theoretical foundation. The objective was to build an empirical model for ion–solvent interactions. With three abstract factors, over 80% of the data were predicted within 30 mV. The factor space seemed reasonably spanned since experimental errors exceeding 30 mV were suspected for only a few of the half-wave potentials.

Howery tested several solvent and ionic parameters by target transformation. Molar enthalpy of vaporization, donor number, and a radial correction term tested moderately well as solvent vectors. A model based on these three solvent factors gave in combination TFA an average error of 28 mV, a quite good solution to the solvent part of the problem. Ionic charge, reciprocal of the crystalline ionic radius, and the ratio of ionic charge to crystalline ionic radius transformed well as ionic vectors. Results from target testing the ionic charge are shown in Table 12.6. The value for lithium ion in the test was deliberately free floated since lithium ion was known to produce high overvoltages in polar solvents. The predicted value of 2.03 implied that the lithium ion behaved like a divalent ion in this polarographic investigation.

Chelate Stability. Realizing that a detailed understanding of the factors affecting chelate stability might lead to guidelines for improving the selectivity of chelate reactions, Duewer and Freiser[26] conducted an abstract factor analysis on the logarithm of the formation constants for 14 diaminetetraacetic-acid ligands and 24 metal ions. Factor analysis is well suited to the study of complex ions, furnishing solutions involving factors for the ligands and for the central metallic ion. The

TABLE 12.6 Target Test of Ionic Charge for Polarographic Half-Wave Potentials[a]

Ion	Test[b]	Prediction
Li^+	—	2.03
Na^+	1.0	1.16
K^+	1.0	1.15
Rb^+	1.0	0.92
Cs^+	1.0	0.87
Mg^{2+}	2.0	2.17
Ca^{2+}	2.0	2.06
Sr^{2+}	2.0	1.76
Ba^{2+}	2.0	1.78

[a]Reprinted with permission from D. G. Howery, *Bull. Chem. Soc. Jpn.*, **45**, 2643 (1972).
[b]Test value for Li^+ was free floated.

logarithm pretreatment was dictated by the well-known proportionality between standard free-energy change and the logarithm of the equilibrium constant.

Four factors accounted for the data within experimental error. To facilitate physical interpretation of the unrotated abstract factors, a simplified three-factor model consisting of an average stability constant, a metal ion factor, and a ligand factor was invoked. The average stability constant corresponded to the first principal eigenvector, the ion factor correlated with the ratio of charge to ionic radius, but the ligand factor could not be identified. Contrary to expectations, the logarithm of the proton association constant for the ligands did not appear to be equivalent to any of the principal factors.

Bond Energy. Bond dissociation energies are among the most fundamental data in chemistry. A bond dissociation energy matrix can be formed by using radicals as both row and column designees, the radicals constituting the chemical groups being joined together by the bond in question. In an effort to study the factors influencing bond formation, Howery and Rubinstein[27] target-factor-analyzed a symmetrical 7×7 matrix of bond dissociation energies of some simple molecules.

Three factors reproduced 96% of the data within the estimated experimental error of 2 kcal. According to the results of uniqueness tests, hydrogen and phenyl were the most atypical radicals. Combinations of typical vectors indicated that hydrogen, isopropyl, and benzyl were the three key radicals, giving a RMS error in combination TFA of 0.78 kcal.

Over 50 basic vectors and three functional forms of each basic vector were target tested. The results indicated that more than 25 test factors were acceptable representations of basic factors. This large number of acceptable factors indicated that many of the factors were interrelated (i.e., were linear functions of each other).

A combination TFA model consisting of the radical-methyl stretching frequency, the radical-hydrogen bond length, and the diagonal of the data matrix produced an RMS error of 0.86 kcal, well within experimental error.

12.6 BIOMEDICAL CHEMISTRY

A variety of applications of factor analysis in biomedical chemistry have been published. Because of the inherent complexity of biomedical data and difficulty in obtaining reproducible, quantitative measurements, these problems are harder to deal with than those discussed in the previous sections of this chapter.

In an early study, Woodbury et al.[28] demonstrated that AFA could be used to predict biochemical data. Several points in the data matrix were deliberately omitted and then were predicted using the following iterative procedure. Arbitrary values were assigned to the missing points. New estimates for the missing points were predicted from the reproduction step of AFA. The new estimates were substituted in the data matrix, reproduction was carried out again, and a second set of estimated values was obtained. This process was repeated until the deleted data were predicted adequately. For a sulfa drug–tissue localization matrix, the predictions were moderately good. For a bacteria–antibody matrix, the predictions were rather poor. A drawback of the approach is the difficulty in determining the proper number of factors to use in the reproduction. In fact, Swain and co-workers[29] have shown that iterative factor analysis can lead to absurd results when points are missing in the data matrix.

Principal factor analysis was used by Sneath[30] to study the relationship between the chemical structure and the biological activity of 20 amino acids. The data matrix was a 20×20 "resemblance" table constructed statistically to take into account 134 attributes of the amino acids. Typical attributes incorporated into the resemblance table included solubility, optical rotation, chromatographic retention, presence or absence of various functional groups, moments of inertia, and the number of lone pairs of electrons. The same amino acids represented both the row designees and the column designees of the data matrix. Each entry in the data matrix was generated by means of a correlation coefficient designed to measure the overall resemblance between each pair of amino acids. Since each amino acid exactly resembles itself, the diagonal elements of the matrix were unities. A resemblance value of zero indicated that the two amino acids had no similarity whatsoever; negative resemblance values indicated opposite characteristics.

Principal factor analysis of the resemblance matrix yielded four eigenvectors with eigenvalues greater than 1, together accounting for 69% of the total variation. By studying the unrotated factor loadings, Sneath related the four factors to the aliphatic character, degree of hydrogenation, aromatic character, and degree of hydrothiolation (involving the abilities of hydroxyl and sulfhydryl groups to form hydrogen bonds), respectively. These four factors were correlated with biological activity and then used to predict the activities of new peptides. The predictions, although not highly accurate, were better than those obtained by chance guessing.

Factor analysis has been used successfully as a preprocessing method for pattern recognition studies involving biological compounds. Chemical pattern recognition techniques, such as the "linear learning machines," attempt to classify a set of compounds into one of two specified categories. For example, an investigator may wish to predict whether or not a compound has a certain biological activity. In order to achieve such a binary classification, a set of linear descriptors representing chemical or physical properties must be found. Descriptors are continually added to the pattern recognition scheme until the classification reaches an acceptable degree of correctness. Ritter and Woodruff[31] pointed out that the number of descriptors employed does not necessarily equal the dimensionality of the factor space and that, in practice, often too many descriptors are used. By factor analyzing a data matrix composed of the descriptors, the true size of the factor space and hence the minimum number of descriptors required can be determined. Thus factor analysis can play a valuable role in pattern recognition studies.

Pattern recognition factor analysis has been used to separate chemical compounds into therapeutic classes on the basis of their molecular structures. The method requires a general molecular descriptor coding that is applicable to the entire set of compounds. The coding must be designed to discriminate between atoms and groups. From the general descriptor a structural feature matrix is generated. Factor analysis of this structural feature matrix yields a minimum set of eigenvectors that can be used for pattern recognition classification, thus reducing the number of required axes to a minimum, simplifying both the classification and its interpretation. A reduction in axes is expected if the structural features are interrelated.

As an example of this technique, Cammarata and Menon[32] coded 13 amino compounds by means of a structural diagram, called a "superstructure," shown in Figure 12.5. Each arbitrary number in this diagram was used to designate a column in the molecule–feature data matrix. The descriptors shown in the footnote of Table 12.7 were proposed to distinguish between the different features of the molecules. The use of 1 or 0 served to indicate the presence or absence of a particular functional group. The use of 2 served to distinguish an aromatic carbon from an aliphatic carbon. Applying the feature descriptors to each of the 13 molecules, the molecule–feature matrix (Table 12.7) was generated. One can easily verify that the thirteenth row of Table 12.7 is a descriptor for benzylamine, $C_6H_5CH_2NH_2$.

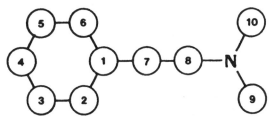

Figure 12.5 Superstructure for 13 amino compounds. [Reprinted with permission from A. Cammarata and G. K. Menon, *J. Med. Chem.*, **19**, 739 (1976).]

TABLE 12.7 Molecule–Feature Data Matrix for Pressor Agents[a]

Compound	X_1	X_2	X_3	X_4	X_5	X_6	X_7	X_8	X_9	X_{10}
					Feature[b]					
1	2	2	2	2	2	2	1	1	0	1
2	1	1	1	1	1	1	1	1	0	1
3	1	1	1	0	1	1	1	1	0	1
4	2	2	2	2	2	2	1	1	1	1
5	1	1	1	0	1	1	0	1	0	0
6	1	1	1	0	0	0	0	1	0	0
7	2	2	2	2	2	2	1	1	0	0
8	1	1	1	0	0	1	0	1	0	0
9	1	1	1	0	0	0	1	1	0	0
10	2	2	2	2	2	2	1	0	0	0
11	2	2	2	2	2	2	0	1	0	1
12	2	2	2	2	2	2	0	0	0	1
13	2	2	2	2	2	2	0	0	0	0

[a]Reprinted with permission from A. Cammarata and G. K. Menon, *J. Med. Chem.*, **19**, 739 (1976).
[b]Based on the following descriptor set:

Feature	Nature	Descriptor
1–6	Aromatic atom	2
	Aliphatic atom	1
	No atom	0
7	CH_2 present	1
	CH_2 absent	0
8	$CHCH_3$	1
	CH_2	0
9,10	CH_3 present	1
	CH_3 absent	0

Factor analysis of the resulting 10×10 correlation matrix yielded two eigenvalues greater than 1, accounting for 79% of the variance. A plot of the abstract row-factor loadings in the two-factor space is shown in Figure 12.6 (which is analogous to Figure 5.1). Each point in the plot is associated with a particular amino compound. The 13 compounds consisted of pressor agents (i.e., agents that cause an increase in blood pressure). Molecules known to be "strong" and "weak" pressor agents are labeled with solid circles and open circles, respectively; the classifications of points labeled with an $\times$ were not known previously. On the diagram the molecules fall into two distinct classes. When matched against known biological responses, the classifications based on factor analysis appeared to be correct.

In another classification study involving 43 compounds, Menon and Cammarata[33] employed coding values based on atom and bond molar refractivities. This coding was designed to take into account differences in "bioisosterism" among the

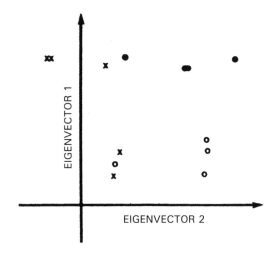

Figure 12.6 Two-dimensional factor space showing weak (open circles), strong (closed circles), and undesignated (crosses) pressor agents. [Reprinted with permission from A. Cammarata and G. K. Menon, *J. Med. Chem.*, **19**, 739 (1976).]

functional groups, rather than simply account for the presence or absence of a functional group as in the previous example. Bioisosterism refers to those atoms, ions, or groups that have identical peripheral layers of electrons. The compounds included antihistamines, antidepressants, antipsychotics, anticholinergics, analgesics, and anti-Parkinsonian agents. The 43 × 8 pharmacophore–feature data matrix yielded four eigenvalues greater than 1, which accounted for 79% of the variance. Although the four-dimensional factor space could not be plotted, projections onto three-dimensional subspaces showed clusterings of points that were associated with various kinds of therapeutic activity.

To further simplify pattern recognition studies, Menon and Cammarata[33] used factor analysis in a two-step procedure. In the first step, factor analysis was used to separate an entire set of 39 drugs into broad clusters. Submatrices based on molecules in the smaller clusters were then factor analyzed separately. Because the small clusters required fewer discriminators, the results were easier to interpret.

Weiner and Weiner[34] used target factor analysis to examine the structure–activity relationships of 16 diphenylaminophenol drugs characterized by 11 biological tests on mice. Eight factors were deemed necessary to reproduce the data satisfactorily. The factor-size decision was somewhat arbitrary, since several of the biological tests were qualitative in nature. Uniqueness values and clusterings based on the results of uniqueness tests were obtained for the drugs and, what was even more intriguing, for the biological tests. Interrelating one biological test with key sets of others was also achieved.

Several structural vectors and a few vectors based on the results of the uniqueness tests were target tested. For example, the presence of a ring linked to the nitrogen atom in the drug structure was shown to be a factor. Selected results for this test are shown in Table 12.8. A binary classification, separating the drugs into those having the ring linkage and those not having the linkage, was indicated. The large value predicted for the last drug appears to be related to the double ring in that compound. An attempt to predict the activity of new drugs using combination TFA based on a key set of eight typical biological test vectors was not very successful. Little loss in predictive ability of the known drugs was observed when the ring linkage factor shown in Table 12.8 was substituted for one of the key biological test vectors.

In addition to the problems discussed above, applications of AFA to medicinal chemistry include studies of survival times for patients having cirrhosis of the liver using clinical and biological tests,[35] electroencephalograms of patients given or not given a drug,[36] effects of drugs on the membrane potential of brain cells,[37] and formulations of drug tablets.[38] Data matrices involving patients versus diagnostic tests and drugs versus evaluative tests are often obtained in medical and pharmaceutical research. Factor analysis should furnish useful classifications in such studies.

Prospective users of factor analysis in biomedical research should apply the technique with caution. Since many kinds of medical data may not obey the factor analytical model, the criteria for factor analyzability described in Chapter 4 should always be met. Malinowski,[39] using the theory of errors described in Section 4.3.2, found that the matrix employed by Weiner and Weiner[34] did not appear to have a good factor analytic solution. Howery and Gelbfish[40] investigated a data matrix concerning the concentrations of several components in the blood sera of patients having heart problems. Since no factor compression was observed, factor analysis was deemed inappropriate in that problem.

TABLE 12.8 Uniqueness Test for Ring Linkage to Nitrogen Atom in Drugs[a]

Drug Substructure[b]	Test[c]	Prediction
$CH_2CH(CH_3)NCH_3$	0.0	0.01
$CH(CH_3)CH_2N(CH(CH_3)_2)_2$	0.0	0.00
$CH_2CH(CH_3)N(CH_3)CH_2C_6H_5$	(0.0)	0.09
$CH(CH_2CH_3)CH_2\underline{N(CH_2)_3CH_2}$	1.0	0.91
$CH_2CH(CH_3)CH_2\underline{N(CH_2)_3CH_2}$	(1.0)	1.00
$CH(C_6H_5)CH_2\underline{N(CH_2)_3CH_2}$	1.0	1.05
$CH_2\underline{(CH_2)_3N(CH_2)_3CH_2}$	(1.0)	1.0
$\underline{CHCH_2N\underline{CH_2CH_2}CHCH_2CH_2}$	—	1.16

[a]Reprinted with permission from M. L. Weiner and P. H. Weiner, *J. Med. Chem.*, **16**, 655 (1973).
[b]Structural component A̲N̲R̲R̲' of the complete molecule $(C_6H_5)_2C(OH)\underline{ANRR}' \cdot HX$.
[c]Known points in parentheses; dashed test value was free floated.

Urinanalysis. There is an imperative need to develop efficient methods for detecting and classifying drug-induced toxic lesions. Holmes and co-workers[41] applied principal component analysis (PCA) to classify urine samples obtained from rats treated with various nephrotoxins (glomerular, papillary, and proximal tubular). The urine samples were buffered at pH 7.4 and centrifuged to remove any precipitate before ^{1}H NMR analysis. The water signal was suppressed by a NOESY pulse sequence. The free induction decay signal was averaged 64 times and zero filled. All spectra were phase corrected.

In preparation for PCA, each spectrum was reduced to digital form by dividing the spectrum into 250 chemical shift regions and integrating the signals within each region. A typical spectrum, before and after processing, is illustrated in Figure 12.7. Plots of the first three principal components produced clusters that corresponded to the toxicity sites in the kidneys. Detailed examination of the eigenvectors revealed the spectral regions characteristic of the different metabolites.

Cystic Fibrosis. A genetic algorithm (GA) has been incorporated into PCA by Lavine and associates.[42] The power of this methodology is illustrated in their investigation of cystic fibrosis (CF) detection using pyrolysis gas–liquid chromatographs of cultured skin fibroblasts. Pyrochromatograms (PyGC) yield chemical fingerprints that are amenable to this technique.

Cultured skin fibroblasts were obtained from 24 cystic fibrosis patients and 22 normal volunteers. Replicate experiments were performed on each sample, yielding 73 CF and 60 normal PyGCs. Typical 700°C pyrochromatograms of CF and normal human skin fibroblast are presented in Figure 12.8. The retention times were standardized with a peak-matching program based on marker peaks. The areas were then adjusted by a scaling factor based on the areas of the reference peaks.

Initially, each PyGC was represented by 84 retention time windows selected by a peak-matching program. Prior to PCA each chromatogram was normalized and autoscaled. This was done to give each chromatogram equal weight in the analysis. The classification scheme was based on the first two factors obtained from PCA. The genetic algorithm was designed to act as a fitness filter, deleting those peaks that did not help in the desired classification. Figures 12.9 and 12.10 are biplots of the first two factors, before and after applying the genetic algorithm. Ten standardized retention time windows (1, 12, 15, 27, 39, 41, 46, 67, 73, and 78) were retained by the genetic algorithm after 100 generations. The separation into two distinct groups is clearly shown in Figure 12.10.

Magnetic Resonance Images of the Human Brain. In 1998 Antalek and co-workers[43,44] recognized that DECRA (Section 6.3.5) could be used to analyze magnetic resonance images of human brain tissues. This is possible because the resonance signal, S, of a standard spin-echo imaging sequence is a function the relaxation times, T_1 and T_2 as well as the echo time, TE, and the repetition time, TR:

$$S = k \sum_i \sum_j \rho_{ij}(1 - e^{-TR,T_{1i}})e^{-TE/T_{2j}} \qquad (12.5)$$

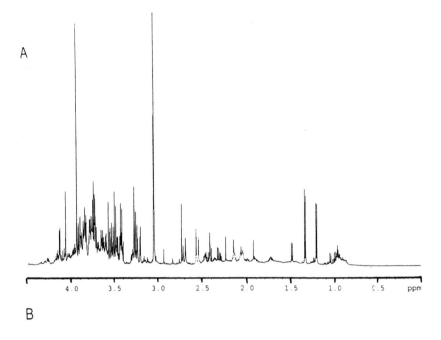

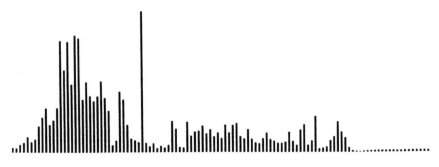

Figure 12.7 600-MHz ^{1}H NMM spectrum obtained from a rat treated with hexachloro-1,3-butadiene (A) and the corresponding data reduced spectrum (B). [Reprinted with permission from E. Holmes, J. K. Nicholson, A. W. Nicholls, J. C. Lindon, S. C. Connor, S. Polley, and J. Connelly, *Chemom. Intell. Lab. Syst.*, **44**, 245 (1998). Copyright 1998 © Elsevier Science.]

Subscripts i and j indicate the unique spin states of the different types of hydrogen atoms in brain tissue. Keeping *TE* constant while varying *TR* converts (12.5) into a sum of single exponential terms ideally suited for DECRA extraction of T_{1i}. On the other hand, keeping *TR* constant while varying *TE* converts (12.5) into a form suitable for determining T_{2j}.

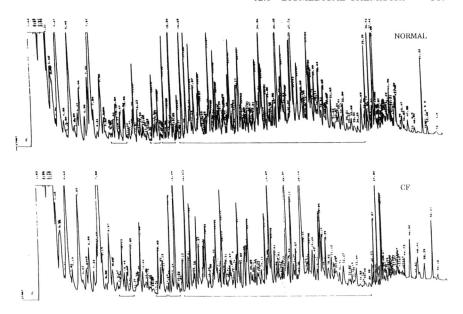

Figure 12.8   Pyrochromatograms (700°C) of CF and normal human skin fibroblast samples. [Reprinted with permission from B. K. Lavine, A. Moores, and L. K. Helfend, *J. Anal. Appl. Pyrol.*, **50**, 47 (1999). Copyright 1999 © Elsevier Science.]

Axial magnetic images of the brain of a healthy 42-year-old male volunteer were recorded with a GE Signa whole body imager. The image plane passed through the lateral ventricles. Each slice formed a 256 × 256 pixel image, which unfolded into a 65,536-element array. Two sets of images of this 5-mm slice were obtained, one set (consisting of 15 images) with *TR* varying from 200 to 3000 ms in 200-ms intervals. The other set (consisting of 14 images) with *TE* varying from 15 to 210 ms in 15-ms intervals.

In preparation for DECRA, each unfolded image was arranged as a column into its respective *TE* or *TR* matrix. DECRA analysis of the *TE* matrix yielded three first-order exponential decay profiles. From the half-lives of these curves, the T_2 relaxation times were determined to be 22, 64, and 290 ms. DECRA analysis of the *TR* matrix yielded only two exponential profiles with T_1 relaxation times equal to 0.92 and 7.0 s. In Figure 12.11 the resolved component images corresponding to these T_2 values are labeled *a*, *b*, and *c*, and those corresponding to the T_1 values are labeled *d* and *e*.

Images produced by DECRA analysis differ significantly from images obtained from other fitting procedures. Other procedures yield images that are linear combinations of exponential time constants for various proton environments. DECRA extracts independent time constants and produces images that represent homogeneous environments. For example, Figure 12.11*a* represents the image of protons with a short T_2. This is characteristic of methylene in long alkyl chains in fat and lipid tissue because of restricted rotation. Although the interpretation of the

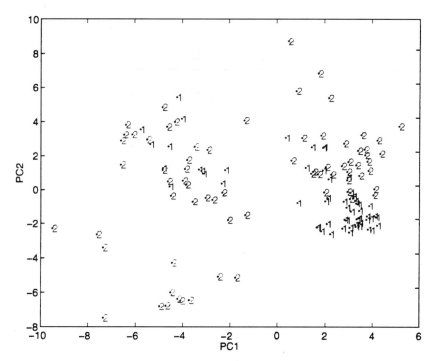

Figure 12.9 Plot of the two largest principal components of the 84 GC peaks. 1, Normal and 2, CF. [Reprinted with permission from B. K. Lavine, A. Moores, and L. K. Helfend, *J. Anal. Appl. Pyrol.*, **50**, 47 (1999). Copyright 1999 © Elsevier Science.]

DECRA images is not complete at this time, the method shows great promise in helping to unravel chemical environments of the living brain.

12.7 FUELS

Paraffins, olefins, naphthenes, and aromatics (PONA) Analysis. Cost-effective methods for analyzing and characterizing gasoline play an important role in industrial trade. Gas chromatography–mass spectra (GC–MS) is commonly used but is difficult and expensive to perform on a routine basis. As an alternative, Zhu and co-workers[45] designed a simple method that employs GC data alone.

Their method is based on factor analyzing the correlation coefficients of three variables: the difference between the retention indices of hydrocarbons on mixed dual stationary columns, at the same temperature, and their temperature coefficients on each column. By subjecting the PFA results to varimax (maximum variance) and promax (maximum projection) rotations (see Section 3.4.2) and plotting the rotated factor scores, they found that the hydrocarbons could be classified into paraffins (P), olefins (O), naphthenes (N), and aromatics (A).

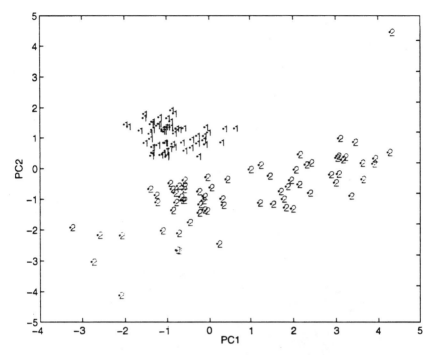

Figure 12.10   Plot of the two largest principal components of the 10 GC peaks identified by the genetic algorithm. 1, Normal and 2, CF. [Reprinted with permission from B. K. Lavine, A. Moores, and L. K. Helfend, *J. Anal. Appl. Pyrol.*, **50**, 47 (1999). Copyright 1999 © Elsevier Science.]

To generate the discrimination plot, the retention indices and temperature coefficients of 191 hydrocarbons measured by Lubeck and Sutton[46] on DB-1 and DB-5 capillary columns were chosen. This data consisted of 72 paraffins, 69 olefins, 36 naphthenes, and 14 aromatics. Two factors accounted for 99.95% of the total variance. Figure 12.12 is a plot of the scores, T_1 and T_2, after the varimax and promax rotations. The PONA groups are separated, the aromatics residing in the first quadrant, the naphthenes in the second quadrant, the paraffins in the lower portions of the second and third quadrants, and the olefins primarily in the fourth quadrant.

A catalytically cracked gasoline was then analyzed by passing it through two capillary columns: OV-1 and SE-54, which are similar to DB-1 and DB-5. The retention indices and temperature coefficients of 58 major peaks were recorded at temperatures identical to those used in the hydrocarbon analysis. The data was factor analyzed as described above, yielding the score shown in Figure 12.13. Based on the known groupings in Figure 12.12, the PONA value of this gasoline sample was determined to be 44.56% P, 26.84% O, 5.25% N, and 8.28% A. This analysis was confirmed by subjecting the sample to GC–MS.

a   b   c

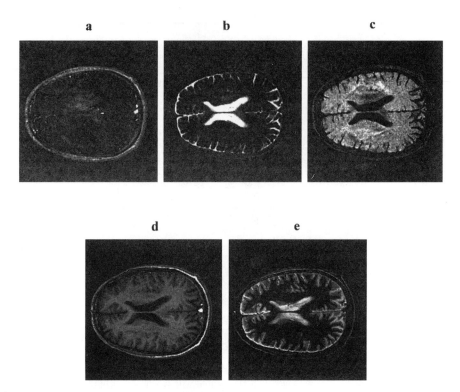

Figure 12.11 DECRA resolved images: (a, b, c) from T_2 data; $(d$ and $e)$ from T_1 data. [Reprinted with permission from W. Windig, B. Antalek, L. J. Sorriero, S. Bijlsma, D. J. Louwerse (AD), and A. K. Smilde, *J. Chemometrics*, **13**, 95 (1999). Copyright 1999 © John Wiley & Sons, Ltd.]

Octane Number. Cooper[47] used PLS to determine gasoline octane numbers from their Raman spectra. Two types of octane numbers were measured using American Society for Testing and Materials (ASTM) procedures, laboratory research octane number (RON), and motor octane number (MON). The pump octane number (PON) is the average of RON and MON. The average standard deviation in the octane measurements of 208 commercial fuel blends was determined to be 0.3 octane unit. The study entailed the Raman regions from 196 to $1851 \, \text{cm}^{-1}$ and 2570 to $3278 \, \text{cm}^{-1}$. There were no spectral lines in the other regions.

Gasoline is known to contain more than 500 chemical components whose concentrations are inherently reflected in their Raman spectra. Different regions are indicative of the vibrational modes of different chemical groups, for example, branched alkanes at 700 to $800 \, \text{cm}^{-1}$, alkanes at 1440 to $1480 \, \text{cm}^{-1}$, olefins at 1630 to $1690 \, \text{cm}^{-1}$, ring-breathing modes at $1000 \, \text{cm}^{-1}$, phenyl C–C stretch at 1570 to $1630 \, \text{cm}^{-1}$, and phenyl C–H stretch above $3000 \, \text{cm}^{-1}$. Partial least squares (PLS) is

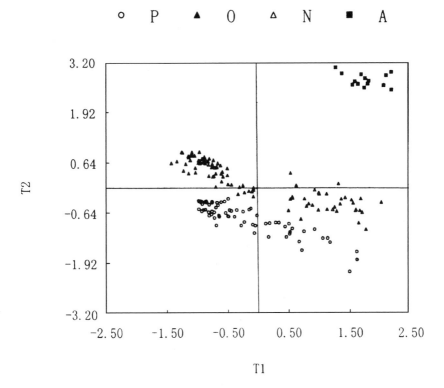

Figure 12.12   Oblique factor score graph of 191 hydrocarbons. [Reprinted with permission from X. Zhu, L. Zhang, X. Che, and L. Wang, *Chemom. Intell. Lab. Syst.*, **45**, 147 (1999). Copyright 1999 © Elsevier Science.]

based on the belief that the octane number and the Raman spectra are functions of the same molecular features, and, therefore, are functions of one other.

Cross-validation PLS indicated that 8 to 9 latent factors were required to correlate the octane numbers with the Raman spectra. The predicted octane numbers of 50 independent samples, not included in the calibration set, are exhibited in Figure 12.14 together with the actual pump octane numbers.

12.8   ENVIRONMENTAL CHEMISTRY

The ability of factor analysis to correlate environmental data has been established. Data matrices typically involve the analytical concentrations of various chemical components at either different sampling stations or sampling times. The objectives are to identify the sources of the components and ultimately to develop a detailed model for the processes responsible for generation and removal of components. Several representative applications are discussed in this section.

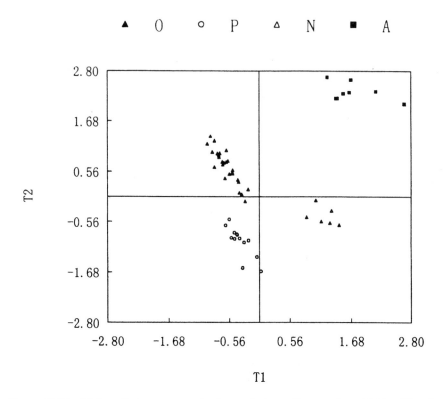

Figure 12.13 Oblique factor score graph of a cracked gasoline sample exhibiting 58 major chromatographic peaks. [Reprinted with permission from X. Zhu, L. Zhang, X. Che, and L. Wang, *Chemom. Intell. Lab. Syst.*, **45**, 147 (1999). Copyright 1999 © Elsevier Science.]

Problems involving elemental analysis of environmental samples are particularly amenable to factor analysis. In such problems the total amount of the ith element in the kth sample, x_{ik}, can be expressed as

$$x_{ik} = \sum_{j=1}^{n} c_{ij} q_{jk} \tag{12.6}$$

Here c_{ij} is the relative concentration of element i in source j, q_{jk} is the amount of source j in the kth sample, and the sum is taken over the n significant sources of the element. Each source acts as a factor; summing over all sources gives the total amount of the element in the sample. The form of (12.6), involving a linear sum of products, indicates that this kind of problem should have a factor analytical solution.

Alpert and Hopke[48] tested (12.6) using artificial data sets. Twenty samples were prepared from known masses of 5 standard materials having known elemental compositions (the sources). Analyses of 37 elements were obtained for each sample using neutron activation analysis. Four criteria were employed to estimate the

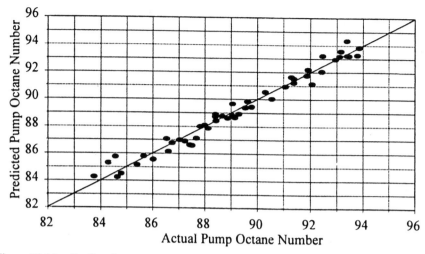

Figure 12.14   Predicted versus actual octane numbers of 50 test samples not included in the PLS model. [Reprinted with permission from J. B. Cooper, *Chemom. Intell. Lab. Syst.*, **46**, 231 (1999). Copyright 1999 © Elsevier Science.]

number of principal factors. The relative magnitude of eigenvalues, the Exner function (see Section 4.3.2), and the factor indicator function each indicated 5 factors, consistent with the number of sources used to prepare the samples, but the chi-squared test (see Section 4.3.1) gave no indication of the number of factors.

Alpert and Hopke[48] felt that target testing should identify the sources since in the artificial data set the known elemental compositions of each source serve as test vectors. The results of target tests showed that the agreement between test and predicted values was good for elements present in high concentrations but poor for trace components, which often tend to distinguish sources.

Factor analysis has been used in conjunction with cluster analysis in several studies of environmental problems. In "hierarchical" cluster analysis, each column designee is represented as a point in a c-dimensional space, where c is the number of columns in the data matrix. A variety of measures of similarity or dissimilarity are then used to classify the designees. The data points are assumed initially to form c separate, single-member clusters. Criteria such as nearest neighbors, farthest neighbors, and average-between-cluster distances are used to form the clusters in a stepwise procedure. The two points calculated to be most similar are combined into the first two-member cluster, which is then treated as a single point in the second stage of comparison with the $c - 1$ remaining points. The two most similar points in the new model are then merged to form a second cluster and the comparison calculation is repeated. At each stage of clustering, the two most similar clusters are combined, irrespective of the number of members in the merged clusters. Ultimately, a single cluster containing all c data points is formed. The connection diagram identifying the data points in each cluster at each stage of clustering is called a *dendogram.* The various clusters calculated from cluster analysis can be compared to those identified via principal or rotated factors.

Water Pollutants. A study of lake sediments by Hopke[49] illustrates the dual factor analysis/cluster analysis approach. A data matrix representing 32 measurements of 79 sediment samples taken from sites throughout Chautauqua Lake, New York, was analyzed. The measurements included 15 elemental analyses, 12 properties related to grain size, water depth above the sample, and the percentages of sand, silt, clay, and organic matter. Both factor analysis and cluster analysis furnished insights into the mechanisms responsible for the distribution of properties. Five principal factors best described the data. Analysis of the loadings in the principal factors and rotated factors (based on a maximum-likelihood method) led to assignments for the processes acting on the sediments. The five factors were related to the coarse-grain source material, the available surface area of the sediments, the glacial till source material, the active sediments due to steam deltas, and wave and current action, respectively. Hierarchical cluster analysis was used to cluster the sample sites. The two largest clusters of sites were shown to represent sites in the center of the lake and sites near shore. The near-shore sites were broken into three subclusters, the first being very sandy and relatively inactive, the second having active sedimentation and high wave action, and the third having intermediate activity.

An abstract factor analysis of Puget Sound, Washington, rainwater, involving concentrations of 16 ions at 22 sampling stations, was conducted by Knudson et al.[50] Three factors accounted for about 70% of the variance in the data. A sea-salt background, a generalized urban source, and an industrial source were interpreted, by use of varimax rotation, to be the three main sources of the ions.

Leaking Fuel. Contamination of water supplies by leaking fuel is a serious problem. The floating fuel layer, collected and analyzed by gas chromatography (GC), is comparably different from the parent fuel, making visual analysis highly subjective. Lavine and co-workers[51–53] have developed a genetic algorithm to classify chemical systems via GC data.

One study[51] focused attention on gas chromatograms from six different types of jet fuels obtained from Wright Patterson Air Force Base and the Mulkilteo Energy Management Laboratory. The fuel samples were divided into a training set (271 quality control samples) and a prediction set (31 weathered samples from well water and soil) as shown in Tables 12.9 and 12.10. The pattern recognition procedure entailed four sequential steps: (1) peak matching, (2) outlier detection and removal, (3) feature selection, and (4) classification.

The peak-matching procedure converted each chromatogram into a vector with 56 elements. Outliers were removed from the training set by application of principal component analysis and a distance test based on Scout.[54] This reduced the training set to 256 GCs with 56 peaks each. A genetic algorithm (GA), based on the principals of natural evolution, was used to develop a fitness function to determine the relative importance of each member. PCA, involving only the first two principal components, was used as a filter for the genetic algorithm, revealing the features

TABLE 12.9   Training Set[a]

Number of Samples	Fuel Type
55	JP-4 (fuel used by USAF fighters)
77	Jet-A (fuel used by civilian airliners)
32	JP-7 (fuel used by SR-71 Reconnaissance planes)
36	JPTS (fuel used by TR-1 and U-2 aircraft)
49	JP-5 (fuel used by Navy fighters)
22	AVGAS (common aviation jet fuel)

[a]Reprinted with permission from B. K. Lavine, A. J. Moores, H. T. Mayfield, and A. Faruque, *Anal. Lett.*, **32(15)**, 2805 (1998) by courtesy of Marcel Dekker, Inc.

(fingerprint patterns) that characterize each fuel. After 100 iterations, the genetic algorithm selected 20 GC peaks. The results are displayed in Figure 12.15. Classification was accomplished by statistical discriminant analysis based on the Mahalanobis distance between a sample and the center of the class (see Table 12.11).

These results, based solely on the neat jet fuels, were used to classify the 31 weathered fuels (the prediction set) listed in Table 12.10. All of the weathered samples were correctly classified, even though they suffered from evaporation of low-molecular-weight components, dissolution of water-soluble compounds, and microbial degradation.

An on-site method for identifying the source of fuel spills was developed by Lavine and associates,[52] eliminating the need for collecting and storing samples. The method uses head space, solid-state microextraction, capillary GC, and pattern recognition factor analysis. The method was developed to identify jet fuel sources from spill samples recovered from subsurface environments on or near military airfields.

To test the suitability of the method, the chromatograms of 180 neat jet fuel samples (Table 12.12) from common aviation fuels were assembled into a data base.

TABLE 12.10   Prediction Set[a]

Number of Samples	Fuel Type
11	JP-4 recovered from soil
8	JP-4 recovered as a neat oily phase
3	JP-4 weathered in a laboratory
5	JP-5 recovered as a neat oily phase
2	JPTS recovered as a neat oily phase
2	AVGA recovered as a neat oily phase

[a]Reprinted with permission from B. K. Lavine, A. J. Moores, H. T. Mayfield, and A. Faruque, *Anal. Lett.*, **32(15)**, 2805 (1998) by courtesy of Marcel Dekker, Inc.

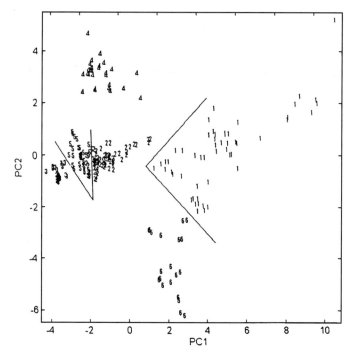

Figure 12.15 Principal component plot of the 20 features selected by the genetic algorithm for 256 neat jet fuel samples: 1 = JP-4, 2 = Jet-A, 3 = JP-7; 4 = JPTS, 5 = JP-5; and 6 = AVGAS. [Reprinted with permission from B. K. Lavine, A. J. Moores, H. T. Mayfield and A. Faruque, *Anal. Lett.* **32(15)**, 2805 (1998) courtesy of Marcel Dekker, Inc.]

TABLE 12.11 Training Set Results[a]

Fuel Type	Number in Class	Right	Wrong	Right (%)
Jp-4	53	53	0	100
Jet-A	74	74	0	100
JP-7	30	30	0	100
JPTS	32	32	0	100
JP-5[b]	47	44	3	93.6
AVGAS	20	20	0	100
Total[c]	256	253	3	98.8

[a]Reprinted with permission from B. K. Lavine, A. J. Moores, H. T. Mayfield, and A. Faruque, *Anal. Lett.*, **32(15)**, 2805 (1998) by courtesy of Marcel Dekker, Inc.
[a]Outliers were removed from the training set.
[b]Misclassified as Jet-A.
[c]Bootstrapped error rate = 0.022 and cross-validated error rate = 0.023.

Five steps were involved in the procedure: (1) peak matching, (2) outlier analysis, (3) feature selection, (4) classification, and (5) prediction. Peak matching involved the construction of a template based on Kovat's retention index. This procedure yielded a reference file containing 84 features for each chromatogram. Twenty-four chromatograms were deleted from the training set because they were classified by SCOUT[54] as outliers. The genetic algorithm selected 22 chromatographic peaks as characteristic of the fuels. These 22 peaks from 156 chromatograms were subjected to PCA. As shown in Figure 12.16, the first two principal components separated the fuels into their appropriate types. SIMCA,[55,56] a PC classification technique that places boundaries on the separations, was then used to classify samples not included in the training study. The results are tabulated in Table 12.12.

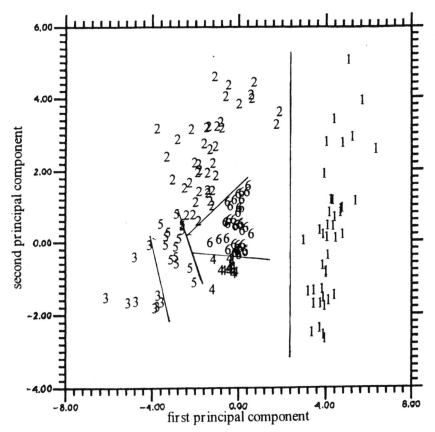

Figure 12.16 Principal component map of 22 GC peaks obtained from 156 neat jet fuel gas chromatograms. 1 = JP-4, 2 = Jet-A, 3 = JP-7, 4 = JPTS, 5 = JP-5, and 6 = JP-8. [Reprinted with permission from B. K. Lavine, J. Ritter, A. J. Moores, M. Wilson, A. Faruque, and H. T. Mayfield, *Anal. Chem.*, **72**, 423 (2000). Copyright 2000 © American Chemical Society.]

TABLE 12.12   Jet Fuels and SIMCA Classification Results[a]

Fuel	No. in Class	Right	Wrong	% Right
JP-4 (fuel used by USAF fighters)	41	41	0	100
Jet-A (fuel used by civilian airliners)	41	38	3	93
JP-7 (fuel used by SR-71 reconnaissance plane)	9	9	0	100
JPTS (fuel used by TR-1 and U-2 aircraft)	12	10	2	83
JP-5 (fuel used by Navy jets)	15	15	0	100
JP-8 (fuel used by USAF fighters in NATO)	38	31	7	82
Total	156	144	12	92

[a]Reprinted with permission from B. K. Lavine, A. J. Ritter, A. J. Moores, M. Wilson, A. Faruque, and H. T. Mayfield, *Anal. Chem.*, **72**, 423 (2000). Copyright 2000 © American Chemical Society.

Herbicides. The introduction of herbicides in surface water by the agricultural industry presents a serious problem in the 11 states that comprise the U.S. Corn Belt. The U.S. Geological Survey in Lawrence, Kansas, analyzed 76 reservoirs in the midwestern United States. Samples were collected at eight time intervals from April 1992 to September 1993. The analysis involved 9 herbicides (alachlor, ametryn, atrazine, cyanazine, metolachlor, metribuzin, prometon, propazine, and simazine), 3 cyanazine derivatives (cyanazine amide, deethylcyanazine, and deethylcyanazine amide), 2 atrazine derivatives (deethylatrazine and deisopropylatrazine), and the ethanesulfonic acid derivative of alachlor. The 9120 analyses fit nicely into a $(15 \times 8 \times 76)$ three-way data cube.

Tauler and co-workers[57] rearranged this data cube into a 15×608 matrix, consisting of 15 row samples and 608 column variables, ideally suited to alternating least squares-multivariate curve resolution (ALS-MCR) analysis via two-way augmented data matrices (Section 7.3). Three principal factors accounted for 96.8% of the variance. Matrices C_{aug} and S, containing the source and composition profiles, were obtained by employing Eqs. (7.5), (7.6), and (7.7) with appropriate constraints and conditions. The results were unique, permitting direct interpretation. This is true because, unlike other methods, the ALS-MCR method does not suffer from rotational ambiguity.

Matrix S is a 15×3 matrix containing the relative concentrations of the 15 herbicides and derivatives for each of the 3 sources. The composition profiles of the 3 sources are shown in Figure 12.17. The first profile is primarily alachlor, the second primarily atrazine, and the third primarily cyanazine and cyanazine amide. Figure 12.18 gives the averaged geographical locations of the 3 sources, averaged over the 8 sample collection periods. The numbers next to the states in the figure refer to specific locations on a geographical map. Among other valuable information, these figures show that there are 2 independent sources for alachlor.

Air Pollution. Hopke et al.[58] were concerned with the concentrations of 18 chemical elements in air particulates collected at 8 sampling stations in the

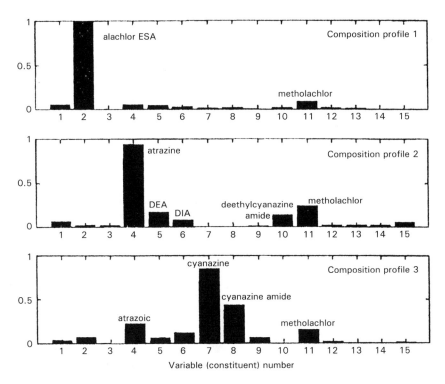

Figure 12.17 Source composition profiles resolved by ALS-MCR. The variable numbers refer to 15 herbicides and derivatives. [Reprinted with permission from R. Tauler, D. Barcelo and E. M. Thurman, *Environ. Sci. Technol.*, **34(16)**, 3307 (2000). Copyright 2000 © American Chemical Society.]

Boston area. Six principal factors accounted for nearly 99% of the variation in the data. Clusterings of elements were identified following varimax rotation. From the clusters, the factors could be associated with various sources of particulate material. The first 4 factors were identified with crustal weathering dust, sea-salt aerosol, combustion of residual fuel, and automotive exhaust, respectively. Hierarchical cluster analysis was employed to further clarify the nature of the factors. Similarities among sites were detected by calculating factor scores for the larger clusters via a special procedure. Several clusters were shown to be strongly associated with particular factors. For example, a cluster that consisted almost entirely of downtown Boston sites was found to have a large score on the factor related by factor analysis to automotive exhaust.

The concentrations of 24 chemical species present in particulate matter collected at 11 locations in the Tucson, Arizona, area of various sampling times over a 1-year period were factor analyzed by Gaarenstroom and co-workers.[59] Species–time data matrices for each location were factor analyzed separately. Between 5 and 8 factors were considered adequate to reproduce the data for the various sampling locations.

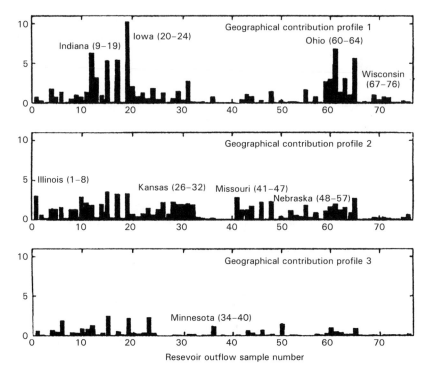

Figure 12.18 Geographical location profiles (averaged over eight sample periods) obtained by ALS-MCR. The reservoir numbers refer to a specific location on a geographical map. [Reprinted with permission from R. Tauler, D. Barcelo, and B. M. Thurman, *Environ. Sci. Technol.*, **34(16)**, 3307 (2000). Copyright 2000 © American Chemical Society.]

By studying the loadings in the varimax-rotated solution, Gaarenstroom et al.[59] attributed the cause of the pollutants to 3 main sources: soil, nonlocal aerosol, and automotive exhaust.

Coal combustion not only introduces large amounts of gaseous pollutants into the atmosphere but also particulates with traces of heavy metals. Using factor analysis. Lu and Zeng[60] have shown that trace elements Cd, Co, Pb, Cu, Ni, and Be in Qingshan bituminous coal are highly correlated with three species of sulfur. More specifically, Cd, Co, Pb, Cu, and Ni are associated with inorganic sulfides, whereas Be is associated with organic sulfides and sulfates. This was done by factor analyzing seven samples with different specific gravities. The elemental analyses were carried out by ICP-AES and GFAAS.

Principal component analysis has aided in controlling chimney emissions.[61] Smoke released from fuel burners produces substantial amounts of submicron particles that can penetrate deeply into the lungs causing severe respiratory ailments. The Bacharach opacity test (ASTM D-2156) is valuable for measuring the amount of blackening caused by a preset volume of smoke that passes through a piece of white filter paper. The Bacharach scale ranges from 0 (white) to 10 (black) in 0.5 unit steps. Some countries, such as Spain, have banned emissions greater than 5 Bacharachs and require corrective action when the index exceeds 4.

The Bacharach opacity is sensitive to a complex variety of operating variables such as the nature of the fuel, temperature, pressure, drop size, viscosity, and density. Prior to PCA no clear-cut correlation between the variables and the Bacharach opacity was found because, primarily, the fuel parameters are highly correlated. Such correlations are easily unraveled by PCA. Four principal factors accounted for 96.63% of the information supplied by nine variables. The scores of the four factors were used as fuel parameters in the ACE[62] (alternating conditional expectations) method to identify the variables that influenced the Bacharach opacity. In particular, the controlling variables involved minimizing the burner-tip viscosity and the SO_2 content in the chimney smoke, using fuel with low coke and asphaltene contents, lowering the smoke temperature, and controlling the O_2 in the combustion chambers and the amount of steam they produce. The PCA–ACE methodology proved to be highly effective in reducing the Bacharach opacity when the fuel burners of a petrochemical plant were adjusted accordingly.

Geladi et al.[63] applied three-way PARAFAC to study Arctic aerosol from Northwest Canada. The method was based on the assumption that the airborne particulate samples are linear sums of contributions from different sources. Samples were collected on Whatman 41 filters on a weekly basis over an 11-year period from 1980 to 1991. The study entailed 24 elements (Table 12.13) quantified by ion chromatography, inductively coupled plasma emission spectroscopy, and instrumental neutron activation analysis. The data were arranged into an $11 \times 52 \times 24$ cube representing years $\times$ weeks $\times$ variables.

Nonnegative constrained PARAFAC with rotation yielded the best solution. Five source factors were found and interpreted as shown in Table 12.13. The yearly variations of the five factors are displayed in Figure 12.19. Such information provides valuable information in environmental studies.

Several other factor analyses of environmental data have been published. Factor analytical investigations of large-scale pollution by Blifford and Meaker,[64] of the

TABLE 12.13   Variables and Factors for Arctic Aerosol from Northwest Canada[a]

Variables				Factors
1. Cl	7. NH_4	13. Zn	19. Si	F1. Photogenic
2. Br	8. K	14. Pb	20. As	F2. Soil
3. NO_3	9. MSA[b]	15. Ca	21. La	F3. Anthropogenic
4. SO_4	10. Mn	16. Ti	22. Sb	F4. Sea salt
5. H	11. V	17. I	23. Sm	F5. Biogenic
6. Na	12. Al	18. In	24. Se	

[a]Reprinted with permission from P. Geladi, Y. Xie, A. Polissar, and P. K. Hopke, *J. Chemometrics*, **12**, 337 (1998). Copyright 1998 © John Wiley & Sons.
[b]MSA≡methane sulfonic acid (CH_3SO_3H).

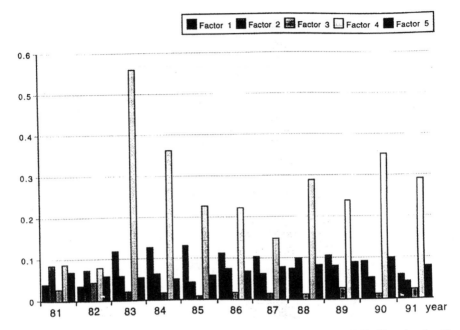

Figure 12.19   Seasonal variation of the five sources listed in Table 12.13. [Reprinted with permission from P. Geladi, Y. Xie, A. Polissar, and P. K. Hopke, *J. Chemometrics*, **12**, 337 (1998). Copyright 1998 © John Wiley & Sons, Ltd.]

effects of meteorological parameters on air pollution by Peterson,[65] of the effects of air pollution on mortality by Lave and Seskin,[66] and of the sources of urban dust by Linton et al.,[67] illustrate the scope of factor analysis environmental problems. Considering the growing need to understand the mechanisms that determine the distributions of pollutants, factor analysis of environmental data should be of even greater value in the future.

12.9   FOOD SCIENCE

Sugar. To improve the quality and process control of sugar beets, a detailed knowledge of the chemical analytes is essential. Direct analysis of raw sugar samples can be achieved by subjecting fluorescence spectra to parallel factor analysis (PARAFAC) (Section 7.4), thus avoiding the need for laborious and time-consuming analysis. Since ~99.99% of sugar is nonfluorescent sucrose, fluorescence spectra focuses on some of the minor constituents.

Such a study has been conducted by Bro.[68] Samples were taken directly from a sugar plant in Scandinavia at various time intervals over a 3-month period. The raw samples (265 in number) were dissolved in water, and the solutions were inserted into a spectrofluorometer. The emission spectrum of each sample was recorded from

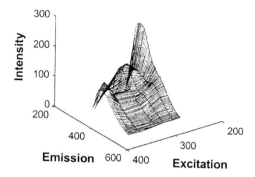

Figure 12.20 Fluorescence emission spectra of a sugar sample. [Reprinted with permission from R. Bro, *Chemom. Intell. Lab. Syst.*, **46**, 133 (1999). Copyright 1999 © Elsevier Science.]

275 to 560 nm at 0.5-nm intervals (571 wavelengths) at 7 excitation wavelengths (230, 240, 255, 290, 305, 325, and 340 nm). A typical spectrum of a sugar sample is shown in Figure 12.20.

The data was assembled into a three-way block ($265 \times 571 \times 7$) appropriate for PARAFAC analysis. The unique advantage of PARAFAC, in comparison to bilinear models, is the fact that it is free from rotational anomalies, yielding unambiguous solutions. The emission spectra of four components (Figure 12.21) were extracted by PARAFAC with nonnegativity and unimodality constraints.

To verify whether or not the correct number of components was determined, a *split-half analysis* was invoked. The stability of the analysis is assured when different subsets of data yield the same results. As seen in Figure 12.21, the emission profiles from the split analyses closely overlap, confirming the results. The results are further substantiated by the fact that the bands 2 and 3 correspond to the

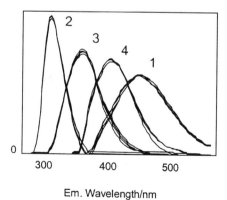

Figure 12.21 Emission spectra of four components extracted directly by PARAFAC. Overlapping spectra are the result of split-half analysis. [Reprinted with permission from R. Bro, *Chemom. Intell. Lab. Syst.*, **46**, 133 (1999). Copyright 1999 © Elsevier Science.]

known emissions of tyrosine and tryptophane. The identities of the components responsible for bands 4 and 1 were not deduced.

In this study the PARAFAC scores represent uncalibrated concentrations as a function of sampling time, based on 8-hour shifts. The scores of the four components are plotted in Figure 12.22. Multiple linear regression indicated that the scores were highly correlated with color and ash content. More specifically, color was predominantly influenced by component 1 and 4. Components 1 and 3 correlated with ash content. Component 2, tyrosine, did not correlate with color and component 4 did not correlate with ash.

Bread. Among other multivariate techniques, PCR and PLS have been used by Kvaal and associates[69] to assess the quality of bread. Textural images of wheat baguettes were recorded and transformed into factors based on sensory perceptions. The factorial design involved four factors: flour quality (four levels), mixing time (two levels), proofing time (two levels), and baking process (two levels). For the PCR and PLS analyses three independent replicated experiments were performed for each of the 32 combinations of design factors, requiring the preparation of 96 baguettes. Half portions of the baguettes were subjected to sensory analysis by humans as well as by instruments.

Although the study focused solely on porosity, 10 trained panelists evaluated the baguettes according to 13 sensory attributes (based on a scale of 1.0 to 9.0): gloss, crust brittleness, porosity, surface cuts, elasticity, odor intensity, fresh smell, flavor intensity, flavor freshness, salt flavor, firmness, juiciness, and crispness of the crust.

True color images of the bread slices were recorded using illumination from four tungsten lamps. Images of four samples spanning the porosity range are illustrated in Figure 12.23. Each image was digitized and converted into a 256×256 gray scale matrix.

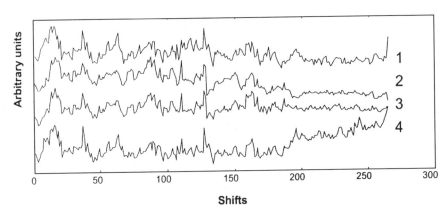

Figure 12.22 Sampling time shifts associated with each of the four components shown in Figure 12.21. Each shift represents an 8-hour period. [Reprinted with permission from R. Bro, *Chemom. Intell. Lab. Syst.*, **46**, 133 (1999). Copyright 1999 © Elsevier Science.]

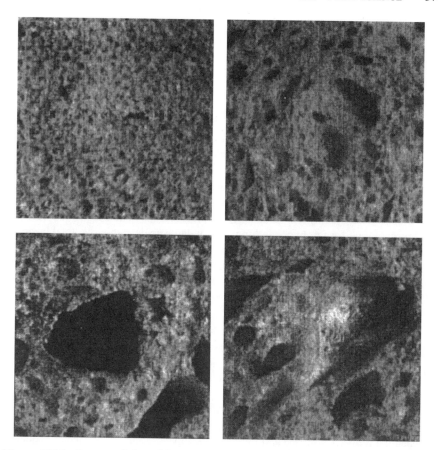

Figure 12.23 Images of four different samples spanning the porosity range for one data block. [Reprinted with permission from K. Kvaal, J. P. Wold, U. G. Indahl, P. Baardseth, and T. Naes, *Chemom. Intell. Lab. Syst.*, **42**, 141 (1998). Copyright 1998 © Elsevier Science.]

The image matrices were further converted into vectors by singular value decomposition (SVD). The singular value spectrum, SV spectra, of an image is defined as the vector containing p singular values in descending order, deleting the remaining set of smaller singular values. Because the singular values are the square-roots of the eigenvalues, the SV spectra is simply

$$\textbf{SV spectra} = [\lambda_1^{1/2}\lambda_2^{1/2} \ldots \lambda_p^{1/2}] \tag{12.7}$$

A data matrix appropriate for PCR and PLS is constructed by inserting these vectors as rows of $\mathbf{X}$ in Eq. (8.6). The dependent vector, $\mathbf{y}$, in (8.6), of course, contains the human sensor porosity assignments. By setting p equal to 256, 100, and 10, three situations were examined, large, medium, and small SV vectors. Based on cross validation, PCR optimally modeled the porosity with 6, 5, and 2 abstract factors, corresponding to the three situations, respectively. PLS required 5, 5 and 1 factors, respectively. Correlations based on p equal to 256 and 100 were found to be greater than 0.8. These correlations were considered to be high because sensory data is very noisy. The correlation, 0.66, based on p equal to 10 was not as good.

Fish. Near-infrared (NIR) spectroscopy has been employed in marine research laboratories for the purpose of distinguishing fish species as well as for developing rapid, accurate, and simple methods of analysis. Solberg[70] generated NIR PCA score plots that distinguished cod and Atlantic salmon. Wold and co-workers[71] found eight NIR wavelengths that gave the best regression coefficients for determining the fat content of fish filet. Nortvedt and co-workers[72] used NIR PLS to determine fat, protein, and dry matter of Atlantic halibut (*Hippoglossus hippoglossus*) filet.

Balladin et al.[73] analyzed 78 samples of [*Scomberomorous brasiliensis* (carite)] fish muscle tissue for fat, hypoxanthine, total volatile acids and bases, total bacterial count, and textural parameters (firmness, hardness, fracturability, cohesiveness, chewiness, and elasticity). The 11×78 data matrix decomposed into five principal factors, accounting for 81% of the total variance. The factors were interpreted as follows: factor 1 (tastiness), factors 2, 3, 5 (texture), and factor 4 (rancidity).

12.10   ADDITIONAL PRACTICAL APPLICATIONS

A hodgepodge of additional practical applications of factor analysis are referenced in this section.

In the chemical industry, abstract factor analysis has been used to analyze a variety of data.[74] For example, in the food and beverage industries, correlations between the properties of rice and the flavor of Japanese sake,[75] between the quality of malt and of beer,[76] between the amino acid composition and varieties of potatoes,[77] and between the amino acid composition and varieties of barley[78] have been determined with factor analysis. Studies of geological data,[79] of the composition of moon rocks,[80] of interacting species in biological systems,[81] of taxonomic classifications of plants,[82] and of anaerobic digestion in ecological problems[83] give but a hint of the potential of factor analysis in sciences closely related to chemistry.

The sheer magnitude of the number of monographs on abstract factor analysis listed in the Bibliography testifies to the leading role of factor analysis in the behavioral sciences. Such is the importance of factor analysis in the social sciences that the method is referred to as the calculus of the social sciences. The many types of applications in a variety of social science disciplines outlined by Rummel[84] include problems of interest to chemists in management and production. Researchers in psychology and education have also benefited greatly from factor analysis. Such applications should be of value to chemists involved in education, counseling, and personnel work.

The overriding objective for writing this book was to equip chemists with the necessary understanding and tools to employ factor analysis in their work. That factor analysis has provided useful results even in exceptionally complicated problems suggests that applications of factor analysis in chemistry are still in an embryonic state. The range of applications discussed in the last five chapters argues well for the continued growth of factor analysis in chemistry. In a broader sense, we can anticipate that the use of diverse chemometric methods will become routine in chemistry. In particular, general procedures for attacking multidimensional problems in chemistry should result from data analyses using factor analysis in conjunction with other multivariate methods such as pattern recognition, cluster analysis, and multiple regression analysis.

REFERENCES

1. M. Charton, *Chem. Tech.*, **1975**, 245 (1975).

2. B. Higman, *Applied Group-Theoretic and Matrix Methods*, Oxford University Press, Oxford, 1955.

3. E. R. Malinowski, Ph.D. thesis, Stevens Institute of Technology, Hoboken, NJ, 1961.

4. H. H. Jaffe, *Chem. Rev.*, **53**, 191 (1953).

5. P. H. Weiner, *J. Am. Chem. Soc.*, **95**, 5845 (1973).

6. J. G. Kirkwood and F. H. Westheimer, *J. Chem. Phys.*, **6**, 513 (1938).

7. S. Wold and M. Sjostrom, *Chem. Ser.*, **2**, 49 (1972).

8. J. B. Justice and T. L. Isenhour, *Anal. Chem.*, **47**, 2286 (1975).

9. R. W. Rozett and E. M. Petersen, *Anal. Chem.*, **47**, 1301 (1975).

10. R. W. Rozett and E. M. Petersen, *Anal. Chem.*, **47**, 2377 (1975).

11. R. W. Rozett and E. M. Petersen, *Anal. Chem.*, **48**, 817 (1976).

12. R. W. Rozett and E. M. Petersen, *Am. Lab.*, **9**(2), 107 (1977).

13. D. R. Burgard, S. P. Perone, and J. L. Weibers, *Anal. Chem.*, **49**, 1444 (1977).

14. S. S. Williams, R. B. Lam, and T. L. Isenhour, *Anal. Chem.*, **55**, 1117 (1983).

15. P. B. Harrington and T. L. Isenhour, *Appl. Spectrosc.*, **41**, 449 (1987).

16. P. B. Harrington and T. L. Isenhour, *Anal. Chem.*, **60**, 2687 (1988).

17. P. C. Gillette and J. L. Koenig, *Appl. Spectrosc.*, **36**, 535 (1982).

18. S. R. Culler, P. C. Gillette, H. Ishida, and J. L. Koenig, *Appl. Spectrosc.*, **38**, 495 (1984).

19. G. Wernimont, *Anal. Chem.*, **39**, 554 (1967).

20. R. N. Carey, S. Wold, and J. O. Westgard, *Anal. Chem.*, **47**, 1824 (1975).

21. J. T. Edward and S. C. Wong, *J. Am. Chem. Soc.*, **99**, 4229 (1977).

22. D. G. Howery and P. Chan, private communication.

23. C. L. deLigny, N. G. van der Veen, and J. C. Van Houweilingen, *Ind. Eng. Chem. Fundam.*, **15**, 336 (1976).

24. W. R. Fawcett and T. M. Krygowski, *Can. J. Chem.*, **54**, 3283 (1976).

25. D. G. Howery, *Bull. Chem. Soc. Jpn.*, **45**, 2643 (1972).

26. D. L. Duewer and H. Freiser, *Anal. Chem.*, **49**, 1940 (1977).

27. D. G. Howery and M. L. Rubinstein, *Can. J. Chem.*, **65**, 1380 (1987).

28. M. A. Woodbury, R. C. Clelland, and R. J. Hickey, *Behav. Sci.*, **8**, 347 (1963).

29. C. G. Swain, H. E. Bryndza, and M. S. Swain, *J. Chem. Inf. Comput. Sci.*, **19**, 19 (1979).

30. P. H. A. Sneath, *J. Theor Biol.*, **12**, 157 (1966).

31. G. L. Ritter and H. B. Woodruff, *Anal. Chem.*, **49**, 2116 (1977).

32. A. Cammarata and G. K. Menon, *J. Med. Chem.*, **19**, 739 (1976).

33. G. K. Menon and A. Cammarata, *J. Pharm. Sci.*, **66**, 304 (1977).

34. M. L. Weiner and P. H. Weiner, *J. Med. Chem.*, **16**, 655 (1973).

35. A. Gauthier, J. Zurli, B. C. Cros, and H. Sarles, *Rev. Eur. Etud. Clin. Biol.* **17**, 574 (1972).

36. J. Farber, J. Tosovsky, and K. Hynck, *Act. Nerv. Super.*, **16**, 258 (1974).

37. E. R. John, *Osnovn. Probl. Elektrofiziol. Golovn. Mozga.*, **1974**, 161 (1974).

38. N. R. Bohidar, F. A. Restaino, and J. B. Schwartz, *J. Pharm. Sci.*, **64**, 966 (1975).

39. E. R. Malinowski, *Anal. Chem.*, **49**, 612 (1977).

40. D. G. Howery and J. Gelbfish, unpublished results.

41. E. Holmes, J. K. Nicholson, A. W. Nicholls, J. C. Lindon, S. C. Connor, S. Polley, and J. Connelly, *Chemometrics Intell. Lab. Syst.*, **44**, 245 (1998).

42. B. K. Lavine, A. Moores, and L. K. Helfend, *J. Anal. Appl. Pyrol.*, **50**, 47 (1999).

43. B. Antalek, J. P. Hornak, and W. Windig, *J. Magn. Reson.*, **132**, 307 (1998).

44. W. Windig, B. Antalek, L. J. Sorriero, S. Bijlsma, D. J. Louwerse (AD), and A. K. Smilde, *J. Chemometrics*, **13**, 95 (1999).

45. X. Zhu, L. Zhang, X. Che, and L. Wang, *Chemometrics Intell. Lab. Syst.*, **45**, 147 (1999).

46. A. J. Lubeck and D. L. Sutton, *J. High Resol. Chromatogr. Comm.*, **6**, 328 (1983); **7**, 542 (1984).

47. J. B. Cooper, *Chemometrics Intell. Lab. Syst.*, **46**, 231 (1999).

48. D. J. Alpert and P. K. Hopke, in *Proceedings of the Conference on Quality Assurance Environmental Measures*, Denver, CO, Nov. 1978, p. 204.

49. P. K. Hopke, *J. Environ. Sci. Health*, **A11**(6), 367 (1976).

50. E. J. Knudson, D. L. Duewer, G. D. Christian, and T. V. Larson, in B. R. Kowalski (Ed.), *Chemometrics: Theory and Applications*, ACS Symp. Ser. 52, American Chemical Society, Washington, DC, 1977, p. 80.

51. B. K. Lavine, A. J. Moores, H. T. Mayfield, and A. Faruque, *Anal. Lett.*, **32**(15), 2805 (1998).

52. B. K. Lavine, J. Ritter, A. J. Moores, M. Wilson, A. Faruque, and H. T. Mayfield, *Anal. Chem.*, **72**, 423 (2000).

53. B. K. Lavine, A. J. Moores, H. T. Mayfield, and A. Faruque, *Microchemical J.*, **61**, 69 (1999).

54. M. A. Stapanian, F. C. Garner, K. E. Fitzgerald, G. T. Flatman, and J. M. Nocerino, *J. Chemometrics*, **7**, 165 (1993).

55. S. Wold. *J. Pattern Recog.*, **8**, 127 (1976).

56. S. Wold and M. Sjostrom, in B. R. Kowalski (Ed.), *Chemometrics: Theory and Applications*, ACS Symposium Series 52, Washington, DC, 1977.

57. R. Tauler, D. Barcelo, and E. M. Thurman, *Environ. Sci. Technol.*, **34**(16), 3307 (2000).

58. P. K. Hopke, E. S. Gladney, G. E. Gordon, W. H. Zoller, and A. G. Jones, *Atmos. Environ.*, **10**, 1015 (1976).

59. P. D. Gaarenstroom, S. P. Perone, and J. L. Moyers, *Environ. Sci. Technol.*, **11**, 795 (1977).

60. X. Lu and H. Zeng, *Chemometrics Intell. Lab. Syst.*, **45**, 311 (1999).

61. M. Blanco, J. Coello, S. Maspoch, and A. R. Puigdomenech, *Chemometrics Intell. Lab. Syst.* **46**, 31 (1999).

62. L. Brieman and J. H. Friedman, *J. Am. Stat. Assoc.*, **80**, 580 (1985).

63. P. Geladi, Y. Xie, A. Polissar, and P. K. Hopke, *J. Chemometrics*, **12**, 337 (1998).

64. I. H. Blifford and G. O. Meaker, *Atmos. Environ.*, **1**, 147 (1967).

65. J. T. Peterson, *Atmos. Environ.*, **4**, 501 (1970); **6**, 433 (1972).

66. L. B. Lave and E. P. Seskin, *Air Pollution and Human Health*, Johns Hopkins University Press, Baltimore, MD, 1977, p. 33.

67. P. W. Linton, D. F. S. Natusch, P. K. Hopke, and R. L. Solomon, in *Proceedings of the 4th Conference on Sensing Environmental Pollutants*, New Orleans, LA, Nov. 1977, p. 221.

68. R. Bro, *Chemometrics Intell. Lab. Syst.*, **46**, 133 (1999).

69. K. Kvaal, J. P. Wold, U. G. Indahl, P. Baardseth, and T. Naes, *Chemometrics Intell. Lab Syst.*, **42**, 141 (1998).

70. C. Solberg, in K. I. Hildrum, T. Isaksson, T. Naes, and A. Tanberg (Eds.), *Near Infrared Spectroscopy. Bridging the Gap Between Data Analysis and NIR Applications*, Ellis Horwood, London, 1992, pp. 223–227.

71. J. P. Wold, T. Jakobsen, and L. Krane, *J. Food Sci.*, **61(1)**, 74 (1996).

72. R. Nortvedt, O. J. Torrissen, and S. Tuene, *Chemometrics Intell. Lab. Syst.*, **42**, 199 (1998).

73. D. A. Balladin, D. Narinesingh, V. A. Stoute, and T. T. Ngo, *Chemometrics Intell. Lab. Syst.*, **40**, 175 (1998).

74. P. H. Weiner, *Chem. Tech.*, **1977**, 321 (1977).

75. K. Yoshizawa, T. Ishikawa, M. Kinoshita, A. Takeda, and I. Fujie, *Nippon Jozo Kyokai Zasshi*, **69**, 581 (1974).

76. L. Reiner and A. Piendle, *Brauwissenschaft*, **27**, 1 (1974).

77. H. Martens, Y. Solberg, L. Roer, and E. Vold, *Potato Res.*, **18**, 515 (1975).

78. H. Martens and K. E. B. Knudsen, *Cereal Chem.*, **57(2)**, 97 (1980).

79. J. Imbrie, Tech. Rep. No. 6, ONR Task No. 389-135, Northwestern University, Evanston, IL, 1963.

80. K. M. Dawson and A. J. Sinclair, *Econ. Geol.*, **69**, 404 (1974).

81. M. E. Magar and P. W. Chuin, *Biophys. Chem.*, **1** 18 (1973).

82. R. R. Sokal and P. H. A. Sneath, *Principles of Numerical Taxonomy*, W. H. Freeman, San Francisco, 1963, Chap. 7.

83. D. F. Toerin, *Water Res.*, **3**, 129 (1969).

84. R. J. Rummel, *Applied Factor Analysis*, Northwestern University Press, Evanston, IL, 1970, Chap. 24.

APPENDIX A

PSEUDOINVERSE

Consider the following multivariate expression concerning vectors $\mathbf{a}$ and $\mathbf{b}$, and matrix $\mathbf{X}$ with elements i, j, and $i \times j$, respectively:

$$\mathbf{a} = \mathbf{Xb} \tag{A.1}$$

Consider further the problem of finding $\mathbf{b}$ from a knowledge of $\mathbf{a}$ and $\mathbf{X}$. Solving this expression for $\mathbf{b}$ yields

$$\mathbf{b} = \mathbf{X}^{-1}\mathbf{a} \tag{A.2}$$

This solution is feasible only if $i = j$ and $\mathbf{X}$ is nonsingular; that is, neither the rows nor the columns of $\mathbf{X}$ are linearly dependent.

For the case where these conditions do not pertain, the problem can be solved by a method of least squares that minimizes the sum of squares of the differences between $\mathbf{a}$ and $\hat{\mathbf{a}}$, the vector calculated by (A.1). The least-squares procedure yields

$$\mathbf{b} = [\mathbf{X}'\mathbf{X}]^{-1}\mathbf{X}'\mathbf{a} \tag{A.3}$$

A derivation of this expression is found in Section 3.4.3, where (A.1) is analogous to (3.108) and (A.3) is analogous to (3.121). Since $\mathbf{X}'\mathbf{X}$ is a square matrix of size $j \times j$, it has an inverse. The combination of matrices on the right of (A.3), because of its similarity to the inverse expressed in (A.2), is known as the "left pseudoinverse" and is given a special symbol,

$$\mathbf{X}^+ = [\mathbf{X}'\mathbf{X}]^{-1}\mathbf{X}' \tag{A.4}$$

Based on this notation, (A.3) can be written in compact form as

$$\mathbf{b} = \mathbf{X}^+\mathbf{a} \tag{A.5}$$

which is analogous to (3.123).

In a similar fashion, for the situation involving the following row vector computation:

$$\mathbf{a}' = \mathbf{b}'\mathbf{X} \tag{A.6}$$

if $\mathbf{a}'$ and $\mathbf{X}$ are known, the least-squares procedure yields

$$\mathbf{b}' = \mathbf{a}'\mathbf{X}^+ \tag{A.7}$$

where

$$\mathbf{X}^+ = \mathbf{X}'[\mathbf{X}\mathbf{X}']^{-1} \tag{A.8}$$

which is known as the "right pseudoinverse." Note that the left pseudoinverse is a $j \times i$ matrix, whereas the right pseudoinverse is an $i \times j$ matrix. In fact, the right pseudoinverse is simply the transpose of the left pseudoinverse.

APPENDIX B

TOOLBOX FOR CHEMICAL FACTOR ANALYSIS

The *Toolbox for Chemical Factor Analysis* is written in MATLAB for use with an IBM PC/XT/AT equipped with MATLAB (or MATLAB for Students), a $3\frac{1}{2}$-inch floppy disk drive and an 80-column printer. (MATLAB is an interactive program system copyrighted by The Mathworks, Inc., Cochituate Place, 24 Prime Park Way, Natick, MA 01760.) The *Toolbox for Chemical Factor Analysis* is designed not only to help you learn the principles of target factor analysis but also to provide the necessary capabilities for tackling real research problems. The programs can be customized very easily to suit your own particular needs and desires. To aid in customization, a glossary of important symbols used in the programs, and literature references are provided.

The *Toolbox for Chemical Factor Analysis* contains sample data taken from the chemical literature, allowing you to get started quickly and easily. The *Instruction Manual* is designed to lead you quickly and carefully through all of the programs. The set of programs provide a wide spectrum of utility and are stored on a single disk as follows:

pfa.m	Principal factor analysis
ccv.m	Complete cross validation
bcv.m	Binary cross validation
uniq.m	Uniqueness test
ttest.m	Target test
tload.m	Target loadings
iksfa.m	Iterative key set factor analysis

spexfa.m	Spectral isolation factor analysis
rafa.m	Rank annihilation factor analysis
efa.m	Evolving factor analysis plot
wfa.m	Window factor analysis
autowfa.m	Automatic window factor analysis
wfax.m	Subroutine for autowfa.m
iksfax.m	Subroutine for autowfa.m
slf.m	Subroutine for pfa.m, uniq.m and ttest.m

Data files stored on the disk are labeled:

chap5.mat	Taken from Chapter 5 in E. R. Malinowski, *Factor Analysis in Chemistry*, John Wiley & Sons, Inc.
glcwlk.mat	Gas–liquid chromatographic retention volumes
pmr.mat	Proton magnetic resonance solvent shifts
ms.mat	Mass spectra of binary mixtures
simms.mat	Simulated mass spectra of unknown mixtures
mixture.mat	UV spectra of unresolved liquid chromatographic fractions
dye.mat	Visible absorbance spectra of a dye at various pH
fabertbl.mat	Table containing Mandel's degrees of freedom

After loading a data matrix into MATLAB, the first step is to perform a principal factor analysis (also known as eigenvector analysis (EVA), principal component analysis (PCA), or singular value decomposition (SVD)). This is done with the **pfa.m** program, which displays a summary of the eigenvalues, the theoretical errors, and the percent significance, information useful for determining the size of the factor space.

By specifying the number of factors, **pfa.m** will compute the abstract row and column matrices as well as the reproduced data matrix, based on the specified number of factors.

To help determine the size of the factor space the **ccv.m** (complete cross validation) or **bcv.m** (binary cross validation) can also be employed.

Typographical errors or outliers, if sufficiently large, can erroneously increase the factor space, thus making the problem more complex than it truly is. Using the SPOIL values or the percent significance levels (%SL) obtained from **uniq.m** the "uniqueness test" you can determine when this occurs and how to locate such errors.

The **ttest.m** (target testing) program is used to determine whether or not a given target vector lies inside the factor space. Each target is tested individually. A SPOIL of three or less, or a %SL(Malinowski) greater than 5 or 10%, is evidence that the target lies inside the factor space. **tload.m** will perform a COMBINATION TRANSFORMATION (transforming the set of eigenvector axes, as best as possible, into the combined set of target axes). The program yields the factor loadings associated with the targets, together with error estimations of the loadings. It also regenerates the data matrix based on the set of targets and their loadings. This is also known as principal component regression (PCR).

iksfa.m (iterative key set factor analysis) is designed to find the key set of "typical" rows and the key set of "typical" columns of the data matrix that best reproduce the data.

rafa.m (rank annihilation factor analysis) will subtract the correct amount of a pure component matrix from a mixture matrix to reduce the rank of the matrix by one unit. The amount subtracted corresponds to the relative amount of component in the mixture. The program uses a single-value analysis technique that accomplishes this task most efficiently. An error analysis is also performed to show the accuracy of the quantification.

spexfa.m (spectral isolation factor analysis) isolates the spectra of the components in a series of related mixtures when the components have unique spectral regions.

efa.m (evolving factor analysis) determines and plots the eigenvalues obtained from forward and backward eigenanalysis of a data matrix obtained from an evolutionary process. An eigenvalue that emerges from the pool of error eigenvalues from the forward analysis indicates the evolution of a component. An eigenvalue that emerges from the pool of error eigenvalues from the backward analysis indicates the disappearance of a component. This information is useful for determining the concentration profiles of the components.

wfa.m (window factor analysis) determines and plots the concentration profile of an *individual* component in an evolutionary mixture, from a knowledge of the window (region of existence along the evolutionary axis). If the spectrum of the component is known, the method can be used quantitatively.

autowfa.m (automatic window factor analysis) attempts to locate the windows of the components present in evolutionary processes. This program makes use of iterative key set factor analysis and window factor analysis. It extracts and plots the concentration profiles of the *chemical* components. The program has not been optimized and some additional refinement may be necessary. This is done by making minor adjustments in the designated windows and running **wfa.m**.

The *Toolbox for Chemical Factor Analysis* is not designed to be an independent, self-contained teaching tool. It presupposes an exposure to the basic philosophical principles and terminology associated with target factor analysis. Copies of the program can be obtained from Applied Chemometrics, Inc., PO Box 100, Sharon, MA 02067, USA, +1 781 784 7700, *info@chemometrics.com, www.chemometrics.com*.

APPENDIX C

MATLAB PROGRAMS

MATLAB is a matrix laboratory, interactive program system (copyrighted by The MathWorks, Inc., Cochituate Place, 24 Prime Park Way, Natick, MA 01760) that is ideally suited to factor analysis. Programs are written almost exactly as they are written mathematically, requiring no matrix dimensioning. MATLAB programs are extremely compact and relatively easy to write, run, and debug. Plotting input and output data in two or three dimensions is spectacularly simple.

To illustrate the power of MATLAB, three FA programs, written by the author are provided in this appendix:

sfa.m   Significance factor analysis—a program designed to help determine the number of significant factors in a matrix

tfa.m   Target factor analysis—a program designed to target test suspected factors

lfa.m   Loading factor analysis—a program designed to calculate factor loadings and errors in the loadings

The programs include references to sections and equations of the text that are pertinent to the particular computation.

```
****************************************************************************************
**       sfa.m                                                                      **
****************************************************************************************
function  [] = sfa(d)
% sfa.m    significant factor anlaysis - a program designed
%                to help determine the number of significant factors in a data matrix.
% sfa(d)
```

```
% d = data matrix
format short e
[r,c] = size(d);
if r < c
    d = d';
    [r,c]= =size(d);
end
[u,s,v] = svd(d,0);
    for j=1:c
        ev(j) = s(j,j) * s(j,j);
        df(j) = (r-j+1)*(c-j+1);
        rev(j) = ev(j) / df(j);
    end
    for k = 1:c-1
        sev(k) = sum(ev(k+1:c));
        sdf(k) = sum(df(k+1:c));
    end
        for i = 1:c-1
            re(i) = sqrt(sev(i) / (r * (c-i)));
            ind(i) = re(i) / (c-i)^2;
        end
semilogy(ind),shg
xlabel('FACTOR LEVEL')
ylabel('IND')
pause
close
clc
[vind,n] = min(ind);
disp(['IND function indicates ',int2str(n),' significant factors (see eq. 4.63).'])
disp(['The real error (RE) is +/-',num2str(re(n)),' (see eq. 4.44).'])
pause
re(c) = NaN; ind(c) = NaN;
  for j = 1:c
        t(j,1) = j;
        t(j,2) = ev(j);
        t(j,3) = re(j);
        t(j,4) = ind(j);
        t(j,5) = ref(j);
  end
  for j = 1:c-1
        f = (sdf(j) * ev(j)) / ((r-j+1) * (c-j+1) * sev(j));
% convert f (see eq. 4.83) into percent significance level
            if j < c
                tt = sqrt(f);
                df = c - j;
                a = tt / sqrt(df);
                b = df / (df + tt * tt);
                im = df - 2;
                jm = df - 2 * fix(df / 2);
                ss = 1;
                cc = 1;
                ks = 2 + jm;
                fk = ks;
```

```
                        if  (im  -  2)  >=  0
                          for  k  =  ks:2:im
                                  cc  =  cc  *  b  *  (fk  -1)  /  fk;
                                  ss  =  ss  +  cc;
                                  fk  =  fk  +2;
                            end
                          end
                      if  (df  -  1)  >  0
                            cl  =  .5  +  (a  *  b  *  ss  +  atan(a))  *  .31831;
                          else
                            cl  =  .5  +  atan(a)  *  .31831;
                          end
                      if  jm  <=  0
                            cl  =  .5  +  .5  *  a  *  sqrt(b)  *  ss;
                          end
                  end
                  sl  =  100  *  (1  -  cl);
                  sl  =  2  *  sl;
                      if  sl  <  le-2,  sl  =  0;  end
                  t(j,6)  =  sl;
      end
t(c,6)  =  NaN;
disp(['SFA  RESULTS  (note  %SL  based  on  eq.  4.83)'])
disp('           n           EV          RE          IND         REV          %SL')
disp(t)
*************************************************************************************
**  tfa.m                                                                         **
*************************************************************************************
function  []  =  tfa(d,n,x)
%  tfa.m     target  factor  analysis  -  a  program  designed  to
%               target  test  suspected  vectors.
%  tfa(d,n,x)
%  d  =  an  (r  x  c)  data  matrix.
%  n  =  number  of  factors  to  be  used  in  the  target  tests.
%  x  =  an  (r  x  m)  matrix  composed  of  m  test  vectors,  each  with  r  elements.
%  No  provisioin  is  made  for  handling  missing  points  (blanks)  in  the  targets.
format  short  e
[rx,nx]  =  size(x);
[r,c]  =  size(d);
if  rx  ~=  r
disp('Target  vectors  must  emulate  columns  of  the  data  matrix!  PROGRAM  ABORTED')
end
          lg  =  r;
          sm  =  c;
      if  r  <  c
          lg  =  c;
          sm  =  r;
          [v,sv,u]  =  svd(d',0);
      else
          [u,s,v]  =  svd(d,0);
      end
       for  j=1:sm
            ev(j)  =  s(j,j)  *  s(j,j);
```

```
                df(j)  =  (r-j+1)*(c-j+1);
                rev(j) = ev(j) / df(j);
                u(:,j) = u(:,j) * s(j,j);
        end
ubar = u(:,1:n);
sev = sum(ev(n+1;sm));
sdf = sum(df(n+1:sm));
re = sqrt(sev / (lg * (sm-n)));
    for j = 1:nx
            t(:,j)  = pinv(ubar) * x(:,j);
            xp(:,j) = ubar * t(:,j);
            dx = xp(:,j) - x(:,j);
            aet(j) = sqrt((dx' * dx) / (rx - n));
            rep(j) = re * norm(t(:,j));
                if rep(j) > aet(j)
                    ret(j)= 0;
                else
                    ret(j) = sqrt(aet(j)^2 - rep(j)^2);
                end
        spoil(j) = ret(j) / rep(j);
        f(j) = (sdf * r * aet(j)^2) / ((r-n+1) * (c-n+1) * sev * t(:,j)' * t(:,j));
    end
clc
df1 = rx -n;
df2 = sm -n;
disp('RESULTS OF TARGET TESTING (see Section 4.6)')
disp(' ')
disp(['F(df1,df2) = F(',int2str(df1),',',int2str(df2),'), (see eq. 4.129)'])
disp(' ')
        for j = 1:nx
                tx(j,1)= j;
                tx(j,2) = aet(j);
                tx(j,3) = rep(j);
                tx(j,4) = ret(j);
                tx(j,5) = spoil(j);
                tx(j,6) = f(j);
        end
disp('       target #      AET        REP        RET        SPOIL        F')
disp(tx)


**********************************************************************************
** lfa.m
**********************************************************************************
function [] = lfa(d,x)
% lfa.m -       loading factor analysis - a program designed to calculate
%               factor loadings and errors in the loadings.
% lfa(d,x)
% d = an (r x c) data matrix.
% x = an (r x n) matrix composed of n test vectors.
% The factor space is assumed to be n dimensional.
% No provision is made for handling missing points (blanks) in the targets.
clf
format short e
```

```
[rx,n]  =  size(x);
[r,c]  =  size(d);
if  rx  ~=  r
disp('Target  vectors  must  emulate  columns  of  the  data  matrix!  PROGRAM  ABORTED')
end
        sm  =  c;
    if  r  <  c
        sm  =  r;
        [v,s,u]  =  svd(d',0);
    else
        [u,s,v]  =  svd(d,0);
    end
        for  j  =  1:sm
            u(:,j)  =  u(:,j)  *  s(j,j);
        end
ubar  =  u(:,1:n);
vbar  =  v(:,1:n);
        for  j  =  1:n
                t(:,j)  =  pinv(ubar)  *  x(:,j);
                xp(:,j)  =  ubar  *  t(:,j);
                dx  =  xp(:,j)  -  x(:,j);
        end
loadings  =  inv(t)  *  vbar';
clf
disp('LOADING  MATRIX  =  Y  (see  eq.3.135)')
disp(loadings)
%  Estimate  the  error  in  the  loadings  (Clifford  method)
e  =  d  -  x  *  loadings;
xx  =  inv(x'  *  x);
        for  j  =  1:c
            v  =  xx  *  (e(:,j)'  *  e(:,j))  /  (rx  -  n);
                for  k  =  1:n
                        loaderr(k,j)  =  sqrt(v(k,k));
                    end
        end
disp('ERRORS  IN  THE  LOADINGS  (Calculational  Method,  see  eq.  4.149)')
disp(loaderr)
```

BIBLIOGRAPHY

Adcock, C. J., *Factorial Analysis for the Non-Mathematician*, Melbourne University Press, Victoria, Australia, 1954.

Ahmavaara, Y., and Markkanen T., *The Unified Factor Model*, Finnish Foundation for Alcoholic Studies, Helsinki, 1958.

Anderson, T. W., *An Introduction to Multivariate-Statistical Analysis*, Wiley, New York, 1958.

Beebe, K. R., Pell, R. J., and Seasboltz, M. B., *Chemometrics: A Practical Guide*, Wiley Interscience, NY, 1998.

Bro, R., *Multi-Way Analysis in the Food Industry: Models, Algorithms and Applications*, Doctoral Thesis, Royal Veterinary and Agricultural University, Denmark, 1998.

Bro, R., Workman, Jr., J. J., Mobley, P. W, and Kowalski, B. R., *Appl. Spectrosc. Rev.*, **32**, 237 (1997).

Burt, C. C., *The Factors of the Mind*, Macmillan, New York, 1941.

Cattell, R. B., *Factor Analysis*, Harper, New York, 1952.

Cattell, R. B., *The Scientific Use of Factor Analysis in Behavioral and Life Sciences*, Plenum Press, New York, 1978.

Child, D., *The Essentials of Factor Analysis*, Holt, Rinehart and Winston, New York, 1970.

Comrey, A. L., *A First Course in Factor Analysis*, Academic Press, Orlando, 1973.

Fruchter, B., *Introduction to Factor Analysis*, rev. ed., Van Nostrand, New York, 1964.

Geladi, P., *J. Chemometrics*, **2**, 231 (1988).

Geladi, P., and Kowalski, B. R., *Anal. Chim. Acta*, **185**, 1 (1986); **185**, 18 (1986).

Gemperline, P. J., *J. Chemometrics*, **3**, 549 (1989).

Gorsuch, R. L., *Factor Analysis*, Saunders. Philadelphia, 1974.

Guertin, W. H., and J. Bailey, *Introduction to Modern Factor Analysis*, Edwards Brothers, Ann Arbor, MI. 1970.

Hamilton, J. C., and Gemperline, P. J., *J. Chemometrics*, **4**, 1 (1990).

Harman, H. H., *Modern Factor Analysis*, rev. 1st ed., University of Chicago Press, Chicago, 1976.

Henrysson, S., *Applicability of Factor Analysis in the Behavioral Sciences*, Almquist & Wiksell, Stockholm, 1960.

Hirsch, R. F. (Ed.), *Statistics*, Franklin Institute Press, Philadelphia, 1978.

Hopke, P. K., *Chemometrics Intell. Lab. Syst.*, **6**, 7 (1989).

Horst, P., *Factor Analysis of Data Matrices*, Holt, Rinehart and Winston, New York, 1965.

Höskuldsson, A., *Prediction Methods in Science and Technology*, Thor Publishing, Arnegaards Alle 7, Copenhagen, Denmark, 1998.

Hotzinger, K. J., and Harman, H. H., *Factor Analysis*, University of Chicago Press, Chicago, 1941.

Howery, D. G., *Am. Lab.*, **8**(2), 14 (1976).

Howery, D. G., in Hirsch, R. F. (Ed.), *Statistics*, Franklin Institute Press, Philadelphia, 1978, p. 185.

Jackson, J. E., *A User's Guide to Principal Components*, Wiley, New York, 1991.

Joreskog, K. G., *Statistical Estimation in Factor Analysis*, Almquist & Wiksell, Stockholm, 1963.

Kalivas, J. H., and Lang, P. M., *Mathematical Analysis of Spectral Orthogonality*, Dekker, New York, 1994.

Karjalainen, E. J., and Karjalainen, U. P., *Data Analysis for Hyphenated Techniques*, Elsevier Science B. V., Amsterdam, 1996.

Kemsley, E. K., *Discriminant Analysis and Class Modeling of Spectroscopic Data*, Wiley, Chichester, 1998.

Kowalski, B. R. (Ed.), *Chemometrics: Theory and Applications*, ACS Symp. Ser., 52, American Chemical Society, Washington, DC, 1977.

Kowalski, B. R. (Ed.), *Chemometrics and Statistics in Chemistry*, Reidel, Dordrecht, Holland, 1983.

Kramer, R., *Chemometric Techniques for Quantitative Analysis*, Marcel Dekker, Inc., New York, 1998.

Lawley, D. N., and A. E. Maxwell, *Factor Analysis as a Statistical Method*, 2nd ed.. Butterworths, London, 1971.

Lawlis, G. F., and D. Cahtfield, *Multivariate Approaches for the Behavioral Sciences*, Texas Tech University Press, Lubbock, TX, 1974.

Livingstone, D., *Data Analysis for Chemists: Applications to QSAR and Chemical Products Design*, Oxford University Press, Oxford, 1995.

Martens, H., and Naes, T., *Multivariate Calibration*, John Wiley and Sons, New York, 1989.

Sharaf, M. A., Illman, D. L., and Kowalski, B. R., *Chemometrics*, Wiley-Interscience, New York, 1986.

Massart, D. L., Vandeginste, B. G. M., Deming, S. N., Michotte, Y., and Kaufman, L., *Chemometrics: A Textbook*, Elsevier Science Publishers, Amsterdam, 1988.

Matthias, O., *Chemometrics*, Wiley-VCH, NY, 1999.

McClure, G. L. (Ed.), *Computerized Quantitative Infrared Analysis*, American Society for Testing and Materials, Philadelphia, 1987.

Mulark, S. A., *The Foundations of Factor Analysis*, McGraw-Hill, New York, 1972.

Pang, Z. X., Si, S. Z., Nie, S. Z., and Zhang, M. Z., *Chemical Factor Analysis*, Publishing House of Science and Technology, University of China, Hehui, 1992.

Rao, C. R., *Linear Statistical Inference and Its Applications*, Wiley, New York, 1965.

Rummel, R. J., *Applied Factor Analysis*, Northwestern University Press, Evanston, IL, 1970.

Spearman, C., *The Abilities of Man*, Macmillan, New York, 1927.

Stephenson, W., *The Study of Behavior*, University of Chicago Press, Chicago, 1953.

Thomson, G., *The Factorial Analysis of Human Ability*, Houghton Mifflin, Boston, 1951.

Thurston, L. L., *The Vectors of the Mind*, University of Chicago Press, Chicago, 1935.

Thurston, L. L., *Multiple-Factor Analysis*, University of Chicago Press, Chicago, 1947.

Van der Geer, J. P., *Introduction to Multivariate Analysis for the Social Sciences*, W. H. Freeman, San Francisco, 1971.

Weiner, P. H., *Chem Tech.*, 1977, 321.

Yu, R. Q., *Introduction to Chemometrics*, Hunan Educational Press House, Changsha, 1991.

AUTHOR INDEX

397

SUBJECT INDEX